MANUFACTURING SCIENCE-I

By:

S.A. RIZVI

Ph.D. (Pursuing) Welding
Mechanical Engineering Department
IIT (BHU), Varanasi
Faculty Member, JMI, New Delhi

Dr. S.P. TEWARI

Ph.D. (Welding)
Prof. Department of Mechanical Engineering
IIT (BHU), Varanasi

WAJAHAT ALI

B.E. (Mechanical), J.M.I.
Mechanical Engineering Department
SCRIET, CCS University, MEERUT (UP)

S.K. KATARIA & SONS®

Publisher of Engineering Books
4885/109, Prakash Mahal, Dr. Subhash Bhargav Lane,
Opposite Delhi Medical Association, Daryaganj, New Delhi–110002
Phone: +91-11-23243489, 43551243; Mobile: +91-9871775858
e-mail: katariabook@gmail.com, katariabooks@yahoo.com
Website: www.skkatariaandsons.com
(e-books, subscriptions and bulk buying also available on our website)

Published by:

S.K. Kataria & Sons®

4885/109, Prakash Mahal, Dr. Subhash Bhargav Lane,
Opposite Delhi Medical Association, Daryaganj, New Delhi-110002
Phone: +91-11-23243489, +91-1143551243; Mobile: +91-9871775858
e-mail: katariabook@gmail.com, katariabooks@yahoo.com

First Edition: 2012; Third Edition: 2015; Reprint: 2023

***Price :* ₹ 395/-**

Laser Type Setting by:

SARA ASSIGNMENTS, Delhi.

PREFACE

This book Manufacturing Science-I has been written to meet the full requirement of **B.Tech** Mechanical Engineering and related branch students. This subject is taught at a professional level as compulsory course in various universities. Any successful Mechanical engineer should be familiar with current and latest (New) Manufacturing Processes.

Best efforts have been made to place the exact course material covering all important topics related with manufacturing.

A set of **Objective Questions,** review questions to reinforce the information within the unit and text provide the basic information to explain the "How as well as why" of different manufacturing processes. **First** unit of this book cover the introduction of **Metal Forming Process. Second** unit cover **Rolling, Forging, Deep drawing** etc. **Third** unit gives the broader idea about the **Sheet Metal Jigs and Fixture, Powder Metallurgy etc.** Last and important **Fifth** unit deals with the **Unconventional Metal Forming Process**; cover the entire **Casting Process.**

It is hoped that this book in its present form shall be preferred by the reader over other books available in the market.

We shall be most grateful to those who will be so kind as to report errors and suggestions for improvement.

S.A. Rizvi
Wajahat Ali

A

ACKNOWLEDGEMENT

I am greatful to my mother (Smt. Amna Begum Rizvi) and all my family members and colleges specially for their encouragement and blessing. I feel greatly indebted to my wife **Saba Zafar Rizvi** and my kids **Aliyan Zehra, Khadija Tul Kubra** and **Syed Mohammad** for their constant support and valuable guidance from time-to-time.

My special thanks are due to **Mr. Sanjeev Kumar Kataria** (MD) of **S.K. Kataria & Sons**, (Publisher) for his encouragement and for the confidence he reposed in this work.

S.A. Rizvi

Dedicated to my wife
Saba Zafar Rizvi

– *S.A. Rizvi*

&

Dedicated to my wife
Dr. Mamata Tewari

– *Dr. S.P. Tewari*

CONTENTS

UNIT-I: METAL FORMING PROCESS (PART-I) 1—44

UNIT-I

METAL FORMING PROCESS
PART-I

CHAPTER 1

METAL FORMING

1.1 INTRODUCTION

Metal forming process is a process in which the metal is converted from one shape into another (final) shape for getting the required shape and size. One of the unique feature of metal forming is that, in metal forming operation no losses (no scrap produce) of raw material take place. In metal forming operation shape of raw material is converted from one shape into another shape by means of the application of external load (mechanical pressure) and plastic deformation of metal.

During the metal forming process, metal flow take place and this flow is known as plastic flow. During metal flow metals grains are changed. If the deformation of metal take place at high temperature, new grains start growing at the location of internal stresses caused in the metal, by slip or twin formation deformation of metal take place at recrystallization temperature stresses induced in this process are higher than yield strength but lower than the fracture (failure) strength of metal.

Following properties of metals play an important role in forming of metals:

(i) Ductility
(ii) Elasticity
(iii) Plasticity
(iv) Strength
(v) Hardness

1.2 UNIQUE FEATURE OF METAL FORMING PROCESS

- Load applied during the metal forming process is greater than the yield load but lower than the fracture load in such a way metal is safe from fracture.
- During the application of load metal deform, deformation may be either elastic or plastic. Plastic deformation caused due to the applied load (stress) and deformation may be by slip or twin.
- During the deformation process the temperature at which the new grains take place is known as recrystallization temperature.
- Deformation basically depends on the **temperature**, **size** and **shape** (design) of workpiece and **type of operation** to be performed.
- Metal forming operation can be carried out in two manness i.e., **hot working** and **cold working** condition.
- A large variety of metal forming process have been developed for various specification, application.

1.2.1 Importance of Manufacturing

Meaning of both words manufacturing or production is the same. Manufacturing/ production referred to convert the raw material into final output (component/ part) by performing some metal forming operation. For civilization of a nation manufacturing play an important role or in other words it can be said that a nation may not be considered as a developed nation until/unless they are not manufacturing the different parts by themselves like, **gear, steel sheets, engine housing, beverage cans** and many more comes in this category. Manufacturing is the biggest segment of Industries. Manufacturing can be performed by use of different machines and fabrication process. Manufacturing is directly related with the economy of a civilized nation, generally manufacturing activity level of a nation shows the living standard of the people of that particular nation. Manufacturing is too much affected by the **product design, selection of material, manufacturing process, planning etc.**

Manufacturing is an activity of combination of **material, tools, machine, power, human** etc. it is a system in which product design is the starting stage and delivery of finished component/part is the last stage. Manufacturing is a complex system as it is composed of many diverse physical and human element some of which are difficult to predict and control such as cost of raw material, market change, human behaviour etc.

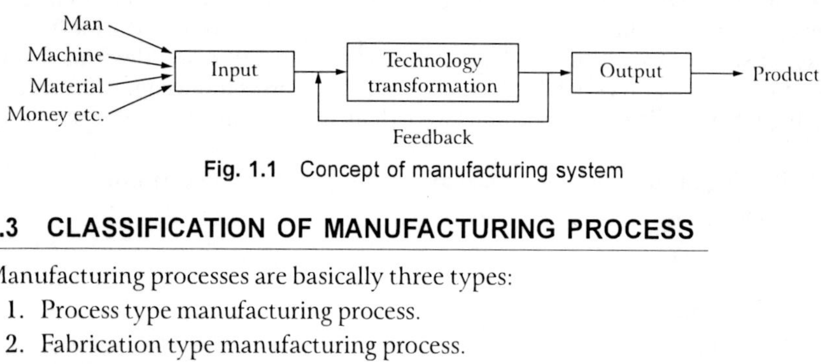

Fig. 1.1 Concept of manufacturing system

1.3 CLASSIFICATION OF MANUFACTURING PROCESS

Manufacturing processes are basically three types:

1. Process type manufacturing process.
2. Fabrication type manufacturing process.
3. Assembly type manufacturing process.

1. Process type manufacturing process

In this type of manufacturing process continuous flow of raw material takes place through a series of step-by-step process to produce a finish product like chemical etc.

2. Fabrication type manufacturing process

It is a secondry manufacturing process. In this type of manufacturing process any parts/components (individual) is produced by a series of operations performed on it. Such as **welding, casting, rolling, unconventional machining** etc.

3. Assembly type manufacturing process

In this type of manufacturing process parts/components are assembled together to get a complete product such as automobile etc.

Further manufacturing processes are classified as:

(i) Metal Joining:

(*a*) Electric arc welding (*b*) Gas welding

(*c*) Resistance welding (*d*) Thermit welding

(*e*) Submerged arc welding (*f*) Soldering

(*g*) Brazing (*h*) Adhesive bonding.

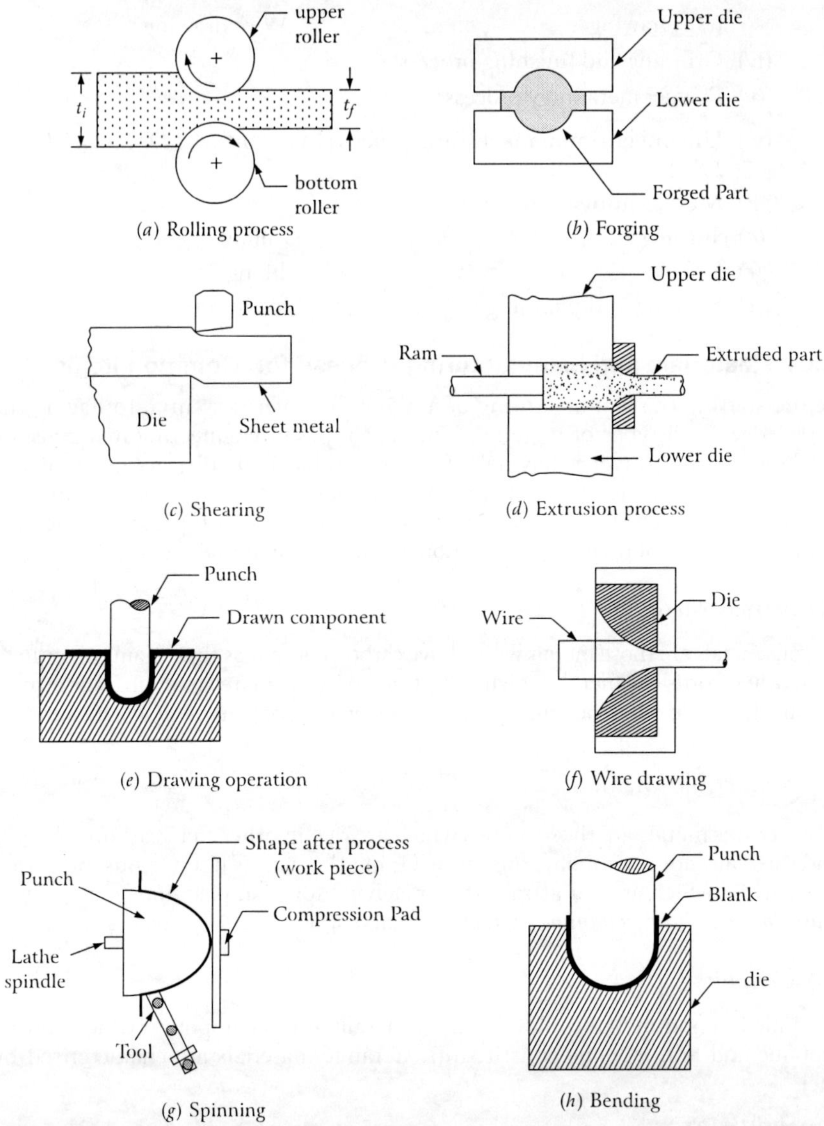

(*a*) Rolling process (*b*) Forging (*c*) Shearing (*d*) Extrusion process (*e*) Drawing operation (*f*) Wire drawing (*g*) Spinning (*h*) Bending

Fig. 1.2 Various metal forming operations

(ii) Casting Process:

(*a*) Sand casting
(*b*) Injection moulding
(*c*) Pressure die casting
(*d*) Hot chamber die casting
(*e*) Investment casting
(*f*) Centrifugal casting.

(iii) Metal Forming:

(*a*) Rolling
(*b*) Forging
(*c*) Extrusion
(*d*) Sheet metal (press work)
(*e*) Drawing
(*f*) Wire drawing.

(iv) Grinding and finishing process

(v) Powder metallurgy process

(vi) Unconventional machining process like — *ECM, EDM, ECG, USM, EBM* etc.

(vii) Metal removal process:

(*a*) Turning
(*b*) Drilling
(*c*) Boring
(*d*) Milling
(*e*) Shaping and planning
(*f*) Slotting etc.

1.3.1 Materials and Manufacturing Process for Common Items

Before starting the manufacturing of any item it is very essential to having the knowledge of all types of material. Different types of engineering materials are available in the universe having different **mechanical**, **physical**, **chemical**, **metallurgical** and **electrical properties** and these engineering material are used to produce different items. It is not possible to discuss all types of engineering materials but out of them some common engineering materials are given below:

(1) Ferrous Metal

Ferrous metals are those metals which have carbon content as their main constituent or in other words the metals which are attracted by magnate are referred as ferrous metals. Example of ferrous metals are *low carbon steel, cast iron, tool and die steel* etc.

(2) Non-ferrous Metal

Non-ferrous metals are those metals which are rich in other than carbon contents and carbon is not their main constitute. Or in other words non-ferrous metals are those metal which are not attracted by magnate, for example– *Aluminium, copper, brass, bronze, Nickel, Titanium, refractory metal, low melting alloys* etc.

(3) Ceramics

Ceramic is a non-metallic material and generally it is a compound which consist metallic and non-metallic constituents. Ceramic materials are characterised by their

⇒ ***brittleness,***

⇒ ***high hardness,***

⇒ ***electrical insulation,***
⇒ ***high melting point,***
⇒ ***chemical inertness.***

Common ceramics materials are ***Glass, refractories, insulating material, abrasive, graphite etc.*** All metal oxides such as ***silicate, carbides*** etc. are also considered as ceramics.

Ceramics are basically of two types—natural ceramics and synthetic ceramics.

(4) Plastics

Plastic is an organic material. They are very attractive organic engineering materials with complex structure extensively used in ***automobiles, electrical appliances, electronic industries*** etc. Plastics have some special properties which make them different from other materials. Plastic basically consists of natural material like *wood, natural rubber, coal, animal fibers* etc.

Plastics are basically of two types

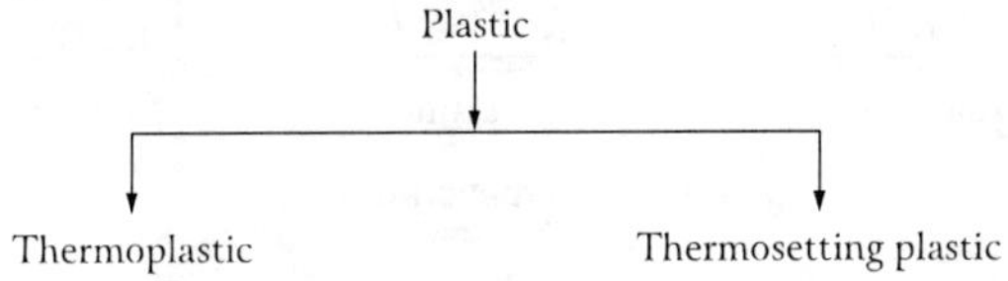

(5) Composite Material

According to ASM composite material is defined as microscopic combination of two or more distinct materials having a recognizable interface between them. Common composite materials are– ***ceramic matrix composite, metal matrix composite*** (MMC), ***polymer matrix composite, fibre reinforced composite, laminar composite*** etc. They have excellent property of metal and non-metal.

(6) New Material

There are some special type of materials for specific items and these new materials have specific properties. Example of new materials are **nano-material, amorphous alloys, super conductors** etc. For manufacturing of any item raw materials, tools, machines, etc. are essential and selection of right material, right machine and right manufacturing process is very essential because finished product basically depends upon the raw material and manufacturing process.

1.4 ECONOMIC AND TECHNOLOGICAL CONSIDERATION IN MANUFACTURING

First and most important process in manufacturing is selection of a material. Before selection a material for producing a finished product. It is very important to known about the material properties and characterstics. Final product (workpiece) is very much affected by the cost of raw material and type of operation to be performed on raw material. For manufacturing of any **item/component/part**, the raw material should be easily available within economy range. If the raw material is not easily available as well as the cost of raw material is very high then it would not be economical to produce an item as the final cost of item would be high and product will be out from customer's purchasing power.

1.5 DEFORMATION OF METAL

[*UPTU 2008-09*]

In metal forming operation, on applying the load (understress) material start to deform. This deformation in metal may be of two types:

- Elastic deformation.
- Plastic deformation.

1.5.1 Elastic Deformation

On applying the load on a metal, metal starts to deform. On removing the applied load if metal regains its original shape and size, it is known as elastic deformation. In a metal, elastic deformation occurs before the yield point.

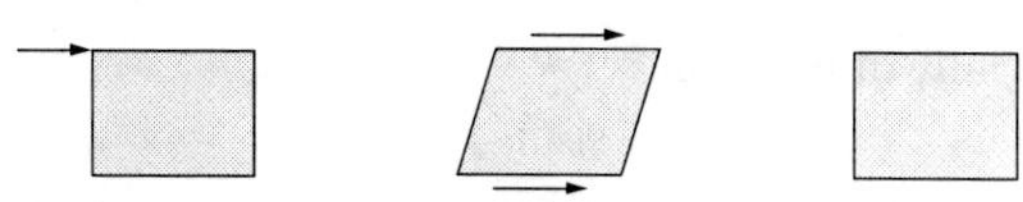

Fig. 1.3 Deformation of material

1.5.2 Plastic Deformation

On applying the load on a metal, metal starts to deform. On removing the applied load if metal does not regains its original shape and size, it is known as plastic deformation. After plastic deformed metal can not be further reused as permanent deformation in work piece had be taken.

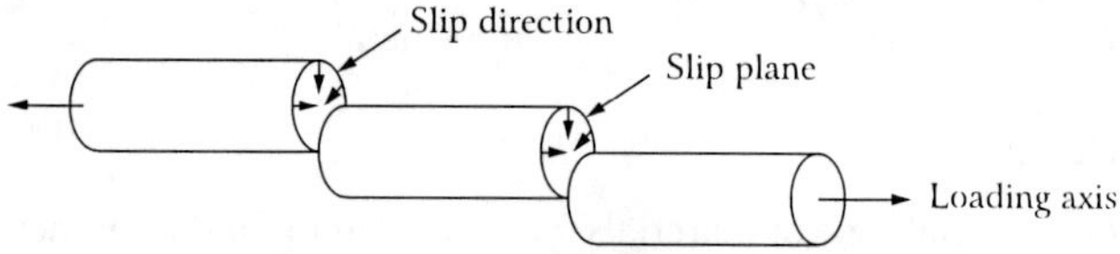

Fig. 1.4 Plastic deformation

Plastic deformation in a metal takes place in the following way:

(*a*) Slip deformation

(*b*) Twin deformation

(*c*) Deviations from regular position of atoms

(*d*) Break down of structure

1.6 FACTORS AFFECTING THE METAL FORMING OPERATIONS

[*UPTU 2008-09*]

Various factors (variable) affect the metal forming operation such as **temperature**, **strain rate** or **speed, friction** etc.

1.6.1 Temperature

Temperature is one of the main variable that affect the metal forming operation.

On increasing the temperature, strength and hardness of metal decreases while ductility increases.

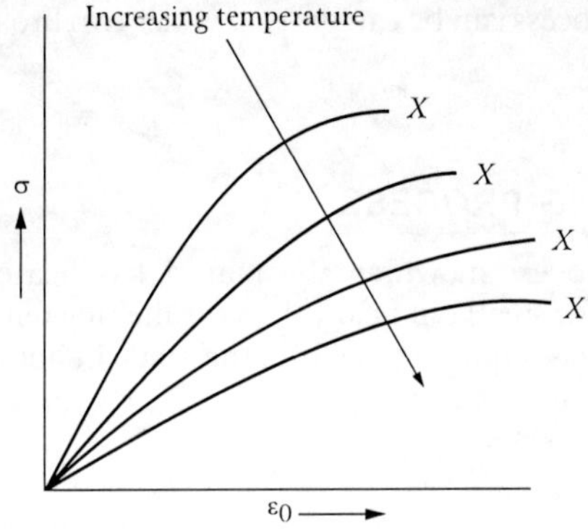

Fig. 1.5 Effect of temperature on stress-strain curve

1.6.2 Strain Rate

Strain rate also affect the metal forming operation. Mathematically it is calculated by

$$\varepsilon = \frac{v}{h}$$

where v = deformation velocity

h = instantaneous length or height of the object (workpiece)

It is observed that effect of strain rate is very small at room temperature or very low temperature.

- On increasing the strain rate, flow stress of metal increases.
- If temperature is considered as a parameter, tensile strength of metal increases with strain rate.
- Due to strain rate, temperature of metal increases.

TABLE 1.1 Typical Deformation Velocity

Process	*Deformation velocity m/sec*
Tensile Test	0.6×10^{-6} to 0.6×10^{-2}
Tube Drawing	0.015 – 0.6
Hammer Forging	2.5 – 10.0
Mechanical Press	0.15 – 1.5

1.6.3 Friction

Friction play an important role in metal forming operation. In metal forming operation if there is intimate contact between the tool and the workpiece, friction produced. This friction not only affects the forming forces (stress) but also affects the deformation rate of metal. Friction is also source of heat. Unless there is no lubrication in between the workpiece and tool, friction will be produced. It can be minimized only by lubrication for cold working process. Coefficient of friction is 0.1 and 0.6 or greater for the hot working process.

Metal forming process can be categorised under two broad group, namely:

(i) Hot forming
(ii) Cold forming

1.7 HOT WORKING PROCESS

Hot working is a process in which the material is heated above its critical temperature, (800°C for steel) but below than the melting temperature of material. Less amount of energy is required to deform the metal in hot working as compared to cold-working process.

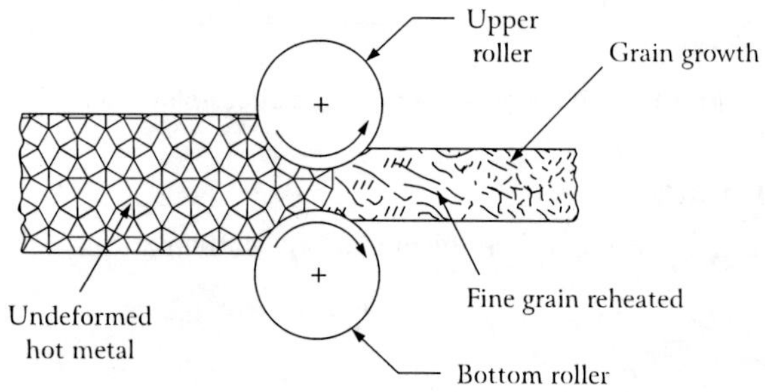

Fig. 1.6 Hot rolling

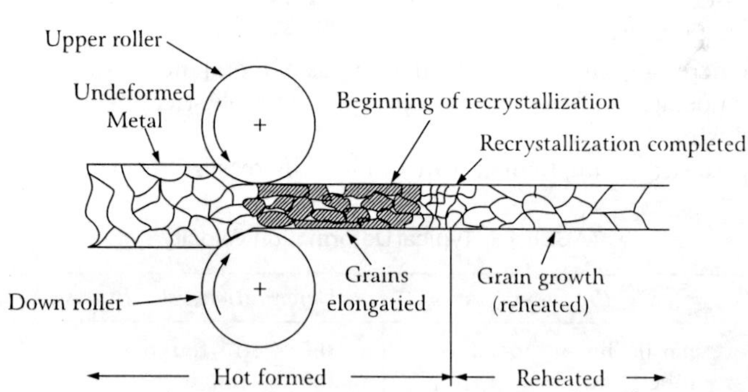

Fig. 1.7 Effect of hot working on the microstructure of cast metals

1.7.1 Advantages of Hot Working [*UPTU 2004-2005*]

(i) No strain hardening takes place as process is carried out at above recrystallization temperature.
(ii) No chance of porosity in metal.
(iii) As the shear stress get reduced at high temperature therefore, less force is required for deformation purpose.

(iv) Ductility and Impact resistance are improved.
(v) Very large workpiece can be deformed with equipment of reasonable size.
(vi) As coarse grains are refined they result in fine grain structure.
(vii) Further stress relieving process is not required.
(viii) Very high reduction is possible without fracture.

1.7.2 Limitations of Hot Working

[UPTU-2004-05]

(i) Due to high temperature tooling, handling of hot metal are very difficult and costly.
(ii) Due to high temperature there is rise in scaling of surface which leads poor surface finish generation.
(iii) Close tolerance on dimensions are not possible.
(iv) Cost of hot working equipments are high and their maintenance is difficult too.
(v) As hot working is carried out at a high temperature therefore brittle material cannot be worked (deformed) easily.
(vi) Metallurgical structure of workpiece may be non-uniform.

Major hot working processes are:

(i) Forming
(ii) Rolling
(iii) Drawing
(iv) Extrusion
(v) Piercing
(vi) Spinning.

1.8 COLD WORKING PROCESS

In cold working process material (object) is heated below its critical temperature. Deformation of metal starts at room temperature. Cold working process basically depends upon the ductility.

1.8.1 Advantages of Cold Working

[UPTU-2004-05]

(i) Mechanical properties such as yield strength and ultimate strength increases.
(ii) As object (job) is heated at below crystallization temperature hence no chance of decarburization of surface.
(iii) Hardness increases but ductility decreases.
(iv) Good quality of surface finish is obtained.
(v) Better chance to control the dimensional tolerance.
(vi) It is far easier to handle cold part and it is also economical for smaller size object.
(vii) Minimum contamination because of low working temperature.
(viii) Suitable for mass production.

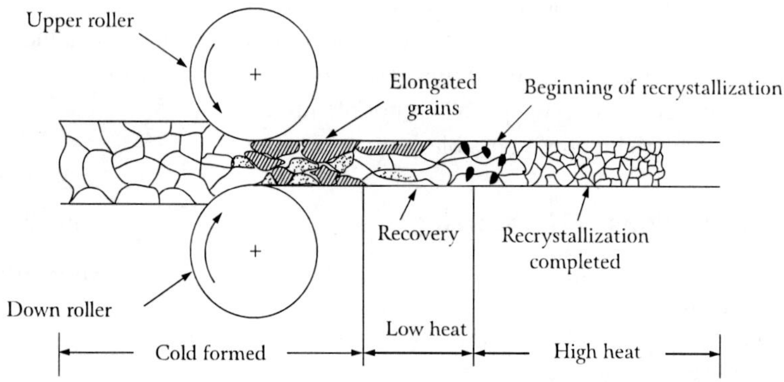

Fig. 1.8 Effect of cold working on the microstructure of cast metals

1.8.2 Limitations of Cold Working [*UPTU-2004-05*]

(i) Only ductile material can be cold worked.
(ii) Residual stresses are produced in cold working which leads to converting material into brittleness.
(iii) Only small size objects are easy to be cold worked.
(iv) Amount of deformation is limited.
(v) More possibility of crack formation.
(vi) Large amount of force is required to perform the cold working process.

Major hot working processes are:

(i) Cold rolling (ii) Cold threading (iii) Cold forging
(iv) Cold drawing (v) Wire drawing (vi) Cold extrusion
(vii) Cold bending (viii) Tube drawing etc.

1.9 RECRYSTALLIZATION TEMPERATURE

Recrystallization temperature is a temperature at which the new grains (atom) start to generate or as per ASM, "recrystallization temperature is defined as minimum temperature at which complete recrystallization of cold worked metal is produced within predetermined time."

TABLE 1.2 Minimum Recrystallization Temperature

	Material	*Recrystallization Temperature °C*
1.	Lead (Pb)	below room temperature
2.	Tin (Sn)	below room temperature
3.	Zinc (Zn)	at room temperature
4.	Aluminium (Al)	150°C
5.	Copper (Cu)	200°C
6.	Iron(Fe)	450°C
7.	Tungsten (W)	1200°C
8.	Tantalum (Ta)	1000°C

1.10 DIFFERENTIATE BETWEEN HOT WORKING AND COLD WORKING

[UPTU 2001-02, 2004-05, 2005-06, 2008-09]

Sl. No.	*Hot Working*	*Cold Working*
1.	In hot working process object is heated above the recrystallization temperature and below the melting point of object.	In cold working process object is heated below recrystallization temperature.
2.	Due to plastic deformation no strain hardening occurs in object.	Strain hardening encountered in the object.
3.	No internal stresses occur in the object.	Internal stresses are encountered in the object.
4.	Less amount of forces are required for deformation.	More amount of forces required for deformation.
5.	Mechanical properties like strength increases and Blow holes and cracks weld up automatically due to heating at very high degree temperature.	Reduction in area, elongation decreases. More chance of crack formation due to low temperature heating.
6.	Light weight equipments and tools are required in hot working process.	Heavy tools and equipments are required.
7.	Very difficult to handle a hot worked object.	Not so difficult to handle a cold worked object.
8.	Closer dimensional tolerance are difficult to be maintained.	Dimensional accuracy can be easily achieved.
9.	Rate of production in hot working is higher.	Lower rate of production in cold working.
10.	It is applicable where high rate of deformation is required and no need of work hardening.	It is preferred where work hardening is required.
11.	Surface finish of hot working parts is relatively poor due to oxidation and scaling.	Cold worked parts carry better surface finish.
12.	Refinement of crystals occurs.	Crystallization does not take place therefore grains are elongated.
13.	Reduction in area, elongation and impact strength value improved.	Cold working decreases the value of elongation, reduction in area and impact strength value.
14.	Surface decarburization in steels likely to occur at high temperature.	Surface decarburization in steel does not occur.

1.11 YIELDING UNDER COMPLEX STRESSES *[UPTU 2006-07, 2008-09]*

It is very important in a metal working process to use the principal stresses. Principal stresses are those stresses which are normal acting on a plane in which there is no shear stress. Such planes are called as Principal planes.

Let σ_1, σ_2 and σ_3 = Principal Stress

Value of these stresses will be three roots of the cubic equation given by determinant for every pair of principal stresses there are two planes of principal shear stresses.

$$\begin{vmatrix} \sigma - \sigma_x & -\tau_{xy} & -\tau_{zx} \\ -\tau_{xy} & \sigma - \sigma_y & -\tau_{zy} \\ -\tau_{xz} & -\tau_{yz} & \sigma - \sigma_z \end{vmatrix} = 0 \qquad \text{for } \sigma_3 = 0$$

$$\sigma_1 = \frac{\sigma_x + \sigma_y}{2} + \frac{1}{2}\sqrt{(\sigma_x - \sigma_y)^2 + 4\tau_{xy}^2} \quad \text{(Maximum)}$$

$$\sigma_2 = \frac{\sigma_x - \sigma_y}{2} - \frac{1}{2}\sqrt{(\sigma_x - \sigma_y)^2 + 4\tau_{xy}^2} \quad \text{(Minimum)}$$

$$\tau_m = \text{Maximum Shear Stress}$$

$$= \frac{\sigma_{max} - \sigma_{min}}{2}$$

Function of yield theories is to show the onset of yielding under complex state of stresses i.e., σ_1, σ_2 and σ_3 and in any metal forming operation, metal is subjected to a complex stress.

1.12 STRESS IN A BODY

Usually two types of stresses act in a body

- Normal Stress
- Shear Stress

Normal Stress

Normal stress is denoted by σ (Sigma a Greek word). Normal stress is the stress which act perpendicular or normal to the surface of body, where tensile stress have positive sign and compressive stress have negative sign.

Shear Stress

Shear stress is denoted by τ (Tau). This type of stress act along the surface, they have positive value when their direction is towards the positive axes.

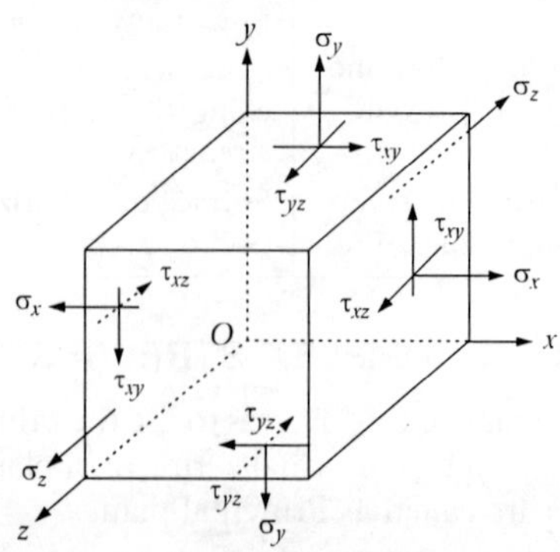

Fig. 1.9 General state of stress at a point

$$\tau_{xy} = \tau_{yx}, \ \tau_{yz} = \tau_{zy}, \tau_{xz} = \tau_{zx}$$

These six components are required to specify the state of stress at a point for ductile material. *[UPTU 2009-10]*

For deformation of ductile materials two main theories are given as –[M. Imp.]

(i) Tresca Theory (Max. Shear Stress Theory),

(ii) Von Mises Theory (Max. Strain Energy Theory)

1.12.1 Tresca Yield Criteria *[UPTU 2010-11]*

Tresca theory is basically related to plastic deformation of a material. According to "Tresca plastic flow occurs in an object when the shear stress reach a maximum value."

If the principal stress at a point are σ_1, σ_2 and σ_3.

Assuming $\sigma_1 > \sigma_2 > \sigma_3$, σ_y = yield stress

then, $\tau_{max} = \dfrac{\sigma_1 - \sigma_3}{2}$ τ_{max} = minimum shear stress

plastic deformation start when $k = \tau_{max}$

$$k = \frac{\sigma_1 - \sigma_3}{2}$$

1.12.2 Von Mises Yield Criteria *[UPTU 2010-11]*

According to Von Mises, "plastic flow start in a material when shear strain energy reach a critical value".

$$U = \frac{1}{6G}\left[(\sigma_1 - \sigma_2)^2 + (\sigma_2 - \sigma_3)^2 + (\sigma_3 - \sigma_1)^2\right]$$

where, U = Shear strain energy/volume

G = Shear modulus of elasticity

$\sigma_1, \sigma_2, \sigma_3$ are the principal stresses

$$(\sigma_1 - \sigma_2)^2 + (\sigma_2 - \sigma_3)^2 + (\sigma_3 - \sigma_1)^2 = 6GC_1 = \text{constant}$$

according to this, plastic flow depends upon all principal stresses.

1.12.3 Relation Between Shear Yield Stress and Tensile Yield Stress

[UPTU 2002-03, 2003-04, 2005-06, 2006-07, 2008-09, 2010-11]

For Tresca's yield criterion:

$\sigma_1 = \sigma_y$ (Tensile yield stress) [for tension]

and $\sigma_2 = \sigma_3 = 0$

in uniaxial $\sigma_1 = \sigma_2 = 0$ [for compression]

$\sigma_3 = -\sigma_y$ σ_y = yield stress in tensile loading

but according Tresca $\dfrac{\sigma_1 - \sigma_3}{2} = k$

$$k = \frac{\sigma_y - 0}{2} \quad \text{(for tension)}$$

$$k = \frac{0 - (-\sigma_y)}{2} \text{(for compression)}$$

$\therefore$ $$k = \frac{\sigma_y}{2}$$

For Von Mises yield criterion:

$$\sigma_1 = \sigma_y, \sigma_2 = \sigma_3 = 0 \qquad \text{[for Tensile Load]}$$

but $$C = (\sigma_1 - \sigma_2)^2 + (\sigma_2 - \sigma_3)^2 + (\sigma_3 - \sigma_1)^2$$

or $$C = (\sigma_y - 0)^2 + (0 - 0)^2 + (0 - \sigma_y)^2$$

$$C = 2\sigma_y^2$$

and $$\sigma_1 = k, \sigma_2 = 0, \sigma_3 = -k \qquad \text{[for Torsion]}$$

again $$C = (\sigma_1 - \sigma_2)^2 + (\sigma_2 - \sigma_3)^2 + (\sigma_3 - \sigma_1)^2$$

$$C = (k - 0)^2 + (0 + k)^2 + (-k - k)^2$$

$$C = 6k^2$$

$$6k^2 = 2\sigma_y^2$$

or $$k = \frac{\sigma_y}{\sqrt{3}}$$

OBJECTIVE QUESTIONS

1. Hot working is carried out at *[UPTU 2009-10]*
   (*a*) below recrystallization temperature
   (*b*) above recrystallization temperature
   (*c*) exactly at recrystallization temperature
   (*d*) none of the above

2. Cold working is carried out at *[UPSC Exam.]*
   (*a*) above recrystallization temperature
   (*b*) below recrystallization temperature
   (*c*) exactly at recrystallization temperature
   (*d*) all of the above

3. Hot working of metals enables
   (*a*) Heavy reduction in porosity of metal
   (*b*) Refinement of its grain structure
   (*c*) Disintegration and distribution of concentrated impurities
   (*d*) All of these

4. Plastic deformation of a metal take, place when
   (*a*) Stress caused in the metal due to external load are much lower than yield load.

(*b*) A metal deforms without external loading.
(*c*) Stress caused in the metal due to external loading reach yield load.
(*d*) None of these

5. Recrystallization temperature for steel is
(*a*) 450°C (*b*) 600°C
(*c*) 800°C (*d*) None of these

6. Cold working produces the following effect.
1. Stress are set up in the metal
2. Grain structure gets distored
3. Strength and hardness of the metal are decreased
4. Surface finish is reduced
which of these statements are correct?
(*a*) 1 and 4 (*b*) 1 and 2
(*c*) 1, 2 and 3 (*d*) 3 and 4. *[IES 2003]*

7. Which one is the correct statement about hot working process
(*a*) Long life of working tools
(*b*) Rapid and economical process
(*c*) Less tooling and handling costs
(*d*) Fine surface finish of the product

8. Cold worked articles are __________ before further making operation.
(*a*) annealed (*b*) carburized
(*c*) spherodized (*d*) normalized

9. In metals subjected to cold working strain hardening effect is due to __________ mechanism
(*a*) twinning (*b*) screw
(*c*) fracture (*d*) dislocation

10. Increasing in hardness due to cold working is called
(*a*) cold hardening (*b*) age hardening
(*c*) hot hardening (*d*) work hardening

ANSWERS

1. (*b*)	2. (*b*)	3. (*d*)	4. (*c*)	5. (*c*)
6. (*b*)	7. (*b*)	8. (*a*)	9. (*d*)	10. (*d*)

REVIEW QUESTIONS

1. Discuss briefly importance of manufacturing in 21st century in India. *[UPTU 2008-09]*
2. What is manufacturing? How will you classify manufacturing process? Give suitable example of products being made by manufacturing process. *[UPTU 2008-09]*
3. Define the Importance of manufacturing.
4. How manufacturing process can be classified?
5. Why material selection and design are important for manufacturing?

6. Classify the metal forming operation in detail.
7. Define the deformation of a metal during the application of the load.
8. How many types of deformation do you know?
9. Differentiate the elastic and plastic deformation.
10. What are the factors affecting the plastic deformation?
11. Describe the theories of plastic deformation (failure). *[UPTU 2004-05]*
12. List the advantages and disadvantages of hot working.
13. Differentiate between the hot working and cold working process. *[UPTU 2001-02, 03-04, 10-11]*
14. Enlist the various (different) cold and hot working process. Compare hot working and cold working process. Indicating advantages and disadvantages of each. *[UPTU 2004-05, 08-09]*
15. What do you understand by yield criterions for ductile materials. Find out the relation between the Von-mises and Tresca yield criteria. *[UPTU 2006-07, 08-09]*
16. Write short notes on:
    (i) Tresca's yield criteria (ii) Von-mises yield criteria. *[UPTU 2010-11]*
17. Describe the elastic and plastic behaviour of metals and also discuss the factors affecting the plastic deformation. *[UPTU 2008-09]*
18. How do you classify metal forming process based on temperature consideration?
19. What is the importance of recrystallization temperature in metal forming?
20. Name a few major hot working processes.

❑❑❑

2 CHAPTER

FORGING

2.1 INTRODUCTION

Forging is another metal forming operation and it is defined as a metal forming operation in which the heated (recrystalization temperature) object (Hot working/ cold working) is kept in between the two dies (Shape of die is replica of component that to be produce) and a compressive force is applied. Due to the application of load, plastic deformation of metal start and object converted from one shape into another shape (final product).

In other words forging is defined as a metal forming process by which metals are plastically deformed to get desired shape and size by applying compressive force.

Forging enhance the quality of steel which becomes stronger after forging. This is the reason to make the parts by forging process which are subjected to heavy load. Forging process is best suited for mass production. Forging process is usually employed for those parts/components which requires high strength, resistance to vibration and uniform properties.

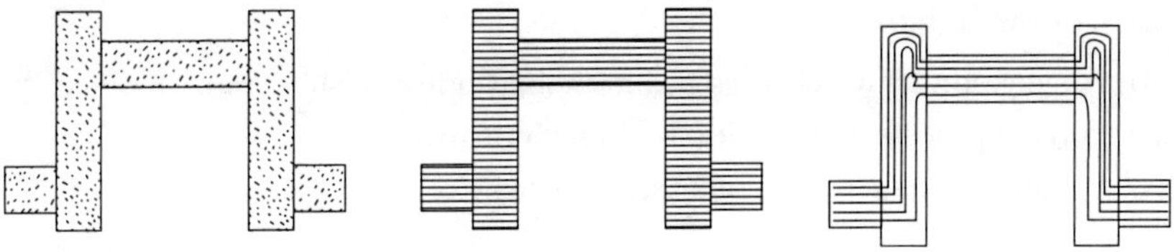

(*a*) Crank shaft produced by casting (no grain-flow)

(*b*) Crank shaft produced by machining (no grain continuity)

(*c*) Crank shaft produced by forging (continuity in grain flow)

Fig. 2.1 Comparison of forging process with other processes

2.1.1 Advantages of Forging Process

(i) Forging process improves the directional strength of components and improves the mechanical properties.

(ii) Forging welds up the previous crack in unidirectional fibres if any.

(iii) Forging process is suitable for ferrous and non-ferrous metals.

(iv) Forging process is best suited for mass production.
(v) Surface finish of part is better by forging as compared to casting process.
(vi) Forging process eliminate the machining process upto some extent hence material saving.
(vii) A large variety of shapes like **round**, **bars**, **shaft**, **rings**, etc. can be easily produced by forging.
(viii) Less chance of material rejections in forging process.
(ix) Forging can be held to within fairly closer dimensional tolerance.

2.1.2 Limitations of Forging

(i) Initial cost of dies are very high.
(ii) Forging process are also very costly.
(iii) More chance of oxidation or scaling of surface in hot forging resulting the poor surface finish.
(iv) Forging are more costly than casting process but components produced by forging are free from porosity.

2.1.3 Application of Forging

Forging process can be performed for following materials.

(i) Low-carbon steel.
(ii) Stainless steel
(iii) Alloys tool steel
(iv) Aluminium and its alloys
(v) Magnesium and its alloys
(vi) Copper and its alloys.
(vii) Forging process widely used in automobile industries-like crank shaft etc.
(viii) Forging process used in ship building industries.
(ix) Forging process used in rail road industries.

2.2 FORGING DEFECT

Forging is a quite better process than other metal forming process but some defects are likely to be produced if proper care is not taken during the process.

Some common forging defects are:

Internal cracks, cold shut, die shift, unfilled section, overheated/burn metal, Scales, Flakes, Improper grain flow, *etc.*

(i) **Internal crack:** This type of defects in metal/components occurs due to high heating and improper striking during the process.

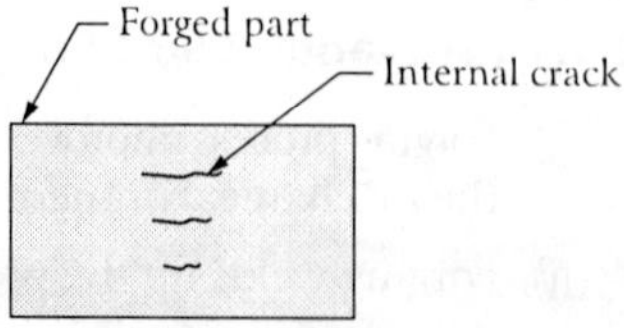

Fig. 2.2 Internal crack

(ii) **Cold shut:** This type of defect occurs only due to improper design of die and occurs at the corner and at right angle of surface.

(iii) **Die shift:** This defect is also known as mismatch and occurs due to misalignment of the upper and lower die. Due to die shift improper shape of item will be formed.

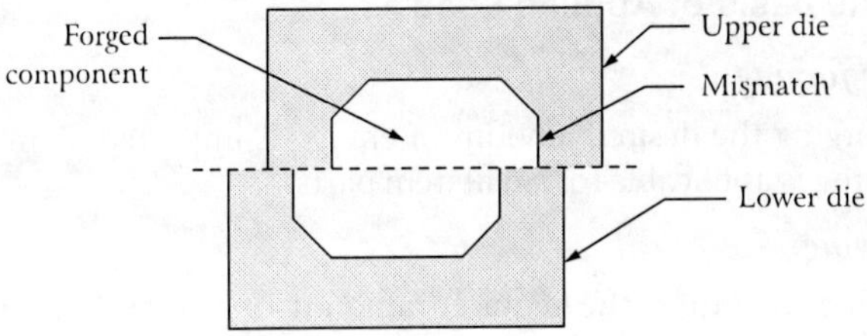

Fig. 2.3 Die shift or mismatch.

(iv) **Unfilled section:** This type of defect occurs due to incomplete fill of the die cavity. It is some what like the misrun of casting defect. It is only due to not proper quantity of filling metal.

(v) **Overheated/Burn metal:** This type of defect occurs due to overheating of metal. In this defect, due to high heat metal burns.

(vi) **Scales:** This type of defect occurs due to hot working condition of metal and produce due to hot metal reaction with air, metal start decarburization.

(vii) **Flakes:** Flakes refer as internal rupture of metal. This occurs due to improper cooling of a forging components.

(viii) **Improper grain flow:** This is only due to improper design of die which makes the flow of metal not flowing the final intended direction.

2.3 CLASSIFICATION OF FORGING PROCESS

Forging process can be classificed on the following basis:

1. On the basis of straining speed
2. On the basis of force applied

2.3.1 On the basis of Straining Speed

In this category forging is classified into two group:

(*a*) low speed forging,
(*b*) high speed forging.

(a) Low speed forging

Most conventional forging process is carried out at slow speed. If the speed of ram is slow then it is referred as slow speed forging. Energy obviously derived by means of large and heavy masses through slow speed forging process, components of superior mechanical properties and maximum reliability can be produced.

(b) High speed forging

If the ram speed of forging press is very high then it is referred as high speed forging. In recent year high speed forging is adopted due to high rate of production. In high speed forging, heavy deformation is possible in a blow. In high speed forging contact between the die and workpiece is very low.

Closer dimensional tolerance and good surface finish can be achieved by high speed forging.

2.3.2 On the basis of Applied Load

(a) Hammer forging

In hammer forging the desired amount of energy is imparted by impulsive load. Hammer forging is applicable for small item parts.

(b) Press forging

In press forging operation the required amount of energy is imparted by the application of squeezing load. Hydraulic press is the example of press forging. It is basically used for forging of heavy parts.

2.4 FORGING DIES

In forging process dies are the main components which play an important role. Shape and size of die is the replica of shape that is to be produced by forging. Two dies are used for this purpose, one is known as upper die and other is referred as lower die. Upper die is associated with ram while lower die is kept on base/anvil.

Classification of Forging Die

Forging die can be classified on the basis of:

1. **Operation based:** (*a*) Drop hammer die, (*b*) Trimming die, (*c*) Press die.
2. **On the basis of manufacturing process:** (*a*) Open die, (*b*) Closed die.
3. **On the basis of number of impression:** (*a*) Single Impression die, (*b*) Multi impression die.

2.4.1 Open die Forging

This process is similar to conventional forging which is done by a black smith. In open die forging process workpiece is kept in between the two dies as shown in Fig. 2.4. In this process upper die is associated with ram, while lower die is attached with anvil. Material is heated above recrystallization temperature in oil fired, gas fired or electric furnace and then hammered the same by a mechanical hammer. Metal deform due to plastic deformation in forging process. In open die forging process more chance of barreling due to frictional forces at the die-workpiece (component) interface. Barreling effect can be reduced by proper lubrication.

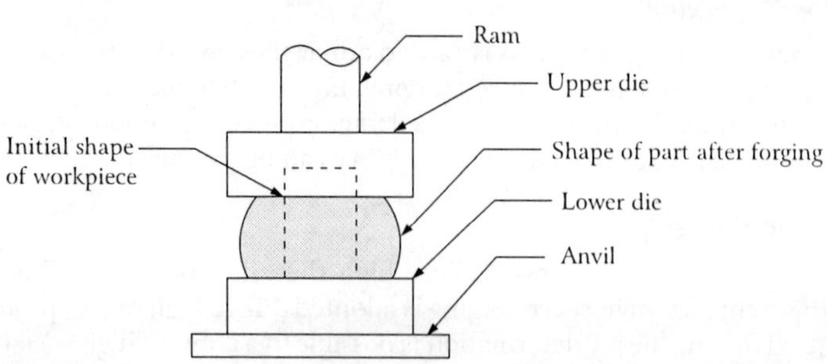

Fig. 2.4(A) Open die forging

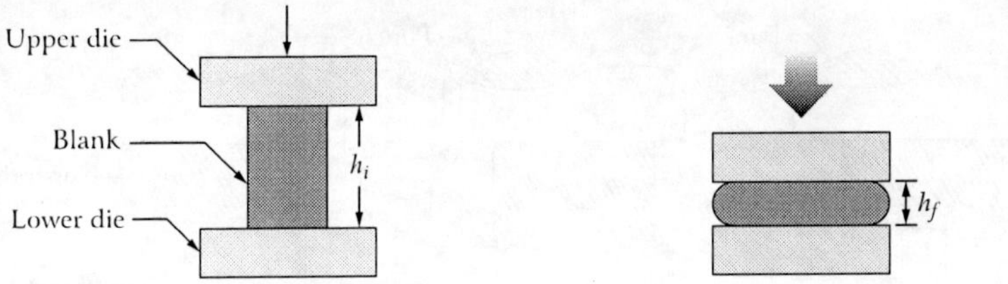

(i) Work piece before applied pressure (ii) Work piece after hammering

Fig. 2.4(B) Forging: Upsetting

2.4.1.1 Advantages of Open Die Forging

(i) Open die forging is a simple process.
(ii) Cost of dies are low.
(iii) Process is suitable only for small quantity.
(iv) Process is suitable for wide range of products size.

2.4.1.2 Limitations of Open Die Forging

(i) Only simple shapes can be produced by this process.
(ii) Skilled operator is required.
(iii) Closer dimensional tolerance cannot be achieved by this process.
(iv) Production rate is very low.
(v) Surface finish is not so good. So machining process is required.
(vi) Material utilization is poor.

2.4.2 Impression Die Forging

Impression die forging is also referred as *closed die forging.* In this process part (workpiece) is deformed in between the two dies. Shape of dies (upper die and lower die) are replica of parts that to be produced by closed die forging process. First of all workpiece is heated up to hot working temperature (recrystalization temperature) and then kept in between the two dies. One die is attached with punch holder and other is attached with anvil (stationary). A high pressure force is applied on upper die by ram and metal deform due to plastic deformation. During impression die forging process no flash is produced as workpiece is completely surrounded in between the dies. No flash is produced during process hence no volume losses open and closed die forging can be carried out at hot or cold state.

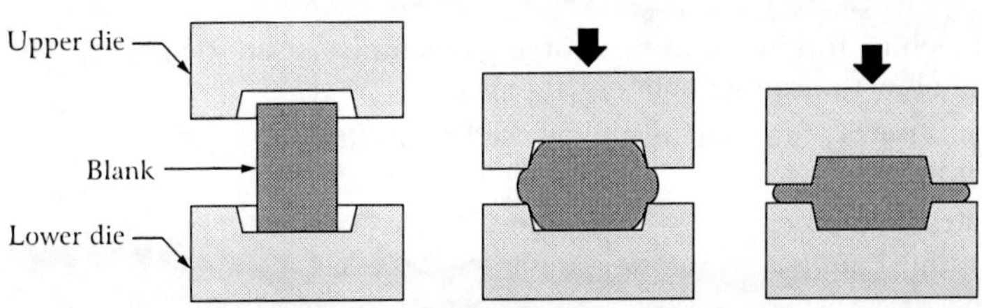

Fig. 2.5(A) Forging: Impression forging

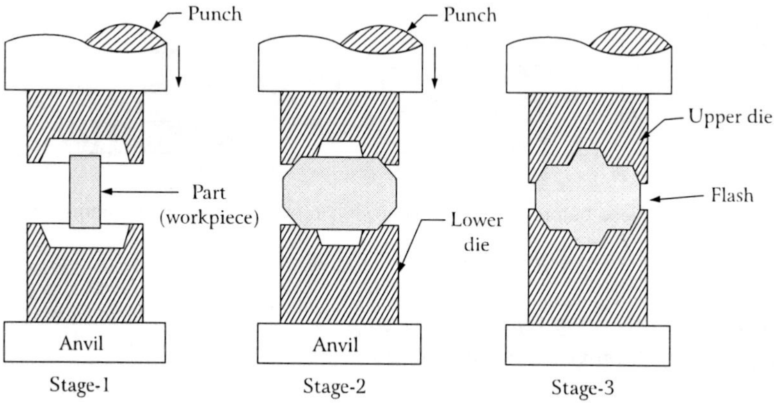

Fig. 2.5(B) Stages in impression die forging

2.4.2.1 *Advantages of Closed Die Forging*

(i) Complex shapes of workpiece can be produced.
(ii) Good utilization of workpiece material.
(iii) Closer dimensional accuracy can be achieved.
(iv) Controlled grain flow of material resulting in improved mechanical properties.
(v) Production rate is higher.
(vi) Less time consuming as compared to open dies forging process.

2.4.2.2 *Limitations of Closed (Impression) Die Forging*

(i) Equipment and tooling costs are high.
(ii) More than one step is required for each forging.
(iii) Separate die is needed for appropriate parts.

Forging temperature range:

Metal	*Temperature*
Steel	1100–1250°C
Aluminium	380–450°C
Coppers it alloys	750–950°C

2.5 FORGING MACHINES

During machine forging process, a forging machine is used to applied continuous and equal blow to produce the forged parts.

Various types of forging machines used are as under:

(*a*) drop hammer
(*b*) steam hammer
(*c*) spring hammer

2.5.1 Drop Hammer

This machine consisting of an anvil or base aligned with a hammer that is raised and then dropped on molten metal, used to forge or stamp the metal resting on the anvil.

In this type of Hammer ram is attached with hard wood board. When rolls are released the ram falls done and thus upper die strike on the red hot workpiece. Hammer weight range vary from 250 kg to 2500 kg and the height from which it falls vary 0.75 m to 1.0 meter and anvil which must absorb the blows usually 20 times are stronger than harmer board generally operated with a prime mover.

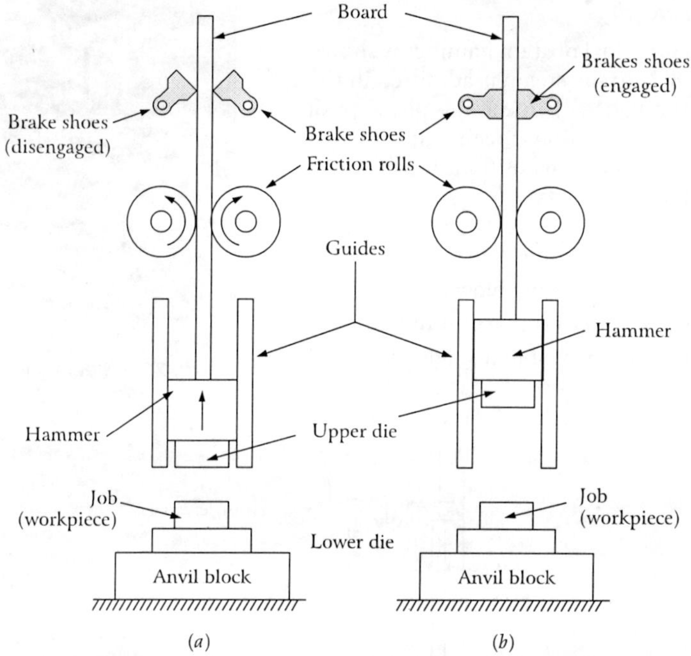

Fig. 2.6 (a) A schematic illustration of the working of a board type drop hammer (a) Board is being raised along with hammer with the help of two rotating friction wheels. Note that the brake shoes are disengaged (i.e., not holding the board), (b) When the hammer is raised to desired height, the brake shoes are pressed against the board which hold it at that height for desired extent of time. When hammer blow is required, the brake shoes are disengaged allowing the board to fall down freely along with the hammer.

2.5.2 Steam Hammer

A steam hammer is a power hammer, driven by steam, used to shape forgings. It consists of a hammer-like piston located within a cylinder. The hammer is raised by the pressure of steam injected into the lower part of a cylinder and falls down with a force by removing the steam. Usually, the hammer is made to fall faster by injecting steam into the upper part of the cylinder. Steam hammers that fall by their own weight are called *steam drop hammers*. Steam hammers vary greatly in weight from 45 kilograms to 90 metric tons.

The steam hammer was invented around 1837 by the Scot James Nasmyth, in Manchester, England and produced in his Patricroft foundry which he built adjacent to the (then new) Liverpool and Manchester Railway and the Bridgewater Canal. From the first, Nasmyth's steam hammer could vary the force of the blow within a very wide range.

Steam hammer be single frame or double frame type. Usually light weight steam hammer are single frame type while heavy weight are double frame.

Fig. 2.7 Steam hammer

Work Principle

Working principal of stam hammer is shown in Fig. 2.8. Initially steam is admitted in the cylinder through valve due to this piston push in downward direction. Steam cut off when piston reach at bottom position. Tup falls over the anvil under gravity.

Advantages of steam hammer

(i) Suitable for heavy blow.
(ii) Very suitable for mass production.
(iii) Shapes achieved in a single blow.

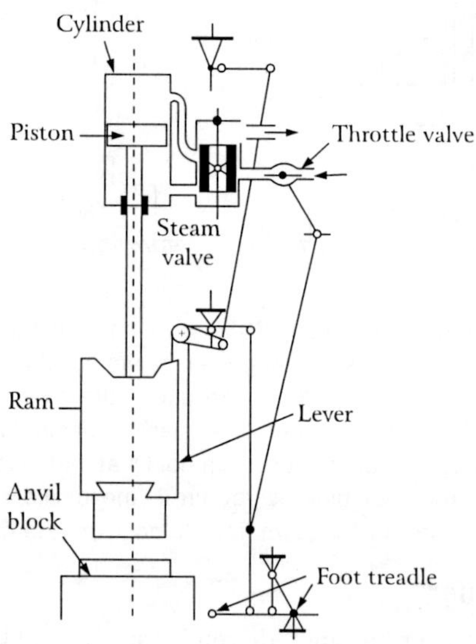

Fig. 2.8 Essentials of steam hammer

2.5.3 Spring Hammer

A spring hammer is a mechanical hammer that is usually powered by the coil and recoil of a spring loaded handle. It can also be powered through the use of *compressed air*. Compressed air is most commonly used in modern mechanical spring hammer designs.

Earlier spring hammers required the user to engage the device with an initial motion. This motion caused the spring mechanism the hammer was attached to at the handle to begin the coil-recoil action. Due to the fact that each one of these actions had an apex point, or point where the force is at its greatest, a table or hammering block was typically placed either just at or after the point of apex in the hammer's radius. This mechanism allowed the user to work more quickly on time-sensitive projects, such as hammering an iron pan while it was still hot enough to be formed.

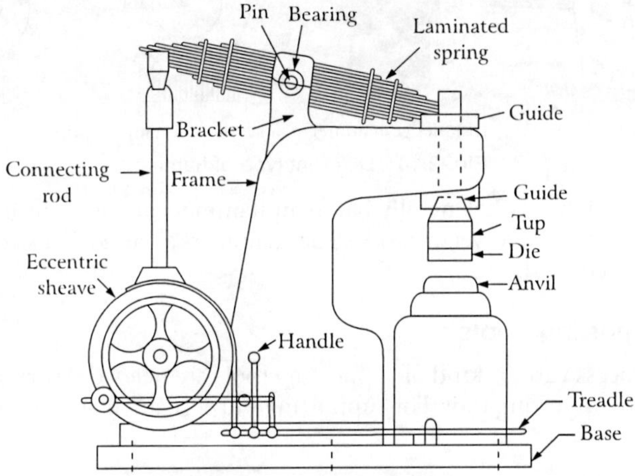

Fig. 2.9 A spring hammer

2.5.4 Hydraulic Hammer

This type of hammer is usually used for heavy and bulky casting purpose. In this ram is operated (raised and downed) by large piston which is operated with a viscous fluid. A large steam piston actuates a small fluid piston which transmit a high pressure to ram piston. Hydraulic hammer capacity vary from 200 to 15000 tones.

Advantages of hydraulic hammer

(i) Suitable for mass production

(ii) Suitable for heavy blow

(iii) Workpiece can be forged in a single stroke.

2.6 FORGING TOOLS

Various type of tools are used during the hand forging operations. They are known as **striking tools, supporting tools** and **holding tools** and they are given below:

2.6.1 Striking Tools

A tool which basically used for applying the load by hand is known as striking tool. In this category different types of hammers are used as per the need. Different types of hammers are **Ball Pein hammer, Sledge hammer, Cross Pein hammer, Double ended hammer, Straight Pein hammer etc.**

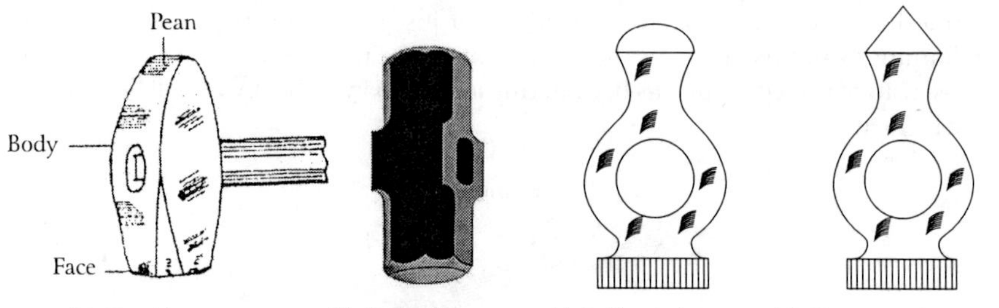

(*a*) Hand hammer (*b*) Sledge hammer (*c*) Ball pein hammer (*d*) Cross pein hammer

Fig. 2.10 Different type of hammers

For light weight work generally ball pein hammers are used during the hand forging where as for heavy weight work sledge hammers are used. For sledge hammer heavy blow is required.

2.6.2 Supporting Tools

In forging process various kind of supporting tools are needed. Workpiece is kept firmly on the supporting tool. For supporting purpose anvil and swage block are used.

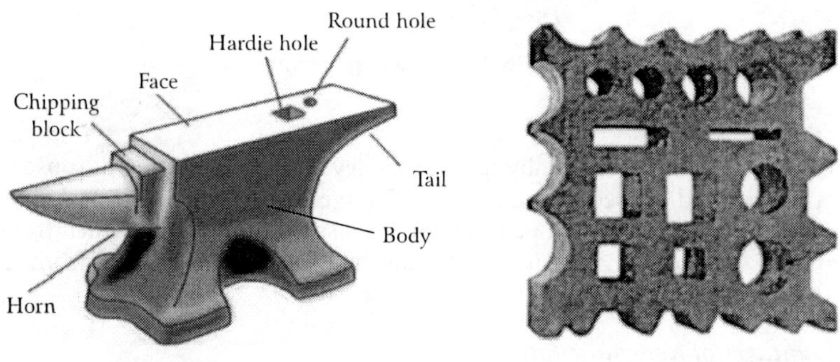

(*a*) Anvil (*b*) Swage block

Fig. 2.11 Supporting tools

(*a*) *Anvil:* Anvil is one of the most important tool used for supporting purpose. Unique feature of anvil is its shape. It consist a horn which is used for converting workpiece into round shape. Anvil also consist two holes of **round** and **rectangular** shape, rectangular shape is known as **hardie hole** whereas round shape is known as **pritchel hole**. Anvil is made out of cast iron, wrought iron and mild steel and it is usually available from 10 kg to 500 kg range.

(*b*) *Swage block:* Swage block is also made by cast iron. It is available in square shape having different grooves like **round shape, rectangular shape, Triangular shapes, half round shapes** etc.

2.6.3 Holding Tools

During forging operation as workpiece is in hot condition, hence the holding of workpiece is required for holding purpose, various types of Tongs are used.

Tongs:

Tong is a device which is used to hold firmly hot workpiece. It may be different types such as **chisel tong, pick up tongs, square hollow tong, close flat tong** etc.

Chiesel tong

Close flat tong

Pick up tong

Pincer tong

Round hollow tong

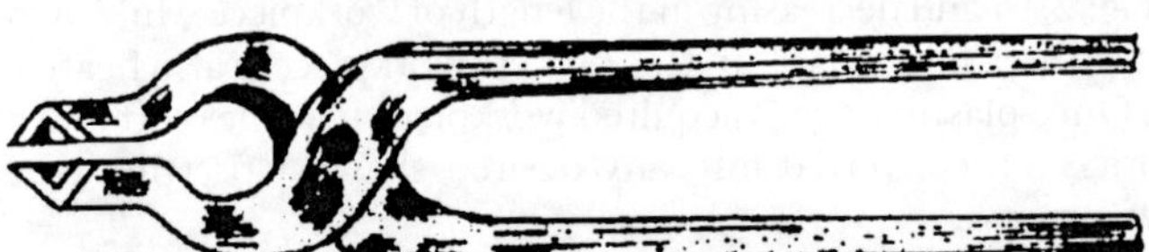

Square hollow tong

Fig. 2.12 Various types of Tongs

2.7 SMITH FORGING OPERATION

Usually six kinds of operations can be performed by Smith forging. Operations that can be performed by forging operations are:

(*a*) Drawing process

(*b*) Upsetting process

(*c*) Punching process

(*d*) Cutting operation

(*e*) Bending.

2.7.1 Drawing Process

Drawing is one of the oldest process performed by forging. In drawing process heated workpiece is kept in between the two die as shown in Fig. 2.13 and hammering is performed on workpiece. In drawing process length of workpiece is increased while cross-section area decreases. Hammering is performed by hand hammer or by any mechanical device. In drawing process a set of fuller is used. In this process workpiece is heated upto plastic stage after that same is kept on the anvil and start hammering (compressive force).

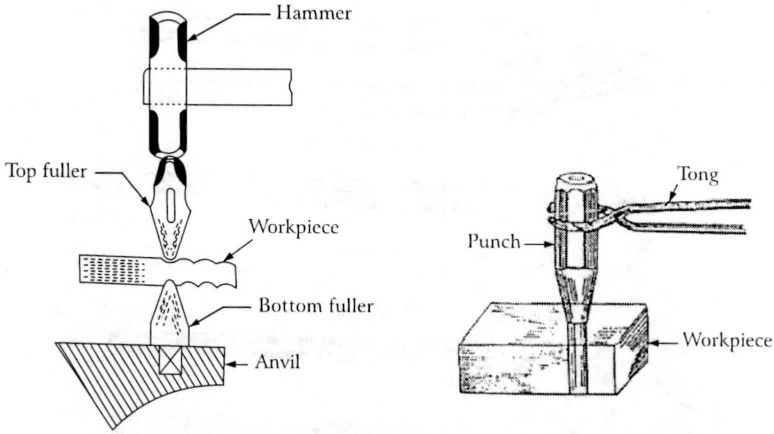

(*a*) Drawing process (*b*) Punching process

Fig. 2.13 Various forging operation

2.7.2 Upsetting Process

In upsetting process hammering (compressive force) is applied on heated workpiece as shown in Fig. 2.14 and decreasing in the length of workpiece while increasing in the cross-section area takes place. In this process workpiece is also heated up to its plastic stage. Once plastic stage is acquired workpiece becomes soft and by a light blow work price can converted into any desired shape. Upsetting is reverse of drawing down process.

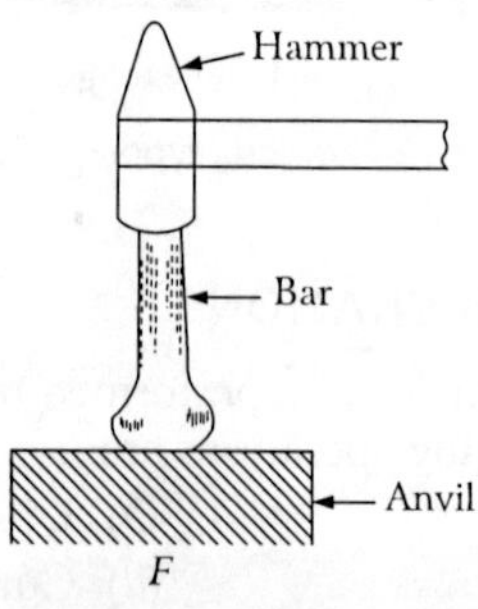

Fig. 2.14 Upsetting process

2.7.3 Bending Process

In bending process heated workpiece is bent at an angle around a straight axis where the natural plane lies by applying compressive force (hammering). It is very common forging operation.

2.7.4 Cutting Process

In cutting process, heated workpiece is cuted with the help of a chisel by applying the compressive force. In this operation chisel is held by tong firmly and compressive force applied due to sharp edges of chisel workpiece cuted into two parts. In cutting process hot or cold chisels can be used. Proper care should requires while cutting operation more power and time is required in cold cutting as compared to hot cutting. Cold chisel is preferred for thin section ($t < 20$ mm). Temperature usually required for hot cutting lies 850°C to 1000°C.

2.8 ANALYSIS OF OPEN DIE FORGING PROCESS

In forging process material deform plastically from one shape into another shape. In forging process flow of metal basically depends upon the following factors:

(i) type of material
(ii) tool geometry
(iii) friction
(iv) thermal conditions
(v) characteristics of the stack material.

Slab method is one of the simplest method to determine the load and stress in forging and other bulk deformation process. In this a slab of infinitesimal thickness is considered and forces are brought in equilibrium. In this method workpiece is compressed/upset or forged in between the two parallel, flat over heating die.

2.8.1 Forging of Strip

Figure 2.15 shows a strip forging by an open die forging method. Workpiece material is not free to flow in the direction perpendicular to the plane of page.

σ_x = Horizontal stress
τ_s = Shear stress
h = Height
B = Width
L = Length
μ = Coefficient of friction

Assumptions

(i) Material of workpiece should be rigid and uniform deformation should be is take place.
(ii) Coefficient of friction is same between the dies and workpiece material.
(iii) Width of block is constant.
(iv) Forging forces reach the maximum value at the end of the process.

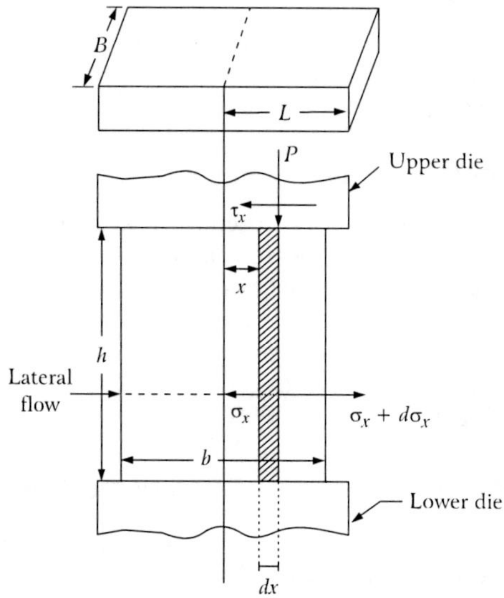

Fig. 2.15 Force balance in an open die forging of a strip

2.8.2 Plain Strain Forging

Figure 2.15 shows the view of open die forging process. Workpiece is not free to flow in the direction perpendicular to the plain of the page as die comes in contact with each other, there is a lateral flow of the workpiece friction, shear stress is produced at die contact surface due to the presence of friction, the horizontal stress σ_x, varies along the length of workpiece. This process is occurring in plastic state hence stresses are not very high.

Equilibrium all forces:

$$(\sigma_x + d\sigma_x)Bh - \sigma_x Bh - 2\tau_s\, dx \times B = 0$$

or
$$\frac{d\sigma_x}{dx} - \frac{2\tau_s}{h} = 0$$

According to friction law of Calumn

$$\tau_s = \mu P$$

where μ = coefficient of friction

$$\frac{d\sigma_x}{dx} - \frac{2\mu P}{h} = 0$$

According to Tresca's theory

$$\sigma_1 - \sigma_3 = \sigma_0 = 2k$$

or
$$\sigma_x + P = \sigma_0 \quad (P \text{ is for compressive})$$

$$\frac{d\sigma_x}{dx} = \frac{-dP}{dx}$$

or $$\frac{-dP}{dx} - \frac{2\mu P}{h} = 0$$

or $$\frac{dP}{dx} = \frac{-2\mu P}{h}$$

or $$\frac{dP}{P} = \frac{-2\mu}{h}dx \quad \text{on integrating both sides, we get}$$

$$\log_e P = \frac{-2\mu}{h}x + c,$$

Now, $x = \frac{b}{2}$ and $\sigma_x = 0$ then $P = 2k$

$$\log_e 2k = \frac{-2\mu}{h}\frac{b}{2} + c$$

or $$c = \log_e 2k + \frac{2\mu}{h}\frac{b}{2} \qquad c = \text{Integration}$$

$$\log_e 2k = \frac{-2\mu}{h}\left(x - \frac{b}{2}\right)$$

or $$P = 2k\, e$$

$$\frac{P}{2k} = \frac{P}{\sigma_y} = e^{\frac{-2\mu}{h}\left(x - \frac{b}{2}\right)}$$

and $$\sigma_x = 2k - P\, 2k\left\{1 - e^{\frac{-2\mu}{h}\left(x - \frac{b}{2}\right)}\right\}$$

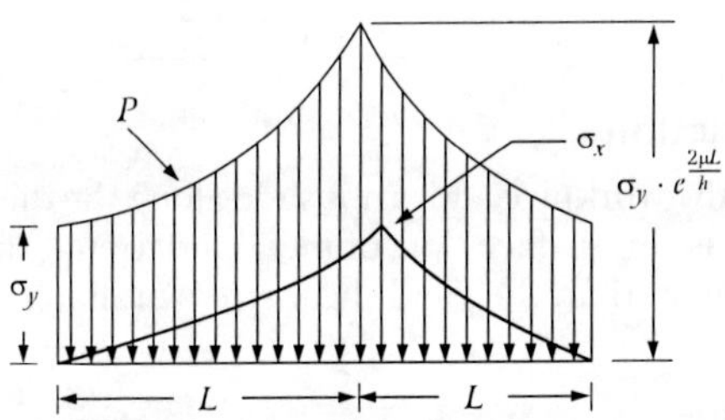

Fig. 2.16 Illustration of the friction hill in plane strain compression for sliding friction

$$\sigma_{x,\,\max} = 2k\left\{1 - e^{\frac{2\mu}{h}\cdot\frac{b}{2}}\right\}$$

Total forging load $$F = 2w\int_0^{b/2} P\,dx$$

Average pressure $$P_a = \frac{F}{2w\left(\frac{b}{2}\right)} = \int_0^{b/2} \frac{P\,dx}{b/2}$$

$$= \frac{\sigma_y}{b/2} \cdot \int_0^{b/2} e^{-\frac{2\mu}{h}\left(x - \frac{b}{2}\right)} \cdot dx$$

$$= \frac{\sigma_y}{b/2} \cdot \frac{h}{2\mu} \left| -e^{\frac{-2\mu}{h}\left(x - \frac{b}{2}\right)} \right|_0^{b/2}$$

$$P_a = \frac{\sigma_y h}{\mu b}\left[-1 + e^{\frac{\mu b}{h}}\right]$$

$$P_a = \frac{\sigma_y h}{\mu b}\left[-1 + 1 + \frac{\mu b}{h} + \frac{1}{2}\left(\frac{\mu b}{h}\right)^2 - \ldots\right]$$

$$= \frac{\sigma_y h}{\mu b}\left[\frac{\mu b}{h} + \frac{1}{2}\left(\frac{\mu b}{h}\right)^2 - \ldots\right]$$

$$= \frac{\sigma_y h}{\mu b} \times \frac{\mu b}{h}\left[1 + \frac{1}{2}\left(\frac{\mu b}{h}\right) - \ldots\right]$$

$$= \sigma_y\left[1 + \frac{1}{2}\frac{\mu b}{h}\right]$$

$$\boxed{P_a = \sigma_y\left\{1 + \frac{1}{2}\frac{\mu b}{h}\right\}}$$

2.8.3 Sticking Friction

For sticking friction the workpiece does not slide along the die face and becomes a part of die and there is sub surface flow of metal. Frictional shear stress at the die interface cannot be more than the yield shear stress of material. Then

According Tresca $\quad \tau_x = \tau_y = \frac{\sigma_y}{2} = k$

$$hd\sigma_x - 2kd\,x = 0$$

as $\quad \sigma_x + P = 2k$

or $\quad d\,\sigma_x = -dP$

$$-\,hdP - 2kdx = 0$$

$$\frac{dP}{2k} = -\frac{dx}{h} \quad \text{on integrating both sides, we get}$$

$$\frac{P}{2k} = \frac{-x}{h} + C$$

where, $\quad C$ = Integrating constant

Now at $x = b/2$ and $P = 2k$ (as $\sigma_x = 0$)

$$C = 1 + \frac{b}{2h} \text{ put the value of } C \text{ in above equation}$$

$$\frac{P}{2k} = \frac{-x}{h} + 1 + \frac{b}{2h}$$

$$\frac{P}{2k} = 1 - \frac{1}{h}\left(x - \frac{b}{2}\right)$$

$$\left(\frac{P}{2k}\right)_{\max} = \left(1 + \frac{b}{2h}\right) \quad \text{(at centre } x = 0\text{)}$$

$$\text{Average pressure } P_a = 2k\left(1 + \frac{b}{4h}\right)$$

2.8.4 Disc Forging or Axi-symmetric Forging

Figure 2.17 shows a circular disc having thickness h and radius R. r, θ, z are the cylinderical co-ordinates of system. Considering a very thin area of dr on which the stresses are acting during forging. Taking the equilibrium of forces in the radial direction.

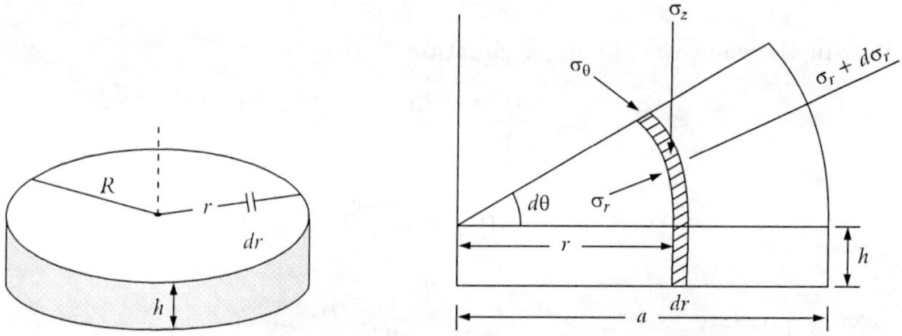

Fig. 2.17 Axi-symmetric forging

$$\sigma_r h d\theta - (\sigma_r + d\sigma_r)\, h \times (r + dr) d\theta + 2\sigma_0 h dr \sin\frac{d\theta}{2} - 2\pi r d\theta dr = 0$$

$$\text{If angle is very small, then } \sin\frac{d\theta}{2} = \frac{d\theta}{2}$$

$$\sigma_r h dr + d\sigma_r \cdot rh - \sigma_\theta h dr - 2\tau r dr = 0$$

As disc is axi-symmetric

$$d\varepsilon_0 = d\varepsilon_r$$

so,

$$\sigma_\theta = \sigma_r$$

$$\frac{d\,\sigma_r}{dr} = \frac{2\tau}{h} = 0$$

According to Von-mises yield criteria

$$\sigma_2 = \sigma_\theta = \sigma_r$$

and $\sigma_3 = -p$ [–ve sign for compressive load]

$$(\sigma_1 - \sigma_2)^2 + (\sigma_2 - \sigma_3)^2 + (\sigma_3 - \sigma_1)^2 = 6k^2$$

$$(\sigma_r - \sigma_\theta)^2 + (\sigma_\theta - P)^2 + (-P - \sigma_r)^2 = 6k^2$$

as $\sigma_r = \sigma_\theta$

$$\sigma_r + P = \sqrt{3}k \quad \text{on differentiation}$$

$$d\sigma_r = -dP$$

$$2\mu P dr + h dP = 0$$

$$\frac{dP}{P} = \frac{-2\mu}{h} dr \quad \text{on Integrating both sides}$$

$$\log_e P = \frac{-2\mu}{h} r + C$$

as $\sigma_r = 0$ and $r = R$

then $P = \sqrt{3}k$

$$C = \log \sqrt{3}k + \frac{2\mu R}{h}$$

Putting the value of C in above equation

$$\log_e \frac{P}{\sqrt{3}k} = \frac{2\mu}{h}(R - r)$$

or
$$P = \sqrt{3}k\, e^{\frac{2\mu}{h}(R-r)}$$

Average pressure
$$P_a = \int_0^R \frac{1}{\pi R^2} \times P \times 2\pi r dr$$

$$= \int_0^R \sqrt{3}k\, e^{\frac{2\mu}{h}(R-r)} \cdot \frac{r}{R^2} dr$$

$$P_a = \frac{1}{2}\left(\frac{h}{\mu R}\right)^2 \sqrt{3}k \left[e^{\frac{2\mu R}{h}} - \frac{2\mu R}{h} - 1 \right]$$

or
$$P_a = \sqrt{3}k \left[1 + \frac{2\mu R}{3h} \right]$$

(i) Sticking friction

$$\tau_x = k$$

$$2kdr - hd\sigma_r = 0$$

$$\sigma_r + P = \sqrt{3}k \quad \text{on differentiation, we get}$$

or
$$d\sigma_r = -dP$$

$$hdP + 2kdr = 0$$

On integration both sides, we get

$$P = \frac{-2k}{h} r + C$$

as $\sigma_r = 0$

when $r = R$

then $P = \sqrt{3}k$

$$\sqrt{3}k = \frac{-2k}{h} r + C$$

or $$C = \sqrt{3}k + \frac{2k}{h} R$$

Putting value of C in above equation, we get

$$P = \sqrt{3}k + \frac{2k}{h} R - \frac{2k}{h} r$$

$$P = \sqrt{3}k + \frac{2k}{h}(R - r)$$

$$\boxed{P_a = \sqrt{3}k + \frac{2kR}{3h}}$$

(ii) Mixed friction condition:

Let r_s = Sticking radius

then $$\tau_x = \mu p = k$$

$$P = \sqrt{3}k\, e^{\frac{2u}{h}(R-r)}$$

$$k = \mu\sqrt{3}\, e^{\frac{2\mu}{h}(R-r_s)}$$

$$e^{\frac{2\mu}{h}(R-r_s)} = \frac{1}{\sqrt{3}\mu} \quad \text{Taking log both sides}$$

or $$r_s = R - \frac{1}{2\mu}\log_e \frac{1}{\sqrt{3}\mu}$$

Pressure at sticking radius is calculated by

$$\left(\frac{P}{\sqrt{3}k}\right)_{r = r_s} = e^{\frac{2\mu}{h}(R-r_s)}$$

Putting $r_s = R - \frac{h}{2\mu}\log_e \frac{1}{\sqrt{3}\mu}$ in above equation, we get

$$\left(\frac{P}{\sqrt{3}k}\right)_{r = r_s} = \frac{1}{\sqrt{3}\mu}$$

$$P = \frac{k}{\mu} = \frac{2k}{h} r_s + C$$

$$= \frac{-2k}{h}\left(R - \frac{h}{2\mu}\log_e \frac{1}{\sqrt{3}\mu}\right) + C$$

or $$C = k\left[\frac{2R}{h} + \frac{1}{\mu}(1 + \log_e \sqrt{3}\mu)\right]$$

Putting this value in above equation

$$P = \frac{2k}{h}(R - r) + \frac{k}{\mu}(1 + \log_e \sqrt{3}\mu)$$

for condition $0 \le r \le r_s$

Example 2.1

A metal component 25 mm × 25 mm × 150 mm long having a yield stress of 7 MPa in the tension, is to pressed between flat dies to a size 6 mm × 100 mm × 150 mm. If μ = 0.2 then calculate the maximum forging load. [*UPTU-2008-09*]

Solution: Assume that it is a case of mixed.

Condition sticking radius x_s is

$$x_s = \frac{b}{2} - \frac{h}{2\mu}\log_e \frac{1}{2\mu}$$

$$= \frac{100}{2} - \frac{6}{2 \times 0.2}\log_e \frac{1}{2 \times 0.2} = 36.26 \text{ mm.}$$

$0 \le x \le 36.26$ and $36.26 \le x \le 50$.

Sliding friction condition exist the above condition yield stress (shear)

$$k = \frac{\sigma_y}{\sqrt{3}} = \frac{7.0}{\sqrt{3}} = 4.04 \text{ MPa}$$

Pressure in the sticking region is governed by equation

$$\frac{P}{2k} = \left[\frac{\left(\frac{b}{2} - x\right)}{h} + \frac{1}{2\mu}\left(1 - \log_e \frac{1}{2\mu}\right)\right]$$

$$P = 2 \times 4.04\left[\frac{\frac{100}{2} - x}{6} + \frac{1}{2 \times 0.2}\left[1 - \log_e \frac{1}{2 \times 0.2}\right]\right]$$

$$P_{\text{sticking}} = 69 - 1.35x \text{ MPa}$$

In sliding region $$\frac{P}{2k} = e^{\frac{-2H}{h}\left(x - \frac{h}{2}\right)}$$

$$P = 2 \times 4.04 \times e^{\frac{2 \times 0.2}{6}\left(x - \frac{100}{2}\right)}$$

$$P_{sliding} = 226.49\, e^{-0.67x}$$

$$F = 2\left[\int_0^{36.26} P_{sticking} + dx \int_{36.26}^{50} P_{sliding} \cdot dx\right]$$

$$F = 2\left[\int_0^{36.26} (69.1.35x)\, dx + \int_{36.26}^{50} 226 \cdot 49\, e^{-0.067x} \cdot dx\right]$$

$$F = 3252.9 \text{ N/mm.}$$

$$\text{Total forging load} = 3252.9 \times 150 = 487935 \text{ N} \qquad \textbf{Ans.}$$

Example 2.2

A block perfectly made by plastic material having yield strength 200 MPa in simple tension with other dimension 200 mm × 100 mm × 125 mm ($b \times h \times w$) at one instant during plane strain forging. Assuming that it is a sticking forging case determine peak pressure.

Solution: Given

$$\sigma_y = 200 \text{ MPa},$$

$$b = 200 \text{ mm}, h = 100 \text{ mm}$$

$$\text{Yield Stress (Shear) } k = \frac{\sigma_y}{\sqrt{3}} = \frac{200}{\sqrt{3}} = 116.5 \text{ MPa.}$$

$$\text{Peak pressure } \left(\frac{P}{2K}\right)_{max} = 1 + \frac{b}{2h},$$

$$P = \left(1 + \frac{200}{2 \times 100}\right) \times 2 \times 116.5$$

$$= 2 \times 2 \times 116.5 = 466 \text{ MPa} \quad \textbf{Ans.}$$

Example 2.3

Determine the forging load at the start and complection of hot forging of a steel billet for the following data:

Length of billet = 2 m, width of billet = 0.9 m and thickness = 0.1 m, tool bite = 0.5 m; σ_y = 60 MPa at start and σ_y = 180 MPa at ending of forging, reduction in forging = 50%, assume it is a sticking case of forging.

Solution:

$$P_a = 2k\left(1 + \frac{L}{2h}\right)$$

or

$$P_a = \sigma_y\left(1 + \frac{L}{2h}\right)$$

Now bite of tool, $b = 2L$

∴

$$L = \frac{b}{2}$$

$$P_a = \sigma_y\left(1 + \frac{b}{4h}\right), \quad \text{forging area} = B \times 2 \times l = B \times b$$

$$\text{Forge load} = \sigma_y \times b \times B\left(1 + \frac{b}{4h}\right)$$

(i) at starting $\sigma_y = 60$, $B = 0.9$ m, $h = 0.1$ m, $b = 0.3$

$$\text{forging load} = 60 \times 0.9 \times 0.3\left(1 + \frac{0.3}{4 \times 0.1}\right) = 28.35\ \text{MN} \quad \textbf{Ans.}$$

(ii) At ending $\sigma_y = 180$ MPa

OBJECTIVE QUESTIONS

1. Welded tubes and pipes are produced by
   (*a*) Hot working (*b*) Cold working
   (*c*) Roll Piercing (*d*) None of these
2. Large articles are generally Hot forged by
   (*a*) Drop forging (*b*) Machine forging
   (*c*) Hand forging (*d*) None of these
3. High carbon steels and alloys are forged at
   (*a*) 750-1300°C (*b*) 750-1250°C
   (*c*) 800-1150°C (*d*) None of these
4. Machine forging/Hot heating enables
   (*a*) Cheaper maintenance (*b*) Better dimensional control
   (*c*) Higher Productivity (*d*) All of the above
5. A forge weld is made by
   (*a*) Gas welding (*b*) Electric welding
   (*c*) Forge welding (*d*) None of these
6. Following are forgeable materials
   (*a*) Brass, Bronze, Copper (*b*) Wrought Iron
   (*c*) Mild steel and high carbon steel (*d*) All of these
7. Main advantage of Hot forging are
   (*a*) Saving in time, Labour and material (*b*) Refinement of metal structure
   (*c*) Metal becomes stronger (*d*) All of these
8. Following type of fuel is used for forging furnace
   (*a*) Gaseous fuel (*b*) Liquid fuel
   (*c*) Solid fuel (*d*) All of the above
9. Spring hammer, drop hammer, pneumatic hammer and steam hammer are
   (*a*) Forging press (*b*) Forging machine
   (*c*) Power hammer (*d*) All of the above
10. Common forging defects are
    (*a*) Cold shuts and cracks in metal (*b*) Incomplete component
    (*c*) Rat tails and swab (*d*) All of the above
11. Following type of furnace are used for forging purpose
    (*a*) Box type furnace (*b*) Open hearth furnace
    (*c*) Smith forge (*d*) All of the above
12. As a result of cold working
    (*a*) Residual stresses are setup, needing subsequent, heat treatment
    (*b*) Strength and hardness of a metal are increased

(*c*) Better surface finish and dimensional control are achieved
(*d*) All of the above

13. Extrusion process enables to produce
(*a*) Hollow object of uniform sections, like seamless tube
(*b*) Solid object of uniform section, like bars, rod etc.
(*c*) Plain object of uniform thickness like sheet
(*d*) All of the above.

14. Hot extrusion
(*a*) Normally uses a hot billet as its raw material
(*b*) Performed either direct or indirect
(*c*) Can be suit best for non-ferrous metals and alloys
(*d*) All of these

15. Deep drawing operation is carried out in a
(*a*) Compound die (*b*) Progressive die
(*c*) Simple die (*d*) None of these

16. Redrawing an already drawn cup is carried out to
(*a*) reduce its diameter (*b*) increase its depth
(*c*) increases its wall thickness (*d*) none of these

17. Expanding the walls of a drawn cup into different shapes is known as
(*a*) Deep drawing (*b*) Redrawing
(*c*) Bulging (*d*) All of these

18. Match the following criteria of material failure, under biaxial stresses σ_1 and σ_2 and yield stress σ_y, with their corresponding graphic representations: [*GATE 2011*]

P. Maximum-normal stress criterion *L.*

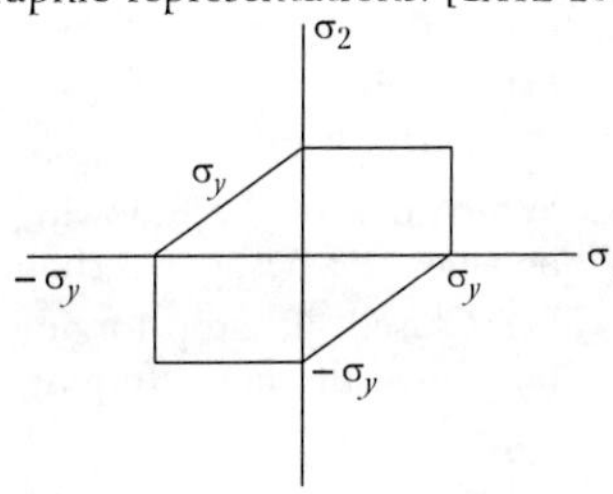

Q. Maximum-distortion energy criterion *M.*

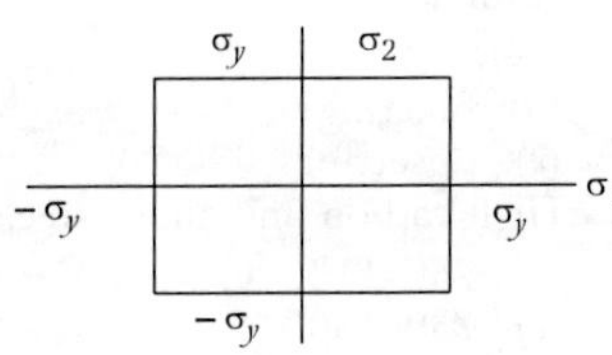

R. Maximum-shear-stress criterion *N.*

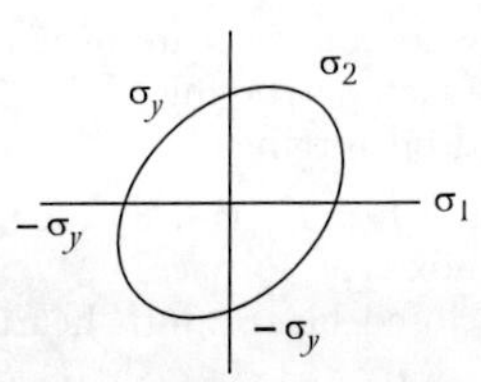

(*a*) *P–M, Q–L, R–N* (*b*) *P–M, Q–N, R–N*
(*c*) *P–N, Q–L, R–M* (*d*) *P–N, Q–M, R–L*

19. Which one of the following is an advantage of forging
(*a*) improved physical property (*b*) good surface finish
(*c*) low tooling cost (*d*) close tolerance [*IES 1996*]

20. Consider the following statements:
1. Forging reduces the grain size of the metal which result in a decrease in strength and toughness.
2. Forged component can be provided with thin section without reducing the strength.

Which of the statements given above is/are correct?
(*a*) 1 only (*b*) 2 only
(*c*) Both 1 and 2 (*d*) Neither 1 nor 2 [*IES 2005*]

21. Match List-I (type of forging) with List-II (operation) and select the correct answer using the code given below the lists. [*IES 2005*]

List-I (Type of forging)	*List-II (Operation)*
A. Drop forging	1. Metal is gripped in the dies and pressure is applied on the heated end.
B. Press forging	2. Squeezing action.
C. Upset forging	3. Metal is placed between rollers and pushed.
D. Roll forging	4. Repeated hammer blows.

Code	A	B	C	D
(*a*)	3	1	2	4
(*b*)	4	1	2	3
(*c*)	4	2	1	3
(*d*)	3	2	1	4

22. Process of removing the burns or flash from a forged component in drop forging is called
(*a*) swaging (*b*) fettling
(*c*) perforating (*d*) trimming [*IES 2005*]

23. Which one of the following processes is most commonly used for the forging of bolt head of hexagonal shape?
(*a*) Closed die drop forging (*b*) Open die progressive forging
(*c*) Open die upset forging (*d*) Closed die press forging [*IES 1998*]

24. A forging method for reducing the diameter of a bar and in the process making it longer is termed as
(*a*) fullering (*b*) punching
(*c*) upsetting (*d*) extruding [*IES 2003*]

25. High carbon and alloys steels are forged at
(*a*) 750–1150 °C (*b*) 800–1150 °C
(*c*) 750–1300 °C (*d*) none of these

26. Large components are usually hot formed by
(*a*) machine forging (*b*) hand forging
(*c*) drop forging (*c*) all of these

27. Type of furnace is used for forging operation
(*a*) box type furnace (*b*) smith forge
(*c*) closed black smith hearth (*d*) all of these

28. A good furnace fuel should
(*a*) burn completely
(*b*) should not be deterimental to furnace efficiently

(*c*) produce 1200°C–1350°C temperature
(*d*) should provide clean fire
(*e*) all of these

29. Unique advantage of hot forging are
(*a*) refinement of metal structure
(*b*) metal becomes stronger
(*c*) saving in time, material and time
(*d*) all of these

30. Which type of fuel is used for forging furnace
(*a*) gaseous fuel
(*b*) liquid fuel
(*c*) solid fuel (coal)
(*d*) all of these

31. Which are forgeable materials
(*a*) mild steel and high carbon steel
(*b*) brass, copper and bronze
(*c*) wrought iron
(*d*) all of these

ANSWERS

1. (*a*)	2. (*a*)	3. (*c*)	4. (*d*)	5. (*c*)
6. (*d*)	7. (*d*)	8. (*d*)	9. (*c*)	10. (*d*)
11. (*d*)	12. (*d*)	13. (*c*)	14. (*d*)	15. (*b*)
16. (*a* & *b*)	17. (*c*)	18. (*c*)	19. (*a*)	20. (*b*)
21. (*c*)	22. (*d*)	23. (*d*)	24. (*a*)	25. (*b*)
26. (*c*)	27. (*c*)	28. (*e*)	29. (*d*)	30. (*d*)
31. (*d*)				

REVIEW QUESTIONS

1. Derive a relation to determine the forging forces for forging of rectangular strip with sticking friction condition. [*UPTU 2006-07*]
2. A thin strip of section $b \times h$ is being cold forged between two parallel over hanging plates, show that the maximum forging pressure is given by:

$$\left(\frac{P}{k}\right)_{max} = \left(1 + \frac{\mu b}{h}\right)$$

3. Derive the equation for the pressure distribution for the forging of rectangular block ($b \times h \times w$) in case of sliding friction.
4. Enlist the various hot and cold working processes. Compare the advantage and disadvantage of hot and cold working processes. [*UPTU 2008-09*]
5. Determine the maximum forging of a metallic compound 25 mm × 25 mm × 150 mm. Yield stress in simple tension is 7 MPa. Component is pressed between flat dies to a size 6 mm × 100 mm × 150 mm. Coefficient of friction μ = 0.20. Consider the mixed friction case. [*UPTU 2008-09*]
6. Compare the forging process with other metal forming process.
7. Define the forging process in detail.
8. Classify the forging process in detail.
9. Define the forging advantages and limitations.
10. Define the various forging defects.

11. Explain the various forging tools with sketch.

12. A metal workpiece has size 200 mm × 100 mm × 150 mm ($b \times h \times w$). Assuming plane strain forging under sticking friction condition, calculate the peak pressure. The material has a yield stress in uniaxial tension of 160 MPa. *[UPTU 2010-11]*

13. Derive the following expression for pressure distribution in forging of a rectangular block of size $b \times h \times w$ under "sticking friction" condition. $\frac{P}{2K} = 1 - \frac{1}{h}\left(x - \frac{b}{2}\right)$, where P is pressure at distance x from centre, K is the shear yield stress of the material of the block, h is height of the block and b is the breadth in the xy plane.

[UPTU 2002-03, 04-05, 05-06]

14. Find the following expression for pressure distribution for sliding friction case for forging of a rectangular block ($b \times h \times w$) and show the variation.

$\frac{P}{2K} = e^{-\frac{2\mu}{h}\left(x - \frac{b}{2}\right)}$ where P is pressure at a distance x from centre. K is shear strength of material and μ is coefficient of sliding friction, h is height and b is width.

[UPTU 2001-02]

❑❑❑

UNIT-II

METAL FORMING PROCESS

PART-II

Chapter 3 : Drawing and Extrusion
Chapter 4 : Rolling

3 CHAPTER

DRAWING AND EXTRUSION

3.1 INTRODUCTION

Drawing is a cold forming process. Basically drawing process consist a die and punch and it deals to produce the shapes having some depth or in other words we can say drawing is a process to produce a container. In drawing process blank convert into final shape due to plastic deformation of metal by applying the compressive forces. Clearance is provided between the die and punch for easy motion of drawn parts. A flat precut metal blank is converted into a hollow vessel without excessive wrinkling, thinning, or fracturing. The various forms produced may be cylindrical or box shaped, with straight or tapered sides or a combination of straight, tapered, and curved sides. The parts may vary from 1/4″ (6 mm) diameter parts or smaller to aircraft or automotive parts large enough to require mechanical handling equipment.

Drawing process is applied to

(i) Ferrous metal and their alloys such as low carbon, high carbon steel.

(ii) Non-ferrous metal and their alloys.

Extrusion is a plastic deformation process in which material is forced under pressure to flow through one or more die orifices to produce products of the desired configuration. This process provides a practical forming method for producing a limitless variety of parallel-surfaced shapes to meet almost any design requirement. Other advantages include improving the microstructure and physical properties of the material, maintaining close tolerances, material conservation, economical production, and increased design flexibility. This process widely used in fabricating non-ferrous metals and their alloys. Now-ever steel and other ferrous alloys can be processed too.

3.2 WIRE DRAWING

Usually wire drawing process is carried out at cold working condition. Wire is produced by cold drawing process. In wire drawing process, wire is pulled through the dies as shown in Fig. 3.1. Dies are always made by hardened as compared to wire material. Usually they are made out of diamond or carbide materials. In wire drawing length increases while there is reduction in cross-sectional area. In wire drawing reduction in each pass through a die range from 10% for steel and 40% for ductile material (like mild steel etc.). If greater reduction in cross-section is required then wire is passed at a time through several dies.

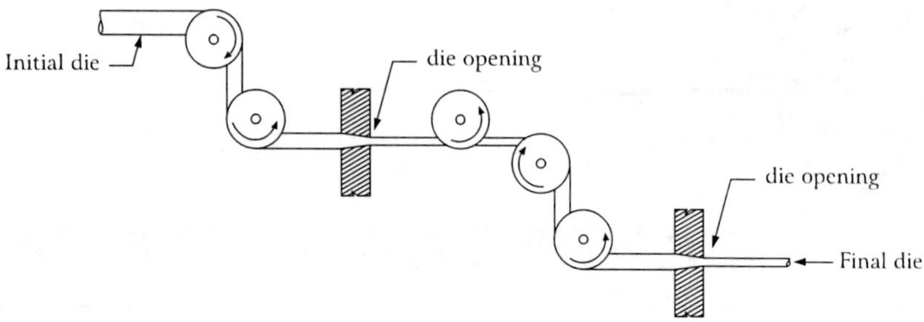

(a)

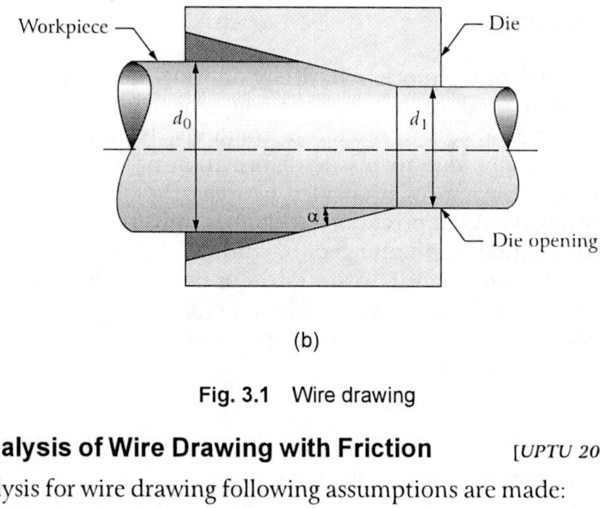

(b)

Fig. 3.1 Wire drawing

3.2.1 Analysis of Wire Drawing with Friction [*UPTU 2003-04, 2008-09*]

Before analysis for wire drawing following assumptions are made:

(i) Material should be uniform.

(ii) Columb's law of friction is assumed to exist within the contact zone.

(iii) Uniform stress should be produced through out all points in a plane normal to the die axis.

(iv) Coefficient of friction should be constant for die and wire material.

(v) Principal stresses are σ_x and $-p$.

(vi) Coefficient of friction μ and semi-cone angle (α) are small.

Figure 3.2 shows a wire drawing operation. In which a wire is passing through a conical die where dia of wire is reducing. Assuming an element of dx length on which the stresses are acting:

(i) due to change in longitudinal stress

$$(\sigma_x + d\sigma_x) \times (D + dD)^2 \times \frac{\pi}{4} - \sigma_x \cdot \frac{\pi D^2}{4}$$

(ii) due to directly die pressure on the circumference $P\left(\pi D \times \frac{dx}{\cos\alpha}\right) \times \sin\alpha$.

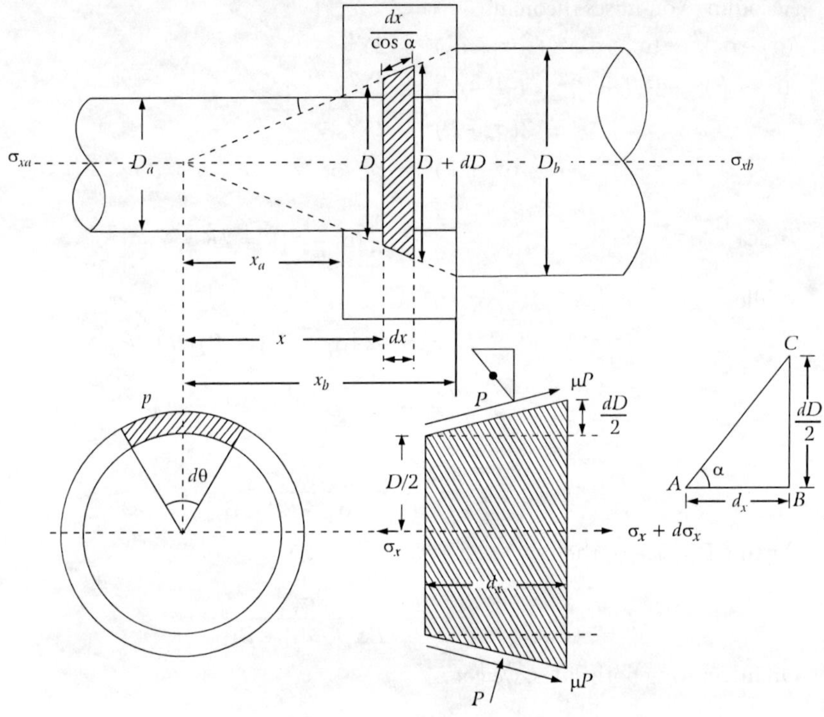

Fig. 3.2 Wire drawing with friction

(iii) due to frictional drag at the circumference $\mu P\left(\pi D \times \frac{dx}{\cos\alpha}\right) \times \cos\alpha$.

For equilibrium condition

$$(\sigma_x + d\sigma_x)(D + dD)^2 \times \frac{\pi}{4} - \sigma_x \cdot \frac{\pi D^2}{4} + P\pi \frac{Ddx}{\cos\alpha} \sin\alpha + \mu P\pi D \frac{dx}{\cos\alpha} \cdot \cos\alpha = 0$$

$$\frac{\sigma_x \cdot D \cdot dD}{2} + \frac{D^2 d\sigma_x}{4} + PDdx \tan\alpha + \mu P\pi Ddx = 0$$

Neglecting the $d\sigma_x .dD$ term as second order.

Since $dD = 2x \tan\alpha$

$$2\sigma_x + D\, d\sigma_x + 2PdD + 2\mu PdD \cot\alpha = 0$$

$$Dd\sigma_x + 2[\sigma_x + P(1 + \mu \cot\alpha)]\, dD = 0$$

Putting $B = \mu \cot\alpha$, we have

$$Dd\sigma_x + 2[\sigma_x + P(1 + B)]\, dD = 0$$

Cylindrical state of stress is assumed

$$\sigma_1 = \sigma_x \;\&\; \sigma_2 = \sigma_3 = \sigma_r = -p$$

According Von-mises theorem

$$(\sigma_1 - \sigma_2)^2 + (\sigma_2 - \sigma_3)^2 + (\sigma_3 - \sigma_1)^2 = 2Y^2$$

$$(\sigma_x + P)^2 + (-P + P)^2 + (-P - \sigma_x)^2 = 2Y^2$$

$$2(\sigma_x + P)^2 = 2Y^2$$

$$(\sigma_x + P)^2 = Y^2 \quad \text{or} \quad \sigma_x + P = Y$$

Tresca $$= \frac{1}{2}(\sigma_1 - \sigma_3)^2 = 2K = Y$$

While $$\sigma_x + P = 2K$$

$$Dd\sigma_x = -2\,[\sigma_x + P(1 + B)]dD$$

$$\frac{D}{dD} = \frac{2[-\sigma_x - (1 + B)]}{d\sigma_x}$$

$$\frac{dD}{D} = \frac{d\sigma_x}{2[-\sigma_x - P(1 - B)]}$$

Putting $P = Y - \sigma_x$ then

$$2\frac{dD}{D} = \frac{d\sigma_x}{B\sigma_x - Y(1 + B)}$$

On integrating both sides, we get

$$2\int\frac{dD}{D} = \int\frac{d\sigma_x}{B\sigma_x - Y(1 + B)}$$

$$2(\log_e D + \log_e C) = \frac{1}{B}\log_e[B\sigma_x - Y(1 + B)]$$

$$2\log_e DC = \frac{1}{B}\log_e[B\sigma_x - Y(1 + B)]$$

$$2B\log_e DC = \log_e[B\sigma_x - Y(1 + B)]$$

$$\log_e (DC)^{2B} = \log_e[B\sigma_x - Y(1 + B)]$$

$$B\sigma_x - Y(1 + B) = (DC)^{2B}$$

where C = integration constant

$$\sigma_x = \frac{DC_1^{2B}}{B} + \frac{Y(1 + B)}{B}$$

$$\sigma_x = \frac{DC_1^{2B}}{B} + \frac{Y(1 + B)}{B} \qquad ...(1)$$

where $C_1 = C^{2B}$ = new constant

for back pull stress $\sigma_x = \sigma_{xb}$ and $D = D_b$

$$\sigma_{xb} = \frac{C_1 D_b^{2B}}{B} + \frac{(1 + B)Y}{B}$$

and $$C_1 = \left[\sigma_{xb} - \frac{Y(1+B)}{B}\right]\frac{B}{D_b^{2B}}$$

Substituting the value of C_1 of Equation (1), we get

$$\frac{\sigma_x}{Y} = \frac{(1+B)}{B}\left[1-\left(\frac{D}{D_b}\right)^{2B}\right]+\frac{\sigma_{xb}}{Y}\left(\frac{D}{D_b}\right)^{2B}$$

Putting $D = D_a$ and $\sigma_x = \sigma_{xa}$ at Die exit

$$\frac{\sigma_{xa}}{Y} = \frac{1+B}{B}\left[1-\left(\frac{D_a}{D_b}\right)^{2B}\right]+\frac{\sigma x_b}{Y}\left[\frac{D_a}{D_b}\right]^{2B}$$

If back full pressure is zero then put $\sigma_{xb} = 0$

$$\frac{\sigma_{xa}}{Y} = \frac{1+B}{B}\left[1-\left(\frac{D_a}{D_b}\right)^{2B}\right]$$

3.2.2 Maximum Reduction for the Case of Wire Drawings

Drawings stress σ_{xa} cannot be more than the tensile yield stress of wire material hence

$$\frac{\sigma_{xa}}{\sigma_y} = 1$$

where, $\sigma_y = 2k$ with back pull zero $\sigma_{xb} = 0$

$$\frac{1+B}{B}\left[1-\left(\frac{D_a}{D_b}\right)^{2B}\right] = 1$$

$$1-\left(\frac{D_a}{D_b}\right)^{2B} = \frac{B}{(1+B)}$$

$$\left(\frac{D_a}{D_b}\right)^{2B} = 1-\frac{B}{1+B}$$

$$= \frac{1+B-B}{1+B} = \frac{1}{1+B}$$

$$\left(\frac{D_a}{D_b}\right)^{2} = \left(\frac{1}{1+B}\right)^{1/B}$$

Reduction, $$R = 1-\left(\frac{D_a}{D_b}\right)^{2} = 1-\left(\frac{1}{1+B}\right)^{1/B}$$

$$R = 1-\left(\frac{1}{1+B}\right)^{1/B}$$

3.3 ANALYSIS OF STRIP DRAWING WITH FRICTION

3.3.1 Assumptions

(i) Uniform stresses are produced at all point of a plane normal to the x-axis.
(ii) Material should deform uniformly and plastically.
(iii) No change in the width of strip.
(iv) Coefficient of friction is same for die and strip material.

Following variables play an important role in drawing operation:

(i) Workpiece material
(ii) Die material
(iii) Lubrication
(iv) Die angle
(v) Speed of drawing

Forces

(*a*) due to change in longitudinal stress

$$(\sigma_x + d\sigma_x) \cdot (h + dh) - \sigma_x dh$$

(*b*) due to die pressure on strip $= 2P \sin\alpha \times \dfrac{dx}{\cos\alpha}$

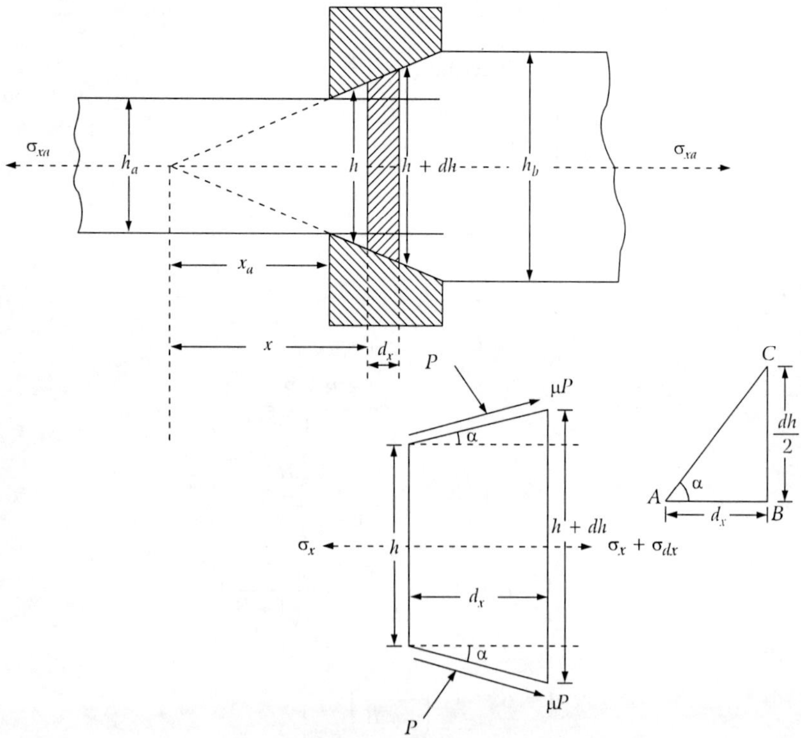

Fig. 3.3 Strip drawing with friction

(c) due to frictional drag $= 2P\cos\alpha \times \dfrac{dx}{\cos\alpha}$

For equilibrium condition

$$(\sigma_x + d\sigma_x)(h + dh) - \sigma_x dh + 2P\sin\alpha \times \frac{dx}{\cos\alpha} + 2P\cos\alpha \frac{dx}{\cos\alpha} = 0$$

$$\sigma_x dh + hd\sigma_x + Pdh + \mu p \frac{dh}{\tan\alpha} = 0$$

Neglecting second order term, i.e., $dh \times d\sigma_x$

From Fig. 3.3, $\tan\alpha = \dfrac{\mu}{B}$

$$B = \frac{\mu}{\tan\alpha}$$

$$\sigma_x dh + hd\sigma_x + Pd\,h + PBdh = 0$$

$$\sigma_x dh + hd\sigma_x + P\,(1 + B)\,dh = 0$$

$$hd\sigma_x + [\sigma_x + P\,(1 + B)]\,dh = 0$$

or

$$hd\sigma_x = [\,\sigma_x + (1 + B)]\,dh$$

$$\frac{dh}{h} = \frac{d\sigma_x}{[\sigma_x + P(1 + B)]} \qquad ...(1)$$

According to Von-mises theorem

$$(\sigma_1 - \sigma_2)^2 + (\sigma_2 - \sigma_3)^2 + (\sigma_3 - \sigma_1)^2 = 2Y^2$$

Put $\sigma_1 = \sigma_x$ and $\sigma_2 = \sigma_3 = -P$

$$(p + \sigma_x)^2 + (-P + P)^2 + (-P + \sigma_x)^2 = 2Y^2$$

$$2(P + \sigma_x)^2 = 2Y^2$$

$$P + \sigma_x = Y$$

According to Tresca's theory $\dfrac{\sigma_1 - \sigma_3}{2} = k$

$$\sigma_x + P = 2k$$

or

$$\sigma_x + P = Y$$

Put $p = (Y - \sigma_x)$ in Eq. (1), we get

$$\frac{d\sigma_x}{\sigma_x.B - y(1 + B)} = \frac{dh}{h}$$

On integrating both sides, we get

$$\frac{1}{B}\log_e[\sigma_x B - (1 + B)y)] = \log_e h + \log_e c = \log_e hc$$

c = Integration constant

$$\log_e[\sigma_x B - Y(1 + B)] = \log\,(hc)^B$$

$$\sigma_x B - Y(1 + B) = (hc)^B$$

or
$$= \sigma_x \frac{(hc)^B}{B} + Y\frac{(1+B)}{B}$$

$$\sigma_x = \frac{h^B C_1}{B} + Y\frac{(1+B)}{B} \qquad \ldots(2)$$

where C_1 = is new integration constant

$$C_1 = C^B$$

$$\sigma_x = \sigma_{xb} \text{ (back pull) and } h = h_b \text{ at die entry}$$

$$\sigma_{xb} = C_1 \frac{h_b^B}{B} + \frac{Y(1+B)}{B}$$

or
$$C_1 = \left[\sigma_{xb} - \frac{Y(1+B)}{B}\right]\frac{B}{h_b^B}$$

putting the value of if C_1 in Eq. (2) we get

$$\frac{\sigma_x}{Y} = \left(\frac{1+B}{B}\right)\left[1 - \left(\frac{h}{h_b}\right)^B\right] + \frac{\sigma_{xb}}{Y}\left(\frac{h}{h_b}\right)^B$$

putting $h = h_a$

$$\frac{\sigma_x}{Y} = \left(\frac{1+B}{B}\right)\left[1 - \left(\frac{h_a}{h_b}\right)^B\right] + \frac{\sigma_{xb}}{Y}\left(\frac{h_a}{h_b}\right)^B$$

putting $\sigma_{xb} = 0$

$$\boxed{\frac{\sigma_x}{Y} = \left(\frac{1+B}{B}\right)\left[1 - \left(\frac{h_a}{h_b}\right)^B\right]} \qquad \boxed{\text{where } Y = 2k}$$

3.4 TUBE DRAWING

3.4.1 Analysis of Tube Drawing

In tube or pipe drawing with a plug or movable mandrel, most of the reduction in tube takes place in wall thickness. Minor reduction take place in inside diameter. Analysis is based on plain strain condition and no hoop strain in the process.

Assumptions

(i) σ_x is a principal stress, die pressure p is assumed same at the both side of tube, i.e., inner and outer.

(ii) uniform state of stress is assumed to exist at all points of a plane normal to the tube axis.

(iii) radial stress $\sigma_r = -p$

(*a*) Axial components of longitudinal stress σ_x

$$(\sigma_x + d\sigma_x)(h + dh)\,\pi D - \sigma_x h\pi D = (\sigma_x dh + h\sigma_x)\,\pi D$$

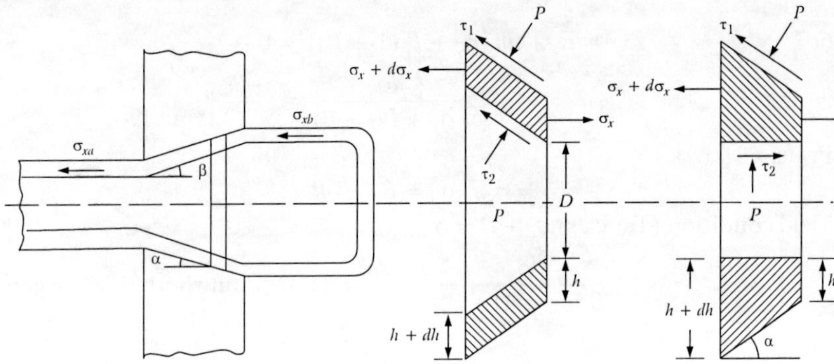

Fig. 3.4 Tube drawing

(*b*) Axial component of normal pressure on the outer die surface

$$\int_0^{2\pi} P\sin\alpha \cdot \frac{dx}{\cos\alpha} \cdot \frac{D}{2} d\theta = P\pi D \tan\alpha \cdot dx$$

α = semi die angle

(*c*) Axial component of the normal pressure on the inner die surface

$$-\int_0^{2\pi} P\sin\beta \cdot \frac{dx}{\cos\beta}\frac{D}{2} d\theta = -P\pi D \tan\beta dx$$

β = Semi = cone angle of Manderal for cylinderical plug β = 0

(*d*) Axial component of the frictional force on the outer die surface

$$\int_0^{2\pi} \mu_1 P\cos\alpha \frac{dx}{\cos\alpha} \cdot \frac{D}{2} d\theta = \mu_1 P\pi D\, dx$$

μ_1 = Coefficient of friction between the tube and die wall

(*e*) Axial component of the frictional force on the inner die surface

$$\int_0^{2\pi} \mu_2 P\cos\beta \frac{dx}{\cos\beta} \cdot \frac{D}{2} d\theta = \mu_2 P \times D\, \pi dx$$

Equilibrium equation will be

$$(\sigma_x \cdot dh + hd\sigma_x)\, \pi D + P\, \pi\, D\, (\tan\alpha - \tan\beta)\, dx + P\, \pi\, D\, (\mu_1 + \mu_2)\, dx = 0$$

decreasing in thickness *dh* as the element moves a distance *dx* is given by

$$dh = dx\, (\tan\alpha - \tan\beta)$$

$$(\sigma_x dh + hd\sigma_x) + Pdh\left[1 + \frac{\mu_1 + \mu_2}{\tan\alpha - \tan\beta}\right] = 0$$

Let $\frac{\mu_1 + \mu_2}{\tan\alpha - \tan\beta} = B$, then

$$(\sigma_x dh + hd\sigma_x) + Pdh\, (1 + B) = 0$$

or $$hd\sigma_x + dh\,[\sigma_x + P\,(1 + B)] = 0$$

$$\frac{d\sigma_x}{\sigma_x + P(1+B)} = \frac{-dh}{h}$$

Principal stress

$$\sigma_1 = \sigma_x, \quad \sigma_2 = \sigma_3 = -P$$

Yield condition (Tresca's) $\sigma_x + P = \sigma_0$

$$\frac{d\sigma_x}{B\sigma_x - \sigma_0(1+B)} = \frac{dh}{h}$$ on integrating both side, we get

$$\int \frac{d\sigma_x}{B\sigma_x - \sigma_0(1+B)} = \int \frac{dh}{h}$$

$$\frac{1}{B}\log_e[B\sigma_x - \sigma_0(1+B)] = \log_e h + \log_e c$$

$$= \log_e hc$$

or $$\log_e[B\sigma_x - \sigma_0\,(1 + B)] = \log_e (hc)^B$$

$$B\sigma_x - \sigma_0\,(1 + B) = (ch)^B$$

$$\sigma_x = \frac{(hc)^B}{B} + \frac{\sigma_0(1+B)}{B}$$

$$= \frac{C_1 h^B}{B} + \frac{\sigma_0(1+B)}{B}$$

where $C_1 = C^B$ = New Integration constant

at entry $h = h_0$ and $\sigma_x = \sigma_{xb}$ (back pull stress)

$$C_1 = \left[\sigma_{xb} + \sigma_0\left(\frac{1+B}{B}\right)\right] \times \frac{B}{h_b^B}$$

Substituting value of C_1 in above equation

$$\frac{\sigma_x}{\sigma_0} = \frac{(1+B)}{B}\left[1 - \left(\frac{h}{h_b}\right)^B\right] + \frac{\sigma_{xb}}{\sigma_0}\left(\frac{h}{h_b}\right)^B$$

draw stress at exit is given by

$$\frac{\sigma_{xa}}{\sigma_0} = \frac{(1+B)}{B}\left[1 - \left(\frac{h_a}{h_b}\right)^B\right] + \frac{\sigma_{xb}}{\sigma_0^2}\left(\frac{h_a}{h_b}\right)^B$$

for zero back pull $\sigma_{xb} = 0$

at exist $h = h_1$

$$\boxed{\sigma_d = \sigma_0 \frac{(1+B)}{B}\left[1 - \left(\frac{h}{h_0}\right)^B\right]} \qquad \sigma_0 = 2k$$

for cylinderical plug = $\beta = 0$

$$\boxed{B = \frac{\mu_1 + \mu_1}{\tan\alpha}}$$

Example 3.1

Compare the forces required to draw 25 mm × 5 mm copper strip to 45% reduction of area with that required for an equal reduction on round bar of the same cross-section area using (i) 12° included angle die, (ii) 30° included angle die, If μ = 0.08 redundent work may be neglected.

Solution: For strip $\dfrac{\sigma_x}{y} = \dfrac{1+B}{B}\left[1-\left(\dfrac{h_a}{h_b}\right)^{2B}\right]$

(i) $2\alpha = 12°$ or $\alpha = 6°$

$$B = \frac{\mu}{\tan\alpha} = \frac{0.08}{\tan 6°} = 0.761$$

(ii) $2\alpha = 30°$, $\alpha = 15°$

$$B = \frac{\mu}{\tan\alpha} = \frac{0.08}{\tan 15°} = 0.298$$

$$h_a = \frac{-5\times 45}{100} + 5 = 2.75, \quad h_b = 5$$

(i) $\dfrac{\sigma_{xa}}{y} = 2.31\,(1 - .634) = 0.844$

(ii) $\dfrac{\sigma_{xa}}{y} = 4.35\,(1 - .836) = 0.7098$

Example 3.2

In a wire drawing operation, initial wire diameter is 5 mm and the final wire diameter is 4.5 mm. Half die angle is 12°. Find the drawing stress considering coefficient of friction μ is 0.2 and *K* = 18 N/mm². Also calculate the maximum reduction.

Solution: Given $\mu = 0.2$, $\alpha = 12°$ [$D_a = 4.5$ mm]

$$\frac{\sigma_{xa}}{2k} = \frac{1+B}{B}\left[1-\left(\frac{D_a}{D_b}\right)^{2B}\right] \qquad D_b = 5 \text{ mm}$$

$$= 2.06\,[(1 - .820)] = 0.370$$

$$B = \left[\frac{\mu}{\tan\alpha} = \frac{0.2}{\tan 12°}\right] = 0.94$$

$$\sigma_{xa} = 0.370 \times 2 \times 18 = 13.33 \text{ N/mm}^2$$ **Ans.**

Maximum reduction

$$R = 1 - \left(\frac{1}{1+B}\right)^{1/B}$$

$$= 1 - \left[\frac{1}{1.94}\right]^{\frac{1}{0.94}} = 1 - 0.494 = 0.5056$$

or $R = 50.56\%$ **Ans.**

Example 3.3

A copper wire is drawn from an initial diameter 10 mm to 8 mm diameter at a speed of 2 m/sec. If the die angle is 14° and the coefficient of friction is 0.1 between copper wire and die and k = 150 MPa then calculate the drawing power.

Solution: $\mu = 0.1$, $2\alpha = 14°$, $\alpha = 7°$, $D_a = 8$ mm, $D_b = 10$ mm

$$B = \frac{\mu}{\tan\alpha} = \frac{0.1}{\tan 7°} = 0.814$$

$$\frac{\sigma_{xa}}{2k} = \left(\frac{1+B}{B}\right)\left[1-\left(\frac{D_a}{D_b}\right)^{2B}\right]$$

$$\sigma_{xa} = 2\times150\times\frac{1.814}{0.814}\times(1-0.695)$$

$$= 2\times150\times\frac{1.814\times0.3046}{0.814} = 203.6 \text{ MPa}$$

drawing force $$F = \sigma_{xa}\times A_a = 203.6\times\frac{\pi}{4}(10)^2 = 15986.1$$

$$\text{Power} = \text{Force} \times \text{velocity}$$

$$= 15986.1 \times 2.0$$

$$= 31972 \text{ watt or } 31.972 \text{ kW}$$ **Ans.**

Example 3.4

Find the force required to draw a wire of size 4 mm from a rod of 5 mm through a conical die of semi-die angle 15° with coefficient of friction 0.05 and $k = 200$ N/mm^2 assume no back tension. *[UPTU 2001-02]*

Solution: $D_a = 4$ mm, $D_b = 5$ mm, $2\alpha = 15°$, $\mu = 0.05$, $k = 200$ N/mm^2, $\alpha = 7.5°$

$$B = \frac{\mu}{\tan\alpha} = \frac{0.05}{\tan 7.5°} = 0.379$$

$$\frac{\sigma_{xa}}{2k} = \left(\frac{1+B}{B}\right)\left[1-\left(\frac{D_a}{D_b}\right)^{2B}\right] + \frac{\sigma_{xb}}{2k}\left(\frac{D_a}{D_b}\right)^{2B}, \quad \sigma_{xb} = 0$$

$$\frac{\sigma_{xa}}{2k} = \frac{1+B}{B}\left[1-\left(\frac{D_a}{D_b}\right)^{2B}\right]$$

$$= \frac{1+0.379}{0.397}\left[1-\left(\frac{4}{5}\right)^{2\times0.379}\right]$$

$$= \frac{1.379}{0.379}\times(1-0.844) = 0.566$$

$$\sigma_{xa} = 2 \times 200 \times 0.566 = 226.47 \text{ N/mm}^2$$

$$\text{force} = \sigma_{xa} \times A_a = 226.47 \times \frac{\pi}{4} \times 5^2 = 4.444 \text{ kN}$$ **Ans.**

Example 3.5

Find out the drawing load for 30% reduction of area 25 mm × 6 mm mild steel strip using tapered die. Take $\alpha = 25°$, $\sigma_y = 25$ kg/mm², $\mu = 0.1$.

Solution: Reduction $= 1 - \frac{h_a}{h_b} = 0.30$, $h_b = 6$ mm.

$$1 - \frac{h_a}{6} = 0.30, \quad \frac{h_a}{6} = 1 - 0.3 = 0.7, \quad h_a = 0.7 \times 6 = 4.2 \text{ mm.}$$

$$\mu = B \tan\alpha \quad \text{or} \quad B = \frac{\mu}{\tan\alpha} = \frac{0.1}{\tan 25°} = 0.21 \qquad \text{as back pull } \sigma_{xb} = 0$$

$$\frac{\sigma_{xa}}{2k} = \left(\frac{1+B}{B}\right)\left[1 - \left(\frac{h_a}{h_b}\right)^B\right]$$

according Von-mises, $k = \frac{\sigma_y}{\sqrt{3}}$ $\quad k = \frac{25}{\sqrt{3}} = 14.43$

$$\sigma_{xa} = 2 \times 14.43 \times \frac{1.21}{0.21}\left[1 - \left(\frac{4.2}{6}\right)^{0.21}\right] = 37.15$$

$$\text{Drawing load} = \sigma_{xa}\,(h_a \times l)$$

$$= 37.15 \times 4.2 \times 25 = 3901 \text{ kg}$$ **Ans.**

Example 3.6

A strip initial width 6.0 mm is drawn by tapered die to a final width of 5.5 mm in a state of plane strain take, $t = 10$ mm, $2\alpha = 20°$, $\mu = 0.05$ and $\sigma_y = 250$ N/mm². Find the stress when (i) $\sigma_{xb} = 0$.

[where, σ_{xb} = back pull]

Solution:

$$B = \frac{\mu}{\tan\alpha} = \frac{0.05}{\tan 10°} = 0.283$$

(i) When $\sigma_{xb} = 0$,

$$\frac{\sigma_{xa}}{2k} = \left(\frac{1+B}{B}\right)\left[1 - \left(\frac{h_a}{h_b}\right)^B\right]$$

but $k = \frac{\sigma_y}{\sqrt{3}} = \frac{250}{\sqrt{3}} = 144.34$

$$\frac{\sigma_{xa}}{288.6} = \frac{1+0.283}{0.283}\left[1 - \left(\frac{5.5}{6.0}\right)^{0.283}\right] = 0.110$$

$$\sigma_{xa} = 0.110 \times 288.6 = 31.82 \text{ N/mm}^2$$ **Ans.**

Example 3.7

A mild steel tube of 20 mm outer diameter having 1.5 mm wall thickness is to be drawn to 15 mm outside diameter of 1.2 mm wall thickness. Find out the drawing load if μ = 0.15 and 0.18 for die and plug respectively. Angle of die and plug are 26° and 18° respectively. σ_y = 150 N/mm². Also calculate the power of motor if V = 0.5 m/sec.

Solution: $\mu_1 = 0.15,\ \ \mu_2 = 0.18,\ \ \alpha = \dfrac{26}{2} = 13°$ and $\beta = \dfrac{18}{2} = 9°,\ \ 2k = 140 \text{ N/mm}^2$

$$B = \frac{\mu_1 + \mu_2}{\tan\alpha + \tan\beta} = \frac{0.15 + 0.18}{\tan 13° + \tan 9°} = \frac{0.33}{0.3892} = 0.8477$$

Now drawing stress,
$$\frac{\sigma_{xa}}{2k} = \frac{1+B}{B}\left[1 - \left(\frac{h_a}{h_b}\right)^B\right]$$

$$\sigma_{xa} = 140 \times \frac{1.8477}{0.8477} \times \left[1 - \left(\frac{1.2}{1.5}\right)^{.8477}\right] = 52.6 \text{ N/mm}^2$$

Area of tube,
$$A_a = \frac{\pi}{4}\left(20^2 - 15^2\right) = 137.4 \text{ mm}^2$$

Drawing load= $\sigma_{xa} \times A_a$ = 52.6 × 137.4 = 7226 N **Ans.**

Motor power, $P = f \times V$ = 7226 × 0.5 = 3613 watt or 3.613 kW **Ans.**

Example 3.8

What is the percentage contribution of friction to the drawing stress at 40% reduction of area using two lubricants if μ = 0.03 and 0.1 and α = 20° and a fixed plug.

Solution:

$$40 = 1 - \frac{h_a}{h_b},\quad \frac{h_a}{h_b} = 1 - 0.4 = 0.6$$

For tube drawing without friction

$$\frac{\sigma_{xa}}{2k} = \log_e\left(\frac{h_b}{h_a}\right)$$

$$\frac{\sigma_{xa}}{2k} = \log_e\left(\frac{1}{0.6}\right) = 0.511$$

When, μ = 0.63 and α = 20°

$2\mu = B\tan\alpha$
$$B = \frac{2\mu}{\tan\alpha} = \frac{2 \times 0.03}{\tan 20°} = 0.165$$

$$\frac{\sigma_{xa}}{2k} = \frac{1+B}{B}\left[1 - \left(\frac{h_a}{h_b}\right)^B\right]$$

$$= \frac{1.165}{0.165}\left[1 - (0.6)^{0.165}\right] = 0.571$$

When μ = 0.1 and α = 20°

$$B = \frac{2\mu}{\tan\alpha} = \frac{2 \times 0.1}{\tan 20°} = 0.549$$

$$\frac{\sigma_{xa}}{2k} = \frac{1+B}{B}\left[1 - \left(\frac{h_a}{h_b}\right)^B\right]$$

$$\frac{\sigma_{xa}}{2k} = \frac{1.55}{0.55} \times \left[1 - (0.6)^{0.55}\right] = 0.690$$

(i) % contribution of μ when μ = 0.03

$$= \frac{0.0571 - 0.511}{0.511} \times 100 = 11.74\% . \quad \textbf{Ans.}$$

(ii) μ = 0.1

$$= \frac{0.69 - 0.511}{0.511} \times 100 = 35.0\% . \quad \textbf{Ans.}$$

Example 3.9

In a wire drawing operation determine the drawing stress and total drawing load. Using following data initial wire diameter = 5 mm. Final wire diameter = 4.5 mm.

Die angle = 18°; dia land = 5 mm; μ = 0.12; yield stress = 25 kg/mm^2

Solution: $r_i = 2.5$ mm, $r_f = 2.25$ mm, $Y = 25$ kg/mm^2, $\alpha = \frac{18}{2} = 9°$

Drawing stress
$$\sigma_x = Y\left(\frac{1+B}{B}\right)\left[1 - \left(\frac{D_f}{D_t}\right)^{2B}\right]$$

$$B = \mu \cot \alpha = 0.12 \cot 9° = 0.75$$

$$\sigma_x = \frac{25(1+0.75)}{0.75}\left[1 - \left(\frac{4.5}{5}\right)^{1.5}\right] = \frac{25 \times 1.75 \times 0.146}{0.75}$$

$$= 8.527 \text{ kg/mm}^2$$

$$\sigma_t = Y - (Y - \sigma_x)\, e^{\frac{-2\mu L}{R_f}}$$

$$= 25 - (25 - 8.527)\, e^{\frac{-2 \times 0.12 \times 2.5}{2.25}}$$

$$= 25 - (16.47) \times 0.766 = 12.38 \text{ kg/mm}^2. \quad \textbf{Ans.}$$

3.5 EXTRUSION

Extrusion is a metal forming process in which the shape of Raw material is converted from one shape into another shape by applying the compressive force. It is process of compressing the metal into a container through die opening. In extrusion process cross-section area reduces to produce the **cylinderical job, hollow tubes** and **irregular shapes** of objects. This process is extensively used to produce the non-ferrous metal and their alloys parts/component. This process is usually carried out at elevated temperature, a lot of hollow and solid small cross-section parts can be fabricated by this process as the die opening is constant during the process hence cross-section of extruded parts is uniform.

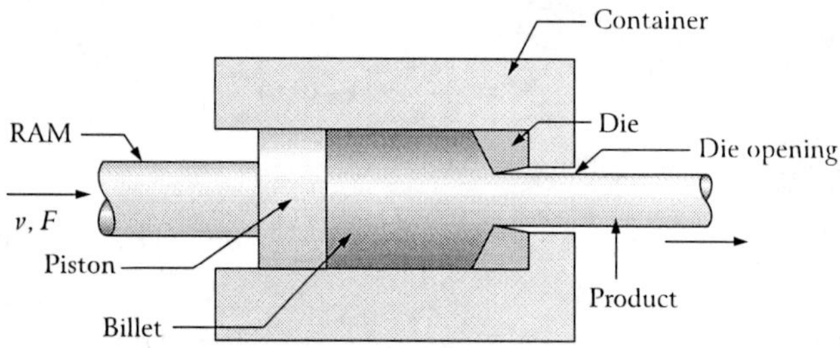

Fig. 3.5 Extrusion

Most important feature of extrusion is that grain flow of material is in the direction most suited to resist imposed stress during the working condition of part.

A very simple and common example of extrusion in daily life ***to come out tooth paste from tooth paste tube*** by applying the compressive force.

3.5.1 Metals can be Extruted

Following metals can be extruted:

- Aluminium and its alloys
- Copper and its alloys
- Tin, *Zn* and their alloys
- Magnesium and its alloys, etc.

3.5.2 Shapes that can be Extruted

Various type of shapes can be produced by the extrusion process.

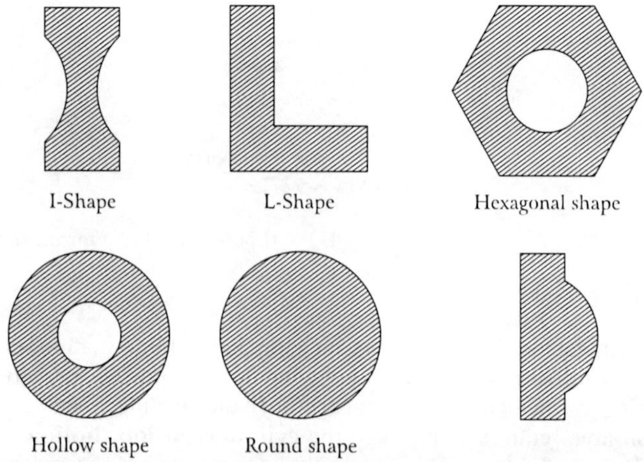

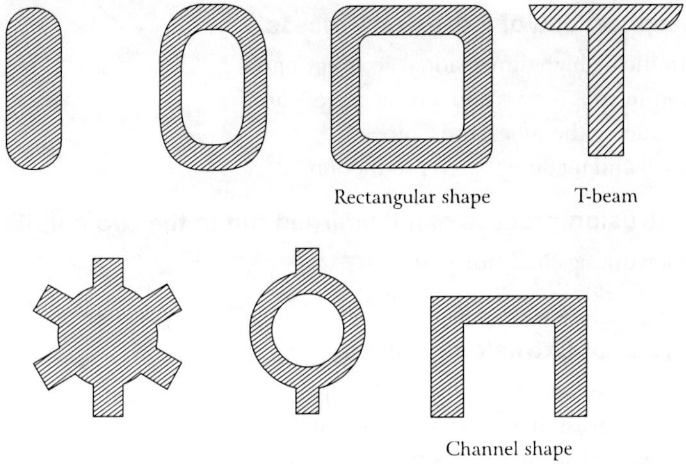

Fig. 3.6 Extruded shapes

Extrusion pressure for a particular material, basically depends upon the *extrusion temperature, extrusion speed, area of reduction*. Very high speed of extrusion cause the generation of overheat which will finally lead to cracks.

3.5.3 Advantages of Extrusion Process

(i) Closer dimensional tolerance is possible by extrusion process.

(ii) Process is suitable for mass production.

(iii) Mostly non-ferrous and ferrous metals can be extruted easily.

(iv) Brittle material can be also extruted.

(v) In some cases extrusion replaces the welding and machining of parts.

(vi) Shapes which are difficult to obtain by other processes can be easily obtained by this process.

(vii) Due to high reduction ratio, metal has excellent transverse flow lines consequently the part has an added strength and secondry processes are made very easy.

(viii) Internal cavity in the shape can be easily produced by using the spider die.

(ix) Rods can be produced from metal having poor ductility.

3.5.4 Limitations of Extrusion Process

(i) Set up cost is very high.

(ii) Die and tooling costs are very high.

(iii) Process is not suitable for batch production or less quantity.

(iv) Speed of process is slow as compared to rolling.

(v) Size of the die is limited.

(vi) In productivity extrusion is much inferior than rolling.

(vii) Process wastage is higher as compared to rolling.

3.5.5 Applications of Extrusion Process

(i) Producing high dimensional accuracy parts.
(ii) Non-ferrous metals and ferrous metal parts.
(iii) Producing the tubes, rods, pipes,
(iv) Small and medium batch production.

3.5.6 Extrusion process can be carried out in the two conditions

(i) Hot working condition of metal.
(ii) Cold working condition of metal.

3.5.7 Types of Extrusion Process

Extrusion processes are basically of four types

(*a*) Direct extrusion or forward extrusion
(*b*) Indirect extrusion or backward extrusion
(*c*) Hydrostatic extrusion
(*d*) Impact extrusion

3.5.7.1 Direct Extrusion

In this process a hot billet to be extuded is placed in a container. At one end there is a provision of end while on the other side there is die opening or die orifice. First of all, pressure (compressive force) is applied through hydraulically operated ram in such a way that metal starts to flow through an orifice and finally acquires the shape of die openings. Figure 3.7(A) shows the direct extrusion process. It is also known *as forward extrusion* as metal flows in the forward direction. To avoid the friction problem lubricants are used.

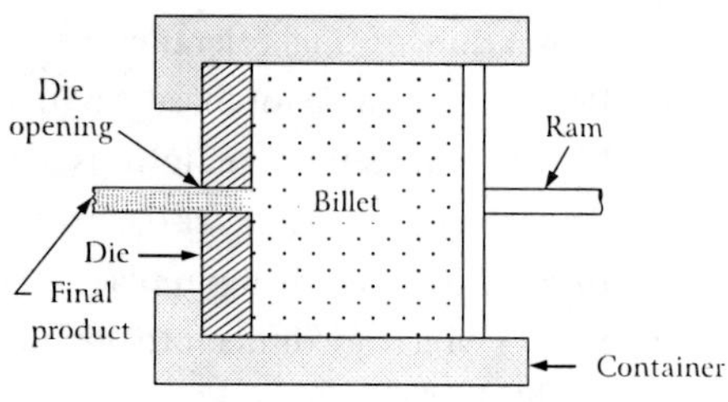

Fig. 3.7(A) Direct extrusion

3.5.7.2 Indirect Extrusion

It is also known as **backward extrusion.** It consist the die chamber in which the hot billet is kept and the force (compressive force) is applied against the back wall of die chamber. Due to pressure finally metal flows through the die opening. Billet in the container remains stationary hence no friction. Also, process is not affected by the length of billet. Main advantages of backward extrusion is that good surface quality can be achieved as there is no heat cracking and the main disadvantage of backward extrusion is that surface defects of the billet would weld up in the final product.

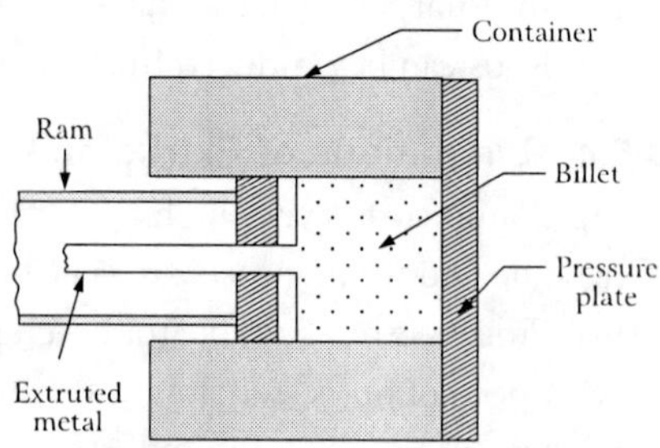

Fig. 3.7(B) Indirect extrusion

Comparison Between Direct and Indirect Extrusion

Sl. No	*Direct or Forward Extrusion*	*Indirect or Backward Extrusion*
1.	In this extrusion, ram moves the billet must slide or shear at the interface between the container and billet. It is very simple process.	In this extrusion, die moves while billet remains stationary. It is some what complicated.
2.	High degree friction forces is required.	Low friction force is required.
3.	High degree extruding forces required; however, mechanically more convenient.	25-30% lesser extruding forces is required as compared to forward extrusion.
4.	Simple type of tools are required.	Due to complex type of tools only limited application.
5.	Scrap or process wastage are very high (15-25% = billet weight).	Scrap or process wastage is very less (5-6% of the billet weight).

3.5.7.3 Hydrostatic Extrusion

A hydrostatic extrusion process is shown by Fig. 3.8. In this process, the pressure is generated through a hydraulic fluid (viscous) surrounded by the billet. Wall friction is eliminated due to the presence of fluid in container. Process have very limited industrial application due to use of special tools/equipments and low production rate. This process is carried out at room temperature.

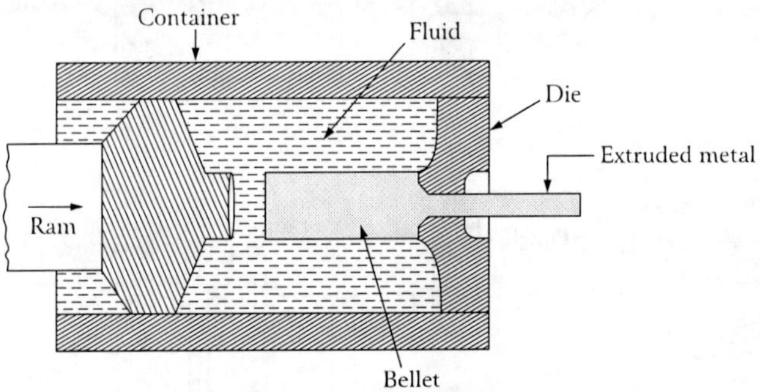

Fig. 3.8 Hydrostatic extrusion

Fluid Used in Hydrostatic Extrusion

- ⇒ SAE-30,
- ⇒ Mineral oil,
- ⇒ Castor oil with 10% alcohol,
- ⇒ Glycerin,
- ⇒ Ethylgol and these oils operate at 1100-3000 N/mm^2 pressure range.

3.5.7.4 Impact Extrusion

This is a direct type of extrusion process and basically carried out by a high speed mechanical press operated by mechanical or any hydraulic system. Punch is the riplica of the shape that to be produced. Punch strikes on the blank at very high speed. Due to high pressure blank extruding upward direction. Process is suitable for soft material like **Aluminium, brass, tin, copper, lead, etc.**

3.5.7.4.1 Advantages of Impact Extrusion

(i) Metal/Blank wastage is very less
(ii) Production speed is very fast
(iii) good dimensional tolerance can be achieved.
(iv) Cheaper in cost.

3.5.7.4.2 Limitations of Impact Extrusion

(i) Process is suitable only for soft material
(ii) complex shapes can not be produced
(iii) only small size of parts can be produced
(iv) Paper clearance is required for moving punch in die.

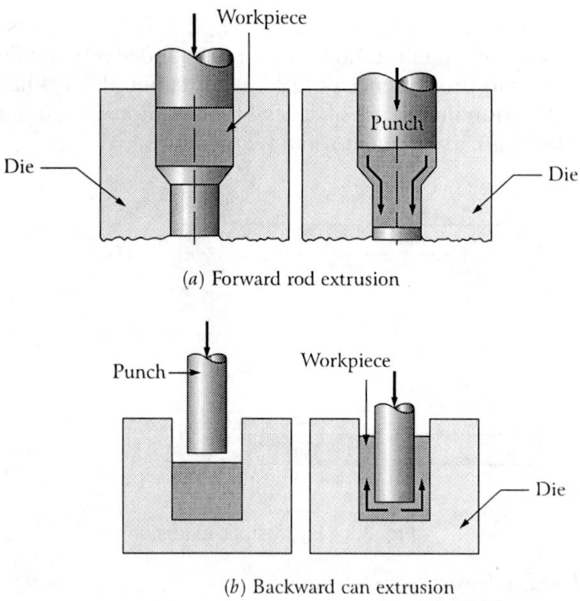

Fig. 3.9 Impact extrusion

3.5.8 Defects in Extrusion

During the extrusion process serveral types of defects are produced which basically depend upon the condition of metal and process variable. Usually three types of extrusion defects are produced:

(i) Extrusion defects
(ii) Surface defect
(iii) Internal defect

3.5.8.1 *Extrusion Defects*

This type of defects also known *as pipe defect* or *fish-tailing* defects and basically produce due to back flow of material, pushing the end face of the billet into the core of product. This type of defect is responsible to weaken the product as surface layer is normally contaminated by oxides.

These type of defects can be reduced by:

(i) Machining the surface before extrusion to eliminate any rusting and impurities.
(ii) By modifying the flow pattern to a less in homogeneous one, such as by controlling friction and minimizing temperature gradient.

3.5.8.2 *Surface Defect*

Due to the friction in extrusion process a very high heat generate which increases the temperature of object and due to high temperature surface cracks may be generated. This type of defect is basically due to hot shortness and are intergranular. They occur specially in *Aluminium, magnesium, zinc* alloys. This type of defect can be reduced by reducing the extrusion speed and temperature.

3.5.8.3 *Internal Defect*

They are also known as centre-cracking, centre burst, arrow head, fracture cracking, in which the extruded product develops cracks in its centre. To reduce this defect, the die angle and impurities in the billet should be reduced.

3.6 ANALYSIS OF EXTRUSION ROD WITH FRICTION

3.6.1 Assumptions

(i) Material is homogeneous, perfectly plastic and isotropic.
(ii) Stress at any point at a plane normal to the axis of extrusion is in uniform state.
(iii) Plane remains plane before and after deformation.

- Total axial stress due to σ_x

$$\sigma_x = (\sigma_x + d\sigma_x)\left(\frac{D + dD^2}{4}\right) \times \pi - \frac{\pi}{4}\sigma_x \times D^2$$

- Total axial stress due to $P = \left(P\pi \times \frac{D\,dx}{\cos\alpha} \times \sin\alpha\right)$

- Total frictional stress at the circumference $= \mu P\left(\pi\frac{D\,dx}{\cos\alpha}\right)\cos\alpha$

for equilibrium condition

$$\frac{\pi}{4}D[2\sigma_x dD + D\sigma_x] + \frac{\pi}{2}pd \times dD + \mu p\pi\, Dd\,x = 0$$

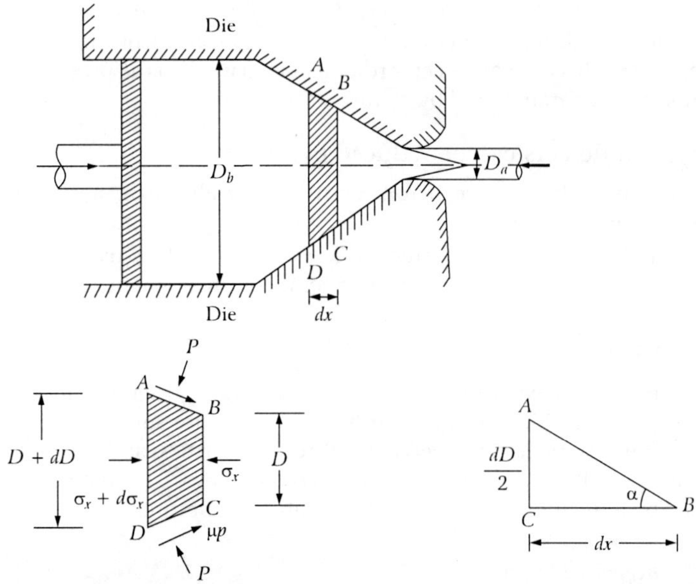

Fig. 3.10 Extrusion of rod with friction

From Fig. 3.10, $dD = 2 \times \tan \alpha \, dx$

$$D d\sigma_x + 2\sigma_x \, dD + 2p \, dD\left(1 + \frac{\mu}{\tan \alpha}\right) = 0$$

According to Tresca's $\quad \sigma_1 = \sigma_3 = \sigma_y = 2k$

$$\sigma_1 = \sigma_x \quad \text{and} \quad \sigma = -p$$

$$\sigma_x + p = 2k$$

or $\quad p = 2k - \sigma_x$

$$B = \frac{\mu}{\tan \alpha} \text{ (assume), then}$$

$$D d\sigma_x + 2\sigma_x \, dD + 2\,(2k - \sigma_x)\,(1 + B)\, dD = 0$$

or $\quad \dfrac{d\sigma_x}{B\sigma_x - 2k(1+B)} = \dfrac{2dD}{D} \quad$ on integrating, we get

$$\frac{1}{B} \log_e[B\sigma_x - 2k\,(1 + B)] = 2 \log_e (DC)$$

where C = constant

$$\log_e [B\sigma_x - 2k\,(1 + B)] = \log_e (DC)^{2B}$$

or $\quad \sigma_x = \dfrac{(DC)^{2B}}{B} + \dfrac{2k(1+B)}{B}$

or $$= \frac{D^{2B}C_1}{B} + \frac{2k(1+B)}{B} \quad \text{...(i)}$$

where C_1 is a new constant and $C_1 = C^{2B}$

at die exit $D = D_a$ and $\sigma_x = \sigma_{xa}$

$$\sigma_{xa} = \frac{C_1 D_a^{2B}}{B} + \frac{2k(1+B)}{B}$$

or $$C_1 = \left[\sigma_{xa} - \frac{2k(1+B)}{B}\right]\frac{D}{D_a^{2B}}$$

Putting the value of C_1 in Equation (i), we get

$$\frac{\sigma_x}{2k} = \left(\frac{1+B}{B}\right)\left[1 - \left(\frac{D}{D_a}\right)^{2B}\right] + \frac{\sigma_{xa}}{2k}\left(\frac{D}{D_a}\right)^{2B}$$

Put $D = D_b$

$$\frac{\sigma_{xb}}{2k} = \left(\frac{1+B}{B}\right)\left[1 - \left(\frac{D_b}{D_a}\right)^{2B}\right] + \left(\frac{\sigma_{xa}}{2k}\right)\left(\frac{D_b}{D_a}\right)^{2B}$$

If front pressure $\sigma_{xa} = 0$, then

$$\frac{\sigma_{xb}}{2k} = \frac{1+B}{B}\left[1 - \left(\frac{D_b}{D_a}\right)^{2B}\right]$$

OBJECTIVE QUESTIONS

1. A brass billet is to be extruded from its initial diameter of 100 mm to a final diameter of 50 mm. Working temperature of 700°C and the extrusion constant is 250 MPa. Forces required for extrusion is [*GATE 2003*]

   (*a*) 5.44 MN (*b*) 2.72 MN
   (c) 1.36 MN (*d*) 0.36 MN.

2. A 4 mm thick sheet is rolled with 300 mm diameter rolls to reduce thickness without any change in its width, the friction coefficient at the work roll interface is 0.1. The minimum possible thickness of the sheet can be produced in a single pass is [*GATE 2006*]

   (*a*) 1.0 mm (*b*) 1.5 mm
   (*c*) 2.5 mm (*d*) 3.7 mm.

3. In a wire drawing operation, diameter of steel wire is reduced from 10 mm to 8 mm. The mean flow stress of the material is 400 mm ideal forces required for drawing (ignoring friction and redundant work) is [*GATE 2006*]

   (*a*) 4.48 kN (*b*) 8.97 kN
   (*c*) 20.11 kN (*d*) 31.41 kN.

4. In open die forging a disc of diameter 200 mm and height 60 mm is compressed without any barreling effect. Final diameter of the disc is 400 mm. The true strain is *[GATE 2007]*
   (*a*) 1.986 (*b*) 1.686
   (*c*) 1.386 (*d*) 0.602.

5. Metal forming means
   (*a*) selective deformation of metal
   (*b*) an intentional plastic deformation of metal
   (*c*) elastic deformation of metal
   (*d*) all of these.

6. Hot working of metal is enable
   (*a*) heavy reduction in porosity of metal
   (*b*) disintegration and distribution of concentrated impurities
   (*c*) refinement of its grain structure
   (*d*) all of these.

7. Plastic deformation of a metal take place
   (*a*) when a metal deforms without external loading
   (*b*) when the stresses caused in the metal due to external loading
   (*c*) when the stresses caused in the metal due to external loading reach yield point
   (*d*) none of these.

8. Recrystallization temperature for steel is
   (*a*) 450°C (*b*) 600°C
   (*c*) 800°C (*d*) none of these.

9. Cutting, setting down and bending are
   (*a*) forging operations (*b*) welding operations
   (*c*) machining operation (*d*) rolling operation.

10. Press forging enables
    (*a*) uniform forging (*b*) faster production
    (*c*) easier maintenance (*d*) all of these.

11. In deep drawing of sheet the value of limiting draw ratio depends upon
    (*a*) thickness of sheet
    (*b*) percentage elongation of sheet metal
    (*c*) yield strength of sheet
    (*d*) type of press used. *[IES 2002]*

12. In wire drawing process, the bright shining surface on the wire is obtained if
    (*a*) no lubricant is used (*b*) thin fluid lubricant is used
    (*c*) solid powder lubricant is used (*d*) thick plate lubricant is used. *[IES 1996]*

13. What does hydrostatic pressure in extrusion process improve?
    (*a*) tensile strength (*b*) ductility
    (*c*) compressive strength (*d*) brittleness. *[IES 2006]*

14. The extrusion process used for the production of tooth paste tube is/are
    1. tube extrusion 2. forward extrusion
    3. impact extrusion.
    Select the correct answer using the code given below
    (*a*) only 1 (*b*) 2 and 3
    (*c*) 1 and 2 (*d*) only 3 *[IES 2002]*

15. Extrusion chamber, die and ram are generally lubricated by
    (*a*) solid lubricant (*b*) kerosene oil
    (*c*) grease (*d*) vegetable oil

16. For extrusion process, metal should be
    (*a*) brittle (*b*) ductile
    (*c*) malleable (*d*) plastic

17. Match the following metal forming processes with their associated stresses in the workpiece.

Metal forming processes	Type of stress
1. Coining	P. Tensile
2. Wire Drawing	Q. Shear
3. Blanking	R. Tensile and compressive
4. Deep Drawing	S. Compressive

(*a*) 1-S, 2-P, 3-Q, 4-R (*b*) 1-S, 2-P, 3-R, 4-Q
(*c*) 1-P, 2-Q, 3-S, 4-R (*d*) 1-P, 2-R, 3-Q, 4-S [*GATE 2012*]

ANSWERS

1. (*b*)	2. (*c*)	3. (*b*)	4. (*c*)	5. (*b*)
6. (*d*)	7. (*c*)	8. (*c*)	9. (*a*)	10. (*d*)
11. (*a*)	12. (*b*)	13. (*b*)	14. (*d*)	15. (*d*)
16. (*d*)	17. (*d*)			

HINTS

Ans. 1: $d_i = 100$ mm

$d_f = 500$ mm

$k = 250 \text{ MPa} = 250 \times 10^6 \text{ N/m}^2$

$$A_i = \frac{\pi}{4}(0.1)^2 = 7.854\times10^{-3}\text{m}^2,$$

$$A_f = \frac{\pi}{4}(0.5)^2 = 1.96\times10^{-3}\text{m}^2$$

Extrusion force
$$P = kA_i \log_e\left(\frac{A_i}{A_f}\right)$$

$$= 250\times10^6\times7.854\times10^{-3}\log_e\left(\frac{7.854\times10^{-3}}{1.96\times10^{-3}}\right)$$

$$= 2.72\times10^6\text{ N} \quad \text{or} \quad 2.72 \text{ MN}$$ Ans.

Ans. 2. $t_i = 4 \text{ mm} = 4\times10^{-3}$ m

$D = 300 \text{ mm} = 0.3$ m

μ = Coefficient of friction at roll work interface

$\mu = 0.1$

$t_f = ?$

We know $t_i - t_f = \mu^2 R$

$$4 \times 10^{-3} - t_f = 0.1 \times 0.1 \times 0.15$$

$$t_f = 0.004 - 0.0015 = 0.0025 \text{ m}$$

or $= 2.5$ mm **Ans.**

Ans. 3. $D_i = 10$ mm

$D_f = 8$ mm

$\sigma_0 = 400$ MPa

Given $\mu = 0$

$\beta = 0$

$$\sigma_d = 2\sigma_0 \log_e\left(\frac{D_i}{D_f}\right) = 2 \times 400 \log_e\left(\frac{10}{8}\right)$$

$$= 178.51 \text{ MPa}$$

$$\text{ideal force} = 178.51 \times 10^6 \text{ N/m}^2 \times \pi r_f^2$$

$$= 178.51 \times 10^6 \times 3.14 \times (0.004)^2$$

$$= 8968 \text{ N} \quad \text{or} \quad 8.968 \text{ kN} \quad \textbf{Ans.}$$

Ans. 4. Since the volume of disc remains constant during the process

Hence $\frac{\pi D_1^2}{4} \times l_1 = \frac{\pi}{4} D_2^2 \times l_2$

or $D^2{}_1 l_1 = D^2{}_2 l_2$

$$200^2 \times 60 = 400^2 \times l_2$$

or $l_2 = 15$

$$\text{change in length} = l_1 - l_2 = 60 - 15 = 45$$

Strain $\varepsilon = \frac{l_1 - l_2}{l_2} = \frac{45}{15} = 3$

True strain $= \log_e (1 + \varepsilon) = \log_e (1 + 3) = \log_e 4 = 1.386$ **Ans.**

REVIEW QUESTIONS

1. What is the difference between the drawing and extrusion process? What type of defect occurs in drawing and extrusion processes. [*UPTU 2006-2007*]
2. Describe the common defects of
   (*a*) Wire drawing
   (*b*) Extrusion [*UPTU 2001-02, 2002-03, 2003-04, 2004-05, 2005-06*]
3. Explain the
   (*a*) forward extrusion (*b*) backward extrusion
   (*c*) impact extrusion (*d*) hydrostatic extrusion.

4. Define extrusion process in detail. Give the advantages, limitations and shapes can be produced by extrusion.
5. Differentiate between Direct and Indirect extrusion.
6. Show that the drawing stress in case of wire drawing with applied back tension σ_{xb} can be expressed by

$$\sigma_{xa} = \sigma_y\left(\frac{1+B}{B}\right)\left[1-\left(\frac{D_a}{D_b}\right)^{2B}\right]+\sigma_{xb}\left(\frac{D_a}{D_b}\right)^{2B}$$

where $B = \mu \cot \alpha$, μ = coefficient of friction, α = semi-die angle, σ_y = yield strength of the wire material in tension, D_a = wire exit diameter, D_b = wire inlet diameter. *[UPTU 2003-04]*

7. Drive the expression for drawings stress σ_{xa} for wire drawing through a conical die of die angle 2α and coefficient of friction is μ as

$$\frac{\sigma_{xa}}{2k} = \frac{1+B}{B}\left[1-\left(\frac{D_a}{D_b}\right)^{2B}\right]$$

where, $B = \mu \cot \alpha$ *[UPTU 2008-09]*

8. Explain
   (i) Forward Extrusion
   (ii) Impact Extrusion. *[UPTU 2009-10]*
9. Give few examples of extruded products. *[UPTU 2002-03, 2005-06]*
10. Why stainless steel can not be extruded?
11. Derive the expression for drawing stress σ_{xa} for wire drawing through a conical die with 2α die angle and μ coefficient of friction as

$$\sigma_{xa} = \frac{1+B}{B}\left[1-\left(\frac{D_a}{D_b}\right)^{2B}\right] \quad \text{where } B = \mu \cot \alpha$$

[UPTU 2008-09]

12. From first principles, derive the formula

$$\frac{\sigma_{xb}}{2K} = \left(\frac{1+B}{B}\right)\left[\left(\frac{D_b}{D_a}\right)^{2b} - 1\right]$$

for extrusion of a wire with friction, where σ_{xb} refer to the stress in wire at inlet to the die. D_b and D_a are the inlet and outlet diameters of the wire $B = \mu \cot \alpha$ (μ = coefficient of friction and α = half die angle) and K = critical stress.

[UPTU 2010-11]

13. Describe the process of wire-drawing. What is the material of drawing die? Why is "in Process" annealing done? Give an idea of wire drawing speed.

[UPTU 2010-11]

❑❑❑

4
CHAPTER

ROLLING

4.1 INTRODUCTION

Rolling is a process in which the metal or alloys is passes between the two rollers which are rotating in the opposite direction and converting shape of object in to another shapes (without metal losses) due to the plastic deformation of metals. Frictional force of Raw material act on the workpiece as it passes through the rollers. Rollers have the cylinderical shapes and plastic deformation of metal take place due to the compressive force.

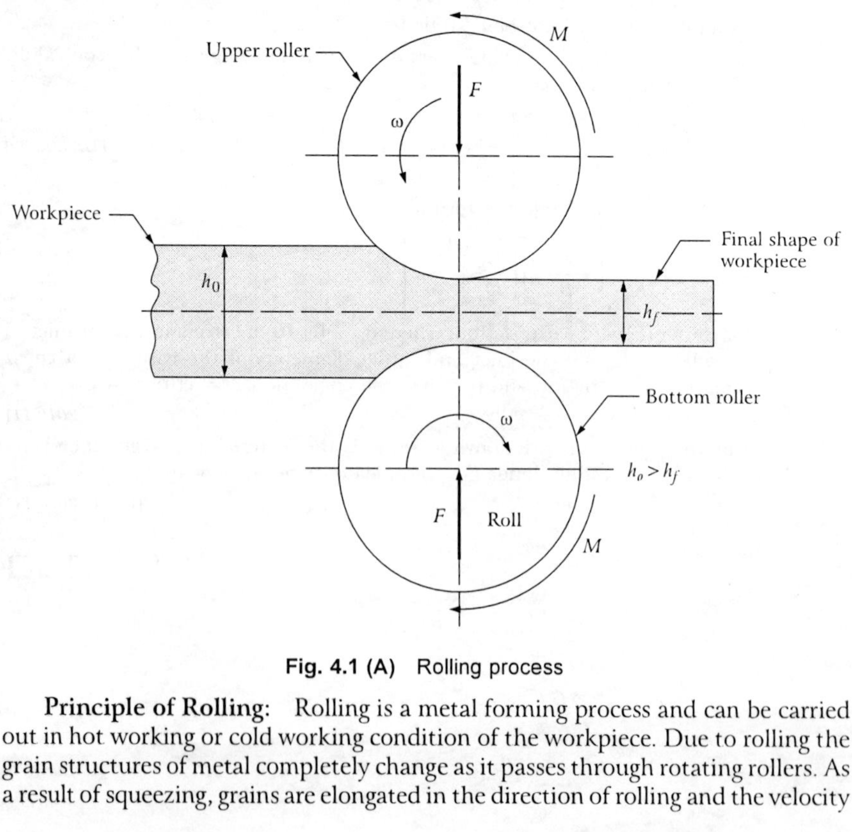

Fig. 4.1 (A) Rolling process

Principle of Rolling: Rolling is a metal forming process and can be carried out in hot working or cold working condition of the workpiece. Due to rolling the grain structures of metal completely change as it passes through rotating rollers. As a result of squeezing, grains are elongated in the direction of rolling and the velocity of metal at exit end is higher than the entry side. Due to plastic deformation grain starts refining. It is only for the case of hot working. For case of cold workingwork pieces are in contact with rollers over a sufficient distance during the process that is represented by arc *AB* in Fig. 4.1(B).

Angle *AOB* is referred as **angle of contact** or maximum angle of bite.

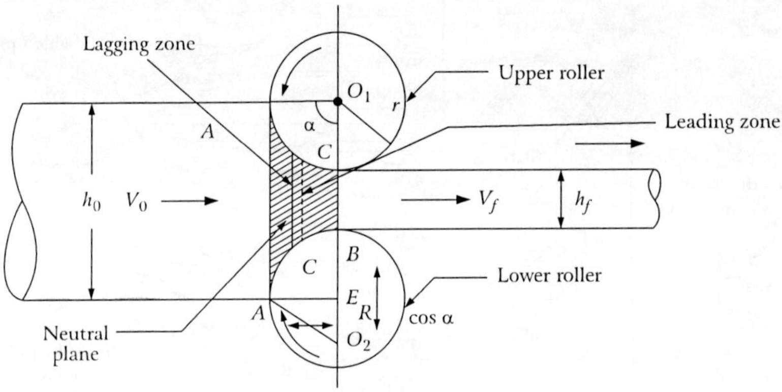

Fig. 4.1 (B) Principle of rolling

Pressure applied on the metal piece by rollers are not uniform throughout. Figure 4.1 (B) shows the typical rolling process, where

h_o = thickness metal sheet before rolling

V_o = velocity at entrance

$c - c$ = neutral point (no slip point)

h_f = final thickness of metal sheet (after rolling)

V_f = final velocity at exit end

$V_f > V_o$ = b_i = bidth of sheet before rolling

b_f = bidth of sheet after rolling

$V_o h_o b_i = V_f h_f h_f$ but $b_i = b_f$

$$V_o h_o = V_f h_f$$

$$V_f = \frac{V_o h_o}{h_f}$$

$$\text{Backward slip} = \frac{V_r - V_o}{V_r}$$

$$\text{Forward slip} = \frac{V_f - V_r}{V_r}$$

Total reduction or draft in rolling $\Delta h = h_o - h_f$

$$= 2\,(R - R\cos\alpha)$$

$$= D(1 - \cos\alpha)$$

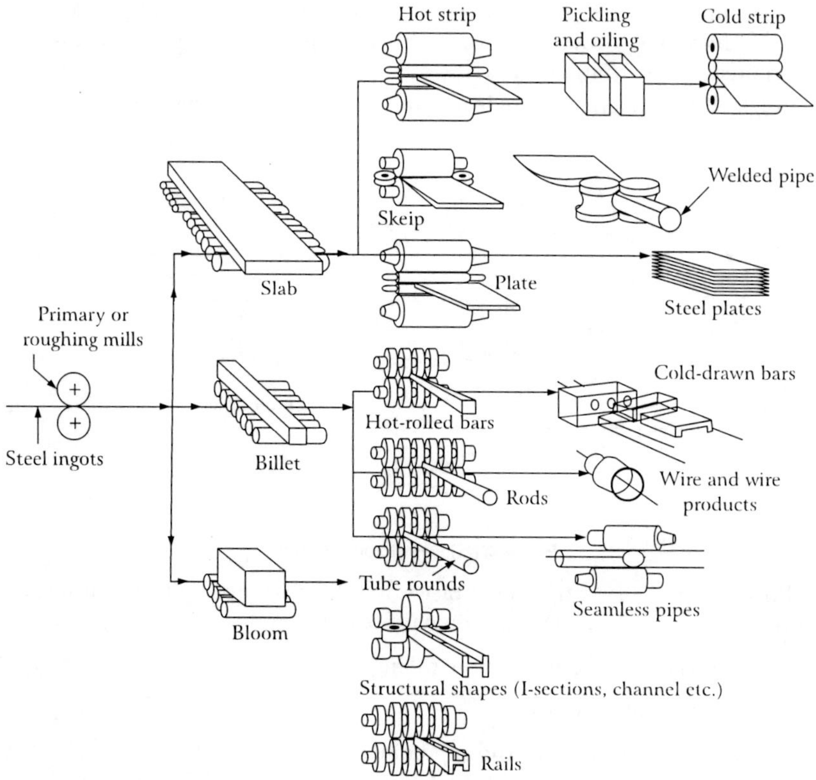

Fig. 4.2 Various flat rolling and shape rolling processes. Hot steel ingots (continuous cast products) are passed through rolls which form the plastic steel into slab, billet and bloom for further processing.

generally reduction in blooming mill is 100 mm and in slab mills around 50 – 60 mm.

Projected length of the arc of contact is

$$l = R \sin \alpha$$

$$l = \sqrt{BC^2 - CE^2}$$

but $$BC = \sqrt{R \Delta h}$$

and $$CE = R(1 - \cos \alpha)$$

$$= 0.5 \, \Delta h$$

$$l = \sqrt{R\Delta h - (0.5\,\Delta h)^2} \quad \text{usually } (0.5\,\Delta h)^2 < \Delta h < R$$

hence $$l = \sqrt{R \Delta h}$$

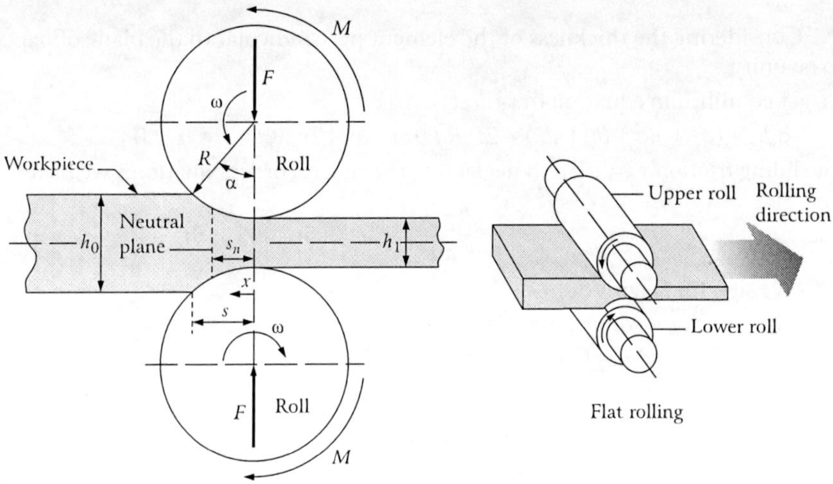

Fig. 4.3 Parameters of float rolling

4.2 ROLLING PRESSURE

4.2.1 Assumptions

Before calculating the rolling pressure, some assumptions are considers and they are followings:

(i) Material is homogeneous throughout.
(ii) Material deform plastically in uniform manner.
(iii) Co-efficient of friction is constant between the roller and metal sheets.
(iv) Rollers are rigid.
(v) Strip is wide compared with its thickness.
(vi) Peripheral velocity of the rolls are uniform i.e., rolls are neither accelerating nor decelerating.
(vii) Elastic deformation of strip is negligible in comparison to plastic deformation.

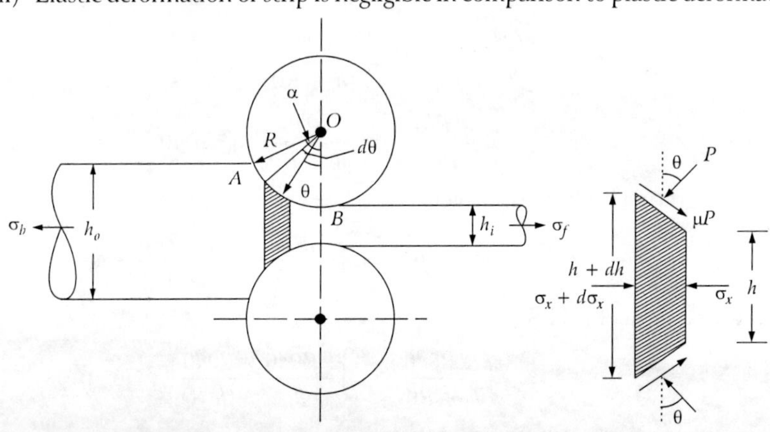

Fig. 4.4 Stress equilibrium of an element in rolling

Considering the thickness of the element perpendicular to the plane of paper to be unity,

we get equilibrium equation in x-direction as

$$-\sigma_x h + (\sigma_x + d\sigma_x)(h + dh) - 2P\,Rd\,\theta \sin\theta + 2\tau_x\,Rd\,\theta\cos\theta = 0$$

for sliding friction $\tau_x = \mu P$ and neglecting the second order equations, we have

$$\frac{d(\sigma_x h)}{d\theta} = 2PR\,(\sin \mp \mu \cos\theta)$$

–ive sign for lagging zone and

+ive sign for leading zone

if $\sin\theta = 0$ and $\cos\theta = 1$, then

$$\frac{d(\sigma_x h)}{d\theta} = 2PR\,(\theta \mp \mu)$$

by Von-Mises equation $\quad P - \sigma_x = \frac{2}{\sqrt{3}}\sigma_0 = \sigma_0'$

$$\frac{d}{d\theta}\left[h(P - \sigma_0')\right] = 2PR\,(\theta \mp \mu)$$

or

$$\frac{d}{d\theta}\left[\sigma_0' h\left(\frac{P}{\sigma_0'} - 1\right)\right] = 2PR\,(\theta \mp \mu)$$

$$\sigma_0' h\frac{d}{d\theta}\left(\frac{P}{\sigma_0'}\right) + \left(\frac{P}{\sigma_0'} - 1\right)\frac{d}{d\theta}(\sigma_0' h) = 2PR\,(\theta \pm r)$$

due to sealed rolling σ'_0 increases as h decreases, thus considering $\sigma'_0 h$ is constant then

$$\frac{(d/d\theta)(P/\sigma_0')}{P/\sigma_0'} = \frac{2R}{h}(\theta \pm \mu)$$

Now $\qquad h = h_1 + 2R\,(1 - \cos\theta)$

or $\qquad h = h_1 + R\theta^2$

$$\frac{(d/d\theta)(P/\sigma_0')}{P/\sigma_0'} = \frac{2R}{h_1 + R\theta_2}(\theta \pm \mu)$$

$$\frac{d(P/\sigma_0')}{P/\sigma_0'} = \frac{2R}{h_1 + R\theta_2}(\theta \pm \mu)d\theta$$

On integrating both sides, we get

$$\log_e (P/\sigma_0') = \int \frac{2R\theta d\theta}{h_1 + R\theta^2} \pm \int \frac{2R\mu}{h_1 + R\theta^2} d\theta$$

Now first term of R.H.S

$$\int \frac{2R\theta d\theta}{h_1 + R\theta^2} = \int \frac{2R\theta d\theta}{h} = \frac{2\theta d\theta}{(h/R)}$$

$$h/R = \frac{h_1}{\theta} + \theta^2$$

Now $$\frac{d}{d\theta}(h/R) = 2\theta \; (h_1 \text{ and } R = \text{constant})$$

$$\int \frac{2\theta d\theta}{h/R} = \log_e \frac{h}{R}$$

Second term of R.H.S. $$\int \frac{2R\mu}{h_1 + R\theta^2} d\theta = \int \frac{2\mu}{\left(\frac{h_1}{R} + \theta^2\right)} d\theta$$

$$= 2\mu \sqrt{\frac{R}{h_1}} \tan^{-1} \sqrt{\frac{R}{h_1}} \cdot \theta$$

Now final equation

$$\log_e \left(\frac{P}{\sigma'_0}\right) = \log_e \left(\frac{h}{R}\right) \pm 2\mu \sqrt{\frac{R}{h_1}} \tan^{-1} \sqrt{\frac{R}{h_1}} \cdot \theta + \log_e C$$

$$P = C\sigma'_0 \left(\frac{h}{R}\right) e^{\pm \mu H}$$

where $$H = 2\sqrt{\frac{R}{h_1}} \tan^{-1} \sqrt{\frac{R}{h_1}} \theta$$

at entry $\theta = \alpha$

(hence $H = H_o$ with θ replaced by μ at exit $\theta = 0$) $\therefore$ $H = H_1 = 0$

Also at entry and exit $\sigma_x = 0$

$$P = \sigma'_0$$

In the lagging zone (entry zone)

$$\sigma'_0 = C\sigma'_0 \left(\frac{h_0}{R}\right) e^{-\mu H_0}$$

$$C = \frac{R}{h_0} e^{\mu H_0}$$

$$P = \sigma'_0 \frac{h}{h_0} e^{\mu (H_0 - H)}$$

In leading zone (exit zone) $C = R/h_1$

$$P = \sigma_0^1 \frac{h}{h_1} e^{\mu H}$$

Maximum Draft:

$$\Delta h = h_0 - h_i$$

$$\Delta h = 2R\,(1 - \cos \alpha)$$

Horizontal component of forces

$$F_x = \mu P \cos \alpha - P \sin \alpha$$

$f_x \geq 0$, $\mu \geq \tan \alpha$

$$\tan \alpha_{max} = \mu$$

$$\cos \alpha_{max} = \frac{1}{\sqrt{1 + \mu_2}}$$

$$\Delta h_{max} = 2R\,(1 - \cos\alpha_{max})$$

$$= 2R\left(1 - \frac{1}{\sqrt{1-\mu^2}}\right)$$

$$= \mu^2 R$$

$$\boxed{\Delta h_{max} = \mu^2 R}$$

4.2.2 Types of Rolling Mills

Rolling mills are usually five types as given below:

- Two-high mill
- Three-high rolls
- Four-high rolls
- Cluster roll
- Planetary mill

4.2.2.1 Two-high Mill

In this type of rolling mill there are two heavy horizontal rolls, placed exactly one over other. Both rollers are rotating in the opposite direction. In two rollers one roller is fixed and the gap between the two rollers can be adjust as per requirement and rollers are supported on a bearing fitted in housing. Workpiece can be rolled by feeding from one direction only.

There is another type of two-high mill which incorporate a drive mechanism that can reverse the direction of rotation of rolls known as Two-high Reversing mill.

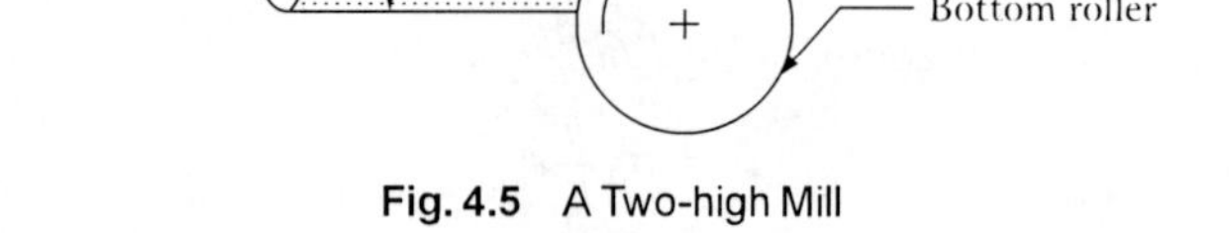

Fig. 4.5 A Two-high Mill

4.2.2.2 Three-high Rolling Mills

There are three rolls in the three rolling mills. Each roller is positioned directly one over the other as shown in Fig. 4.6. Upper and bottom rolls are rotating in the same direction while the middle (centre) roller rotates in the opposite direction of both of these. All the rolls continuously rotate in the fixed direction and direction can not be changed. In three high rolling mills workpiece is fed in one direction only i.e., between the upper and middle rolls and finally pass between the centre rolls and bottom rolls less power is required in these mills. This process is used for **blooming, billet rolling** or **finish rolling**.

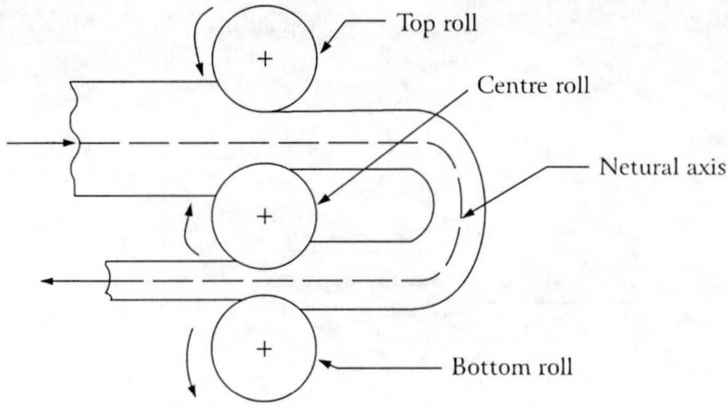

Fig. 4.6 Three-high rolls

4.2.2.3 Four-high Mills

It consist four rolls. Out of four rolls, two rolls are small and two are larger in diameter. Larger diameter roll is known as backup roll while small diameter roll is known as working roll. Main function of larger diameter roll is to prevent the direction of small rolls and the main function of small diameter roll is to concentrate the total rolling pressure over the workpiece.

Application of these mills are subsequent rolling the slab and common products of these mills are hot or cold plate and sheets.

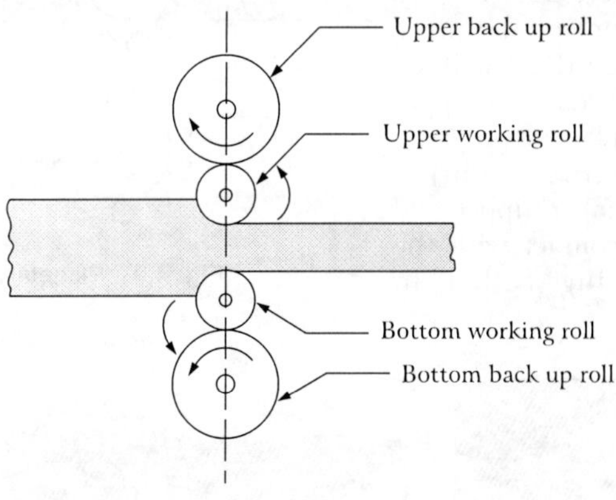

Fig. 4.7 Four-high rolling mill

4.2.2.4 Cluster Rolling Mills

In cluster rolling there are two working rolls of smaller diameter while four or more backup rolls of large diameter is used. Figure 4.8 shows the arrangement of cluster rolling mills and the number of back up rolls may be upto 20 or more depending upon the amount of support needed for the working rolls during process. This type of mills are generally used for cold rolling. Common rolled widths are between 0.66 to 1.5 meter.

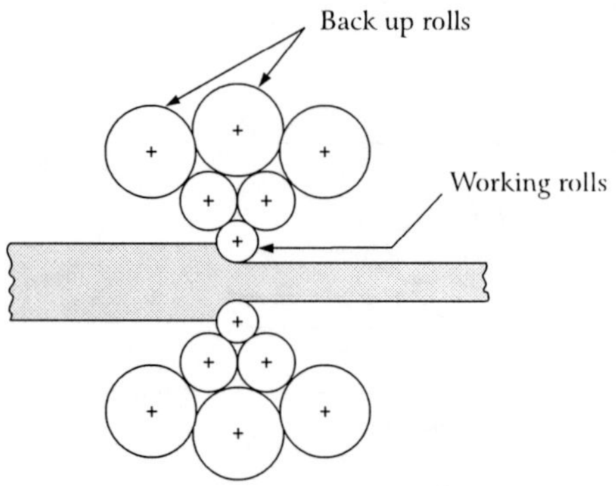

Fig. 4.8 Cluster roll

4.2.2.5 Planetary Rolling Mills

In this type of mills, small diameter rolls are used for rolling purpose. Planetary rolling mills consist of two heavy and large backing rolls as shown in Fig. 4.9 and these backing rolls are surrounded by various small diameter planetary rolls mounted in a cage by the planetary rolling mills. Thickness of sheet can be reduced 25:1 ratio as against in conventional rolling mills. In this process the strip is fed first through the feed rolls as upper and lower cage of planetary rolls are geared together to bring corresponding smaller, upper and lower rolls into contact with the metal strip being called in synchronism.

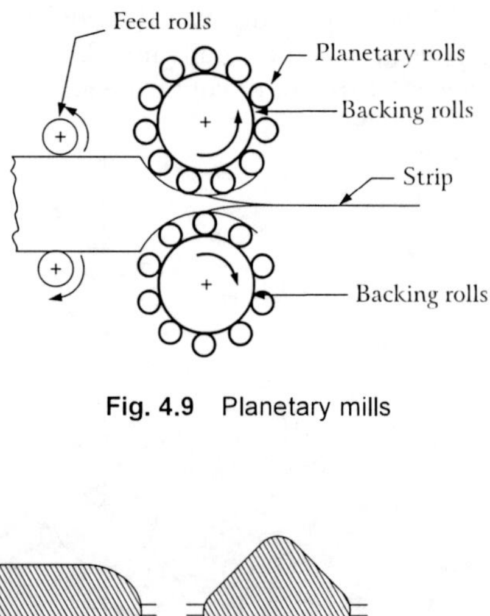

Fig. 4.9 Planetary mills

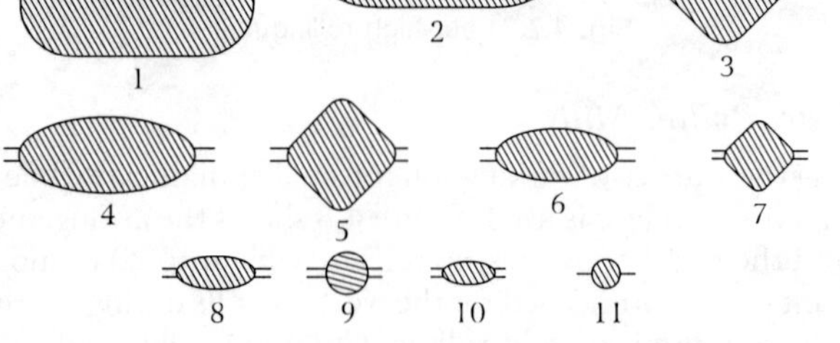

Fig. 4.10 Showing the sequences in reducing a billet of 100 mm × 100 mm to a round bar stock

4.3 ROLLING DEFECTS

[UPTU 2008-09]

During the rolling process various type of defects are produced due to which product or parts can not be further used and these defects are such as:

(i) Wavy edge
(ii) Edge cracks
(iii) Zipper cracks
(iv) Alligatoring defect
(v) Surface defect

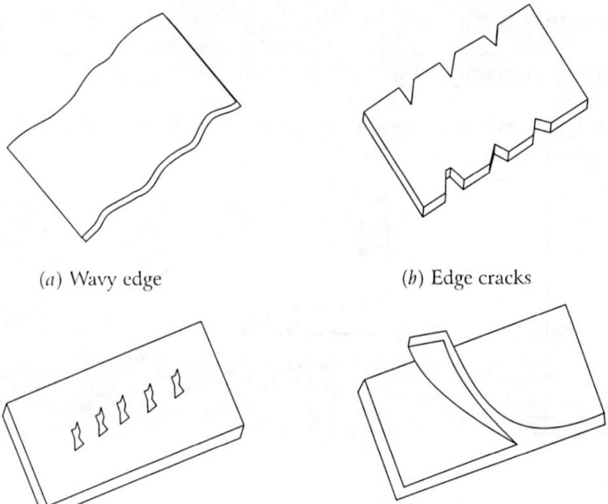

(*a*) Wavy edge (*b*) Edge cracks

(*c*) Zipper cracks (*d*) Alligatoring

Fig. 4.11 Rolling defects

4.3.1 Wavy Edge Crack

This type of cracks produced by the bending of the rolls. Figure 4.11 (*a*) shows the wavy edge crack as edge of strip elongates more as compared the centre position resulting the edge of strip are thinner than the centre. In this defect strip may buckle.

4.3.2 Edge Crack

This type of cracks are produced at the edge of the strip Fig. 4.11(*b*) shows the edge cracks. This type of cracks is produced due to low ductility of material (strip) as well as not applying the uniform load on strip.

4.3.3 Zipper Cracks

This type of cracks produced at the centre of strip. Fig. 4.11(*c*) shows the zipper cracks. This type of defect also produced due to low ductility of material or manufacturing process.

4.3.4 Alligatoring Defect

This type of defect occur very commonly in rolling process. Figure 4.11(*d*) shows the alligatoring defects and this type of defect basically produce due to in homogeneous deformation of metal during rolling process. As the name depicts this type of the defect are like the shape of alligator.

4.3.5 Surface

This type of defect may be produced due to inclusion, impurities in metal, rust, scales, scratches etc. on the sheets surface. In surface defects, cracks, roll marks, gauges, pits etc. may occur.

4.4 ROLLING PRODUCTS

Through rolling process various types of shapes can be generate.

They are given below:

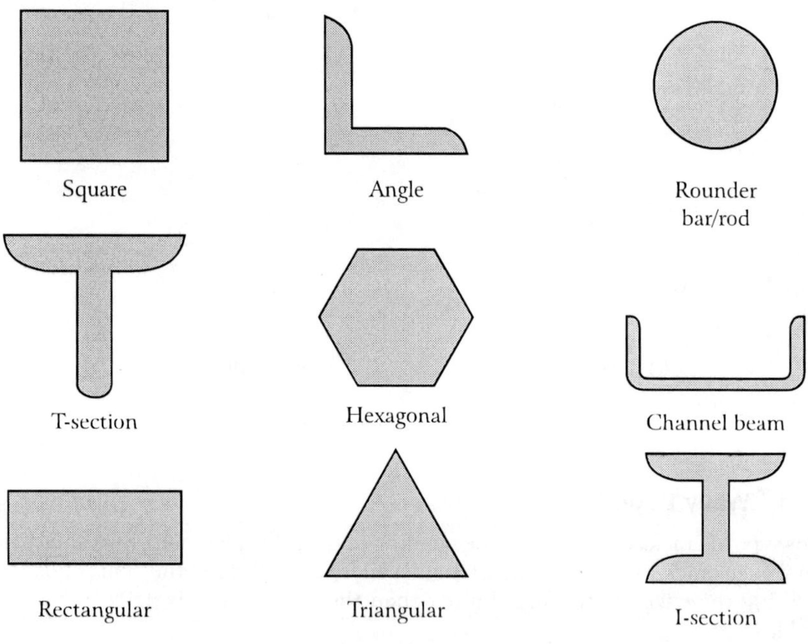

Fig. 4.12 Different shapes produced by rolling process

Example 4.1

A steel plate 25 mm thick is to be rolled to 15 mm in a five high mill having roll diameter 500 mm. If the yield stress is 15 kg/mm^2 then determine:

(i) angle of bite
(ii) coefficient of friction if max reduction possible
(iii) position of natural section
(iv) backward slip
(v) forward slip
(vi) max pressure.

Solution:

t_i = initial plate thickness = 25 mm
t_f = final plate thickness = 15 mm
$\Delta t = t_i - t_f = 25 - 15 = 10$ mm

$$R = \text{radius of roll} = \frac{500}{2} = 250 \text{ mm}$$

(i) Angle of bite

$\Delta t = 2R\,(1 - \cos\alpha)$,

$10 = 2 \times 250\,(1 - \cos\alpha)$

$\cos\alpha = 0.98$ or $\alpha = 11.5°$ **Ans.**

(ii) Coefficient of friction

$\mu = \tan\alpha = \tan 11.5° = 0.203$ **Ans.**

(iii) Position of netural section

$$n = \frac{2\mu L}{\Delta t}$$

$$L = \sqrt{R\,\Delta t} = \sqrt{250 \times 10} = 50 \text{ mm}$$

$$n = \frac{2 \times 0.203 \times 50}{10} = 2.03 \simeq 2$$

(Px) leading = (px) lagging

$$(n+1)\left(\frac{t_x}{t_f}\right)^n - 1 = (n-1)\left(\frac{t_i}{t_f}\right)^n + 1$$

$$(2+1)\left[\left(\frac{t_x}{15}\right)^2 - 1\right] = (2-1)\left[\left(\frac{25}{t_x}\right)^2 + 1\right]$$

$$3\left[\left(\frac{t_x}{15}\right)^2 - 1\right] = \left(\frac{25}{t_x}\right)^2 + 1$$

$t_x = 21.21$ **Ans.**

(iv) Backward slip

$$\frac{V - V_0}{V} = 1 - \frac{V_0}{V}$$

$$= 1 - \frac{t_x}{t_f} = 1 - \frac{21.21}{25} = 0.1514$$

or $= 15.14\%$ **Ans.**

(v) Forward slip

$$\frac{V_1 - V}{V} = \frac{V_1}{V} - 1$$

$$= \frac{t_x}{t_i} - 1 = \frac{21.21}{15} - 1 = 0.414$$

or $= 41.4\%$ **Ans.**

(vi) Max. Pressure $= 2K\left(1 + \dfrac{\mu L}{2h_{max}}\right)$

but $h_{max} = \mu^2 R$ or $D(1 - \cos\alpha)$

$= (0.203)^2 \times 250$

$= 10.302$

$$P_{\max} = 2\times15\left(1+\frac{0.203\times50}{2\times10.302}\right)$$
$$= 30(1 + 0.49)$$
$$= 30 \times 1.49$$
$$= 44.77 \text{ MPa}$$

Example 4.2

(*a*) A 500 mm thick slab is to be colled rolled. The roll diameter is 800 mm and the coefficient of friction is 0.05. Determine the maximum possible reduction.

(*b*) What is the maximum reduction on the same mill for hot rolling when co-efficient of friction is 0.6.

Solution:

(*a*) $(\Delta h)_{\max} = \mu^2 R = (0.05)^2 \times 400 = 1$ mm

[as $D = 800$ mm or $R = \frac{800}{2} = 400$]

(*b*) $(\Delta h)_{\max} = \mu^2 R = (0.6)^2 \times 400 = 144$ mm **Ans.**

Example 4.3

A strip of cross-section 160 mm × 5.0 mm is rolled with 25% reduction of area via 500 mm diameter rolls. Calculate

(*a*) Angle subtended by deformation zone at the roll centre.

(*b*) Final thickness of strip.

(*c*) Location of neutral plane. Take $\mu = 0.2$.

Solution:

(*a*) $\alpha = \sqrt{\frac{\Delta h}{R}} = \sqrt{\frac{h_i - h_f}{R}}$, $\quad h_f = (1 - 0.25) \times 5.0 = 3.75$

$D = 500, R = 250$

$$\alpha = \sqrt{\frac{5-3.75}{250}} = 0.071 \text{ rad} \quad \textbf{Ans.}$$

(*b*) $h_f = 3.75$ mm **Ans.**

(*c*) $$H_i = 2\sqrt{\frac{R}{h_f}}\tan^{-1}\left(\sqrt{\frac{R}{h_f}}\alpha\right) = 2\times\sqrt{\frac{250}{3.75}}\tan^{-1}\left(\sqrt{\frac{250}{3.75}}\times0.71\right)$$

$$= 491.54$$

$$H_n = \frac{1}{2}\left[\frac{1}{\mu}\log_e\frac{h_f}{h_i} + H_i\right] = \frac{1}{2}\left[\frac{1}{0.2}\log_e\frac{3.75}{5.0} + 491.54\right] = 245.1$$

(*d*) $$\theta_n = \sqrt{\frac{h_f}{R}}\tan\left(\sqrt{\frac{h_f}{R}}\cdot\frac{H_n}{2}\right) = \sqrt{\frac{3.75}{250}}\tan\left(\sqrt{\frac{3.75}{250}}\times\frac{245.1}{2}\right)$$

$$= 0.122 \times 0.268 = 0.0327 \text{ rad} \quad \textbf{Ans.}$$

Example 4.4

A mild steel strip is rolled to a 5 mm final thickness with a reduction of 25% roll diameter is 800 mm. Calculate the neutral plane. If μ = 0.4.

Solution:

$$h_i = \frac{h_f}{(1-.25)} = \frac{5}{0.75} = 6.67$$

$$\Delta h = 6.67 - 5 = 1.67 \text{ mm.}, \quad \Delta h = R(1 - \cos\alpha)$$

$$1.67 = 800\,(1 - \cos\alpha), \quad 1 - \cos\alpha = 0.00417,$$

$$\cos\alpha = 0.9958$$

or

$$\alpha = 5.23° \quad \text{or} \quad 0.0916 \text{ rad}$$

$$H_i = 2\sqrt{\frac{R}{h_f}}\tan^{-1}\left(\sqrt{\frac{R}{h_f}} \times \alpha\right)$$

$$= 2 \times \sqrt{\frac{400}{6.67}}\tan^{-1}\left(\sqrt{\frac{400}{6.67}} \times 0.0916\right) = 70.7 \times 7.74 = 547.5$$

$$H_n = \frac{1}{2}\left(H_i - \frac{1}{\mu}\log_e\frac{h_i}{h_f}\right)$$

$$= \frac{1}{2}\left(547.5 - \frac{1}{0.4} \times \log_e\frac{6.67}{5}\right) = 273.38$$

$$\theta_n = \sqrt{\frac{h_f}{R}}\tan\left(\sqrt{\frac{h_f}{R}} \times \frac{H_n}{2}\right)$$

$$= \sqrt{\frac{6.67}{400}}\tan\left(\sqrt{\frac{6.67}{400}} \times 136.69\right)$$

$$= 0.3182 \times 0.129 = 0.041 \text{ rad}$$

or

$$\theta_n = 2.35° \quad \textbf{Ans.}$$

Example 4.5

A mild steel strip is rolled from 4.0 mm to 3.0 mm thickness in a single pass rolling mill. Dia of roller is 280 mm and the width of strip is 400 mm while coefficient of friction is 0.1. Take stress 150 MPa for metal. Calculate the roll separating force assume negligible spreading.

Solution: $F = LWP_{av}$, $W = 400$ mm, $K = 150$ MPa

$$L = \sqrt{R\,\Delta h}\,; \quad \text{and} \quad P_{av} = 2K\left(1 + \frac{\mu L}{2h_{av}}\right)$$

$$D = 280 \text{ mm}, \quad R = \frac{280}{2} = 140$$

$$h_{av} = \frac{h_i + h_f}{2} = \frac{4.0 + 3.0}{2} = 3.5 \text{ mm}$$

$$\Delta h = h_i - h_f = 4.0 - 3.0 = 1.0 \text{ mm}$$

$$L = \sqrt{140 \times 1} = 11.83 \text{ mm}$$

$$P_{av} = 2\times150\left(1+\frac{0.1\times11.83}{2\times3.5}\right) = 300\times1.169 = 350.71 \text{ MPa}$$

$$F = L \times W \times P_{av}$$

$$= 11.83 \times 400 \times 350.7 \times 10^{-3} \times 10^{3} = 1659.5 \text{ kN} \quad \textbf{Ans.}$$

4.5 LUBRICATION IN METAL FORMING PROCESS

Lubrication is very important in metal forming process. When workpiece (raw material) is rolled/passed in between the two rollers/die and due to friction heat produced which results in wearing the die or final products, to avoid this, lubrication in between the dies/rollers is must. Lubrication also reduces the friction and also responsible for smooth flow of metal between dies/rollers. Lubrication protects metal surface to prevent **scratching, blistering** etc.

Lubrication in metal forming process basically depends upon the following factors:

(i) workpiece material,
(ii) speed of metal forming operation,
(iii) pressure at which the metal forming operation performed.

4.5.1 Functions (Characteristics) of Lubrication

(i) To control the friction between the die/roller and workpiece,
(ii) To reduce the wear and tear of die,
(iii) To increase the tool life and die life,
(iv) To control the material flow and there by die filling,
(v) To control the deformation energy and load.
(vi) To control the heat flow from interface.

However if lubrication is not proper then wear and tear of die may take place, lubrication means to make the surface smooth or oily in motion.

The primary function of lubrication is to avoid the metal contact with each other by making a thin film of lubricant on the wall of dies.

Lubrication is done in metal forming process to remove the heat by reducing the friction and to transmit the minor finish surface to the workpiece/die.

4.5.1.1 Advantages of Lubrication in Metal Forming Process

The following advantages are possible by lubrication in metal forming,

(i) Less power requirement is possible by lubrication,
(ii) Lubrication improves the workability,
(iii) Lubrication improves the ability of object to pass through die at high speed without sticking.

4.5.2 Method of Lubrication

Lubrication is expressed by Reynold's equation. Modern lubrication theory includes thermal effect, inertia effect and turbulence effect.

Lubrication are usually two types:

(i) Fluid Lubrication
(ii) Boundry Lubrication

4.5.2.1 *Fluid Lubrication*

Fluid lubrication is also known as hydrodynamic lubrication. It include complete separation of the sliding surface by a thick layer of lubrication in such a way that no direct contact between the two surfaces take place and it is possible by reducing the value of coefficient of friction.

4.5.2.2 *Boundry Lubrication*

Boundry lubrication occurs when the layer of lubricant is only a few molecules thick, boundry lubrication condition is generally produced by adding a relatively small quantity of polar organic component *i.e.,* Fatty acid in a lubricating oil. Boundry lubricant gives best performance when it is solid like WS_2, MOS_2 etc.

There is also another type of lubrication that is known as **mixed lubrication**. In this type of lubrication system metal surfaces are partly in direct contact and partly separated by a liquid film. This type of lubrication system is governed by viscosity of lubricant, load applied for operation, relatively speed of two surfaces.

In general the basic **object** of lubricant is to **reduce the coefficient of friction, rolling pressure, forging load, die wear** and it also **improves the flow of metal**. Lubricant act as a parting compound and thus prevent sticking of the parts in die/rollers etc.

4.5.3 Recommendation of Different Lubrication

Different type of lubrication are used or recommended for different metal forming process.

S.No.	*Type of Lubricants*	*Metal forming Processes*
1.	Colloidal graphide,	Forging
2.	Paraffin	Rolling
3.	Sodium soap in hot solution of calcium stearate, lead.	Cold drawing
4.	Loanolin	Extrusion

OBJECTIVE QUESTIONS

1. The property of material by which it can be rolled into sheets is known as
   (*a*) malleability (*b*) plasticity
   (*c*) ductility (*d*) elasticity

2. It is required to reduce a slab directly to strip in one pass, which of the following rolling mills can do this function?
   (*a*) continuous rolling mill (*b*) two-high mill
   (*c*) three-high mill (*d*) planetary mill

3. Equipment or machinery used to produce I-section or channel, angle, rail section etc. is known as
   (*a*) rolling mills (*b*) hot spinning machine
   (*c*) extrusion process (*d*) continuous casting machine

4. In four-high rolling mill the bigger rollers are referred as
   (*a*) support rolls (*b*) back up rolls
   (*b*) guide rolls (*d*) main rolls

5. In a four high rolling mill the diameter of backing up roll in comparison to diameter of working rolls is
   (*a*) smaller (*b*) bigger
   (*c*) equal (*d*) no such condition

6. Two rolls of a two high mill rotates
   (*a*) Same speed and same direction
   (*b*) Different speed but same direction
   (*c*) Different speed and different direction
   (*d*) Same speed and different direction

7. In a 3-high mills
   (*a*) upper and lower rollers rotate in same direction
   (*b*) upper and lower rollers rotate in opposite direction
   (*c*) middle roller also rotate the direction of upper and lower roller
   (*d*) none of these

8. In a four high rolling mills
   (*a*) Two working roller and two backup rollers
   (*b*) In between there are two working rollers
   (*c*) Backup rollers are placed at top and bottom
   (*d*) None of the above

9. Rolls used on rolling mills contain the following
   (*a*) Neck (*b*) Body
   (*c*) Wabbler (*d*) All of these

10. Blooming mills and stabbing mills are
   (*a*) Secondry mills (*b*) Both rolling mills
   (*c*) Primary mills (*d*) All of the above

11. In a rolling process, sheet of 25 mm thickness is rolled to 20 mm thickness. Roll is of diameter 600 mm and it rotates at 100 r.p.m. roll strip contact length will be *[GATE 2004]*
   (*a*) 5 mm (*b*) 39 mm
   (*c*) 78 mm (*d*) 120 mm.

12. Thickness of a metallic sheet is reduced from an initial value of 16 mm to a final value of 10 mm in one single pass. Rolling with in a pair of cylindrical rollers each diameter is 400 mm, bite angle on degree is *[GATE 2007]*
   (*a*) 5.936 (*b*) 7.936
   (*c*) 8.936 (*d*) 9.936

13. In a single pass rolling operation a 20 mm thick plate with plate width of 100 mm is reduced to 18 mm, roller radius is 250 mm and rotational speed is 10 r.p.m. The average flow stress for the plate material is 300 MPa. Power required for the rolling operation in kW is closet to *[GATE 2008]*
   (*a*) 15.2 (*b*) 18.2
   (*c*) 30.4 (*d*) 45.6
   (*e*) none of these

14. In a single pass rolling process using 410 mm diameter steel rollers, a strip of width 140 mm and thickness 8 mm undergoes 10% reduction of thickness. The angle of bite in radians is *[GATE 2012]*
   (*a*) 0.006 (*b*) 0.031
   (*c*) 0.062 (*d*) 0.600

ANSWERS

1. (*c*)	2. (*d*)	3. (*a*)	4. (*b*)	5. (*b*)
6. (*a*)	7. (*a*)	8. (*a*)	9. (*d*)	10. (*b* and *c*)
11. (*b*)	12. (*d*)	13. (*e*)		

HINTS

Ans. 11: t_i = 25 mm,
t_f = 20 mm,
D = 600 mm

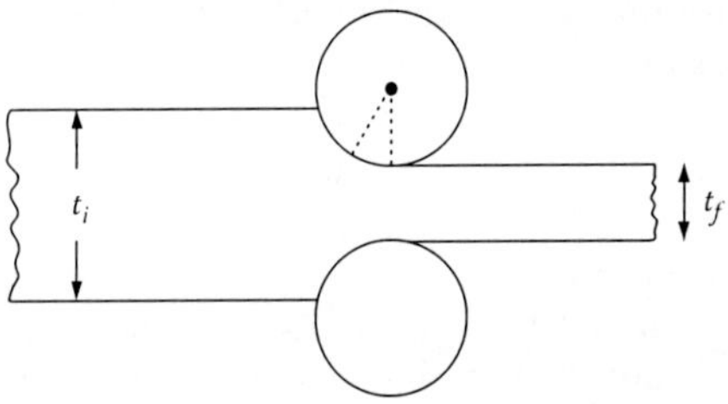

$$R = \frac{600}{2} = 300 \text{ mm}$$

$$N = 100 \text{ r.p.m.}$$

Angle subtended by deformation zene at the roll centre

$$\theta = \sqrt{\frac{t_i - t_f}{R}} = \sqrt{\frac{25-20}{300}} = 0.129 \text{ rad.}$$

Roll strip contact length = $\theta \times R$
= 0.129 × 300 = 38.72 mm **Ans.**

Ans. 12. $\tan\alpha = \sqrt{\frac{t_i - t_f}{R}}$

where t_i = 16 mm
t_f = 10 mm
D = 400 mm or R = 200 mm

$$\tan\alpha = \sqrt{\frac{16-10}{200}} = \sqrt{\frac{6}{200}} = \sqrt{0.03}$$

$$\tan\alpha = 0.173 \quad \text{or} \quad \alpha = 9.826 \quad \textbf{Ans.}$$

Ans. 13. t_i = 20 mm
b = 100 mm
t_f = 18 mm
R = 250 mm
N = 10 r.p.m
σ_0 = 300 MPa
P = ?

$$\text{Load} = \sigma_0 \times L \times N$$

when $$L = \sqrt{R\Delta h}$$

$$= 300 \times 10^6 \sqrt{0.25 \times (0.02 - 0.018)} \times 0.1 = 670 \text{ N}$$

$$\text{Power} = \text{Force} \times \text{Velocity}$$

$$= 670 \times \frac{10 \times 3.14 \times 2}{60} \times 0.25 = 175.3 \text{ kW} \quad \textbf{Ans.}$$

REVIEW QUESTIONS

1. Define the rolling process in detail. *[UPTU 2004-05]*
2. Discuss the mechanism and working principle of rolling.
3. Define the various type of rolling mills.
4. Define the following
   (*i*) Two-high roll mills (*ii*) Three-high roll mills
   (*iii*) Four-high mills (*iv*) Cluster rolling mills
   (*v*) Planetary rolling mills
5. Prove that maximum possible draft in case of strip rolling can be expressed as $(h_b - h_a)_{max} = \mu^2 R$ if the friction condition for included entry of the workpiece is satisfied. Where h_b and h_a are the thickness of strip before and after rolling, respectively.
   μ = coefficient of friction between roll and strip, R = radius of roll. *[UPTU 2005, 08-09]*
6. Describe common rolling defects. Also give their remedies. *[UPTU 2001-02, 2002-03, 2003-04, 2004-05]*
7. Draw the figure of component produce by rolling. *[UPTU 2002-03, 2005-06]*
8. A flat strip of thickness h_1 = 5 mm is to be reduced to h_2 = 4 mm. Taking μ = 0.1 and considering this friction to be enough for rolling. Find the roll radius R also
   (*i*) roll pressure = $2K\left(1 + \frac{\mu L}{(h_1 + h_2)}\right)$, L = roll contact length, (*ii*) Power required for rolling. Take roll rpm as 6 and b = 20 mm and shear strength for strip $K = 200 \text{ N/mm}^2$. *[UPTU 2005-06]*
9. What do you mean by friction hill rolling. Drive a relation to determine roll separating forces. *[UPTU 2006-07]*
10. Describe the various rolling defects and their remedies. *[UPTU 2001, 02, 02-03, 03-04, 04-05, 10-11]*
11. What is a deformation zone in rolling process. Explain the concept of area of contact and bite angle in rolling process.
12. State the factors that affecting the rolling process.
13. Explain the term **backward slip, forward slip, no slip.**
14. Explain the terms:
    (*i*) Bite angle, (*ii*) Draught, (*iii*) No slip angle, (*iv*) Friction angle.
15. Calculate the bite angle when rolling 15 mm thick plates using rolls of 400 mm diameter. Final thickness of plate is 12 mm. *[UPTU 2010-11]*

❑❑❑

UNIT-III

SHEET METAL WORKING

Chapter 5 : Presses

5

CHAPTER

PRESSES

5.1 INTRODUCTION

Press is a machine tool which generally used for performing the cold working operations. Mostly presses are used for mass production purpose of identical components quickly and accurately from sheet. Press working is a metal forming operation in which a set of die and cutting tool (punch) is used to perform the operation and the shape and size of tool and die is the replica of shape that is to be produced.

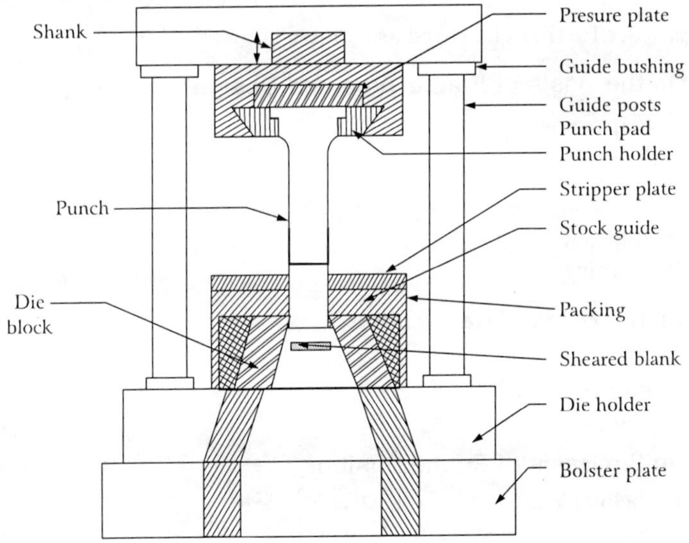

Fig. 5.1 Main components of a press

Large number industries involving the use of presses like **automobile, aircraft, electrical appliances, coining, kitchen, notched, motor laminations, jewellery etc.**

Presses are of different types such as hand operated or motorised press as per size of component.

Press basically consist the following components:

(i) **Rigid Frame:** Every press consist a rigid frame and entire frame weight is **borne** by the bed.

(ii) **Bed:** Every press have a rigid bed. Usually it is made out of cast iron. Entire weight of press body is borne by bed and bed is mounted by the means of some fasteners. Cast iron is used as bed material because it can absorb all vibration it self.

(iii) **Ram:** Ram take the motion in up ward and down ward position and **punch/ cutting** tool is mounted with Ram. As Ram take up and down motion tool also moves up and down.

(iv) **Die:** Die is mounted with the bed and it is stationary. Die assembly comprises
(*a*) Female member or a die mounted and bolted to the bolster plate.
(*b*) Male part or punch mounted with ram, for reciprocating up and down.

(v) Press accessories and auxiliary equipment which helps in conducting the press operation effectively.

5.2 CLASSIFICATION OF PRESSES

It is not easy task to classify the presses on the basis of operation performed on them. As most of them are capable to performing the same operation, however, presses are classified into two groups in a broad manner.

1. **Manual operated press:** hand press, fly press, ball press.
2. **Power operated:** hydraulic, mechanical, pneumatic etc.

Presses are further classified as follows:

5.2.1 On the Basis of Frame Type and Design

(i) Gap frame
(ii) Straight side
(iii) Inclinable
(iv) Open end
(v) Adjustable bed
(vi) Horning.

5.2.2 On the Basis of Ram Action

(i) single action
(ii) double action
(iii) triple action.

5.2.3 On the Basis of Frame Position

(i) inclinable
(ii) horizontal
(iii) vertical
(iv) inclined.

5.2.4 On the Basis of Method of Power Transmission by Motor to Crankshaft

(i) Direct
(ii) Non-geared fly wheel
(iii) Single geared
(iv) Double geared
(v) Multiple geared

5.2.5 On the Basis of Mechanism used to Applying Power to Ram

(i) Crank
(ii) Cam
(iii) Rack and Pinion
(iv) Toggle
(v) Eccentric
(vi) Hydraulic
(vii) Screw
(viii) Pneumatic

5.2.6 On the Basis of Number of Drive Gear

(i) Single drive
(ii) Twin drive
(iii) Quadrable drive

5.2.7 On the Basis of Number of Crank Shaft Used

(i) Single
(ii) Double

5.2.8 On the Basis of for which Purpose used

(i) Shearing
(ii) Punching
(iii) Seaming
(iv) Coining
(v) Extruding
(vi) Straightening
(vii) Forging

5.3 SELECTION OF A PRESS

Before starting the press work it is very essential to select a right press for right work. For example, it is very economical to select a press which perform a complete operation in a single stroke rather than selecting the different (more than one press) press. Design of press tool should be simple. Hydraulic presses are used for forging and coining operation where a comparatively slow stroke is advisable.

Before selecting a press the following points are consider:

(*a*) Forces required to cut the metal
(*b*) Stroke length
(*c*) Method of feeding and size of sheet blank
(*d*) Size and type of die
(*e*) Shut height
(*f*) Type of operation

5.4 DIE AND PUNCH ASSEMBLY

Die assembly is the most important part of a press tool. Die and punch assembly shown in Fig. 5.2. It consist a **die block, die retainer punch, stipper plate** or **pressure plate, punch retainer/pad**, vertical guide post, pushing. Die and punch are made by die steel (hardend material) and Punch is the replica of shape that to be produced. In die and punch assembly punch is in solid form where as die have hollow (cavity) shape.

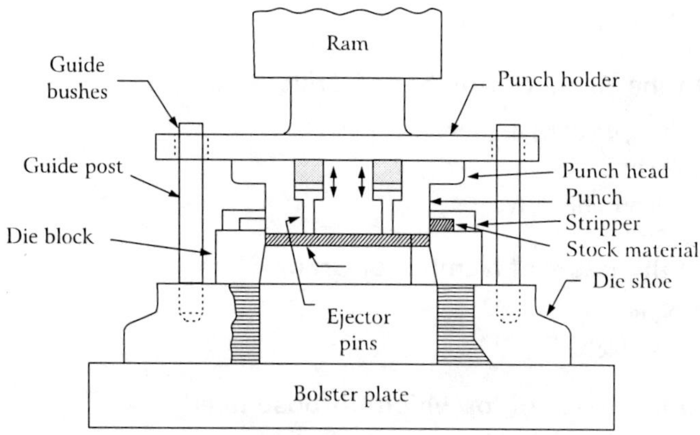

Fig. 5.2 Details of a die set (Die and Punch Assembly)

5.4.1 Main Parts of a Press

5.4.1.1 Die block

It is die cavity which cut the blank to required size and shape from sheet (Raw material). It also imparts the desired profile to drawn part. Die block is made out of hardend steel usually die steel.

5.4.1.2 Stripper

It is mounted on the die in different ways depending upon the types and design of die. It helps in freeing the punch from scrap in the return stroke as the later tries to cling to the former.

5.4.1.3 Die shoe and die retainer

It holds the die block at proper position with regard to the Punch. It is also known as lower shoe of die set.

5.4.1.4 Punch

Punch is the main element of the die assembly set. Basic function of Punch is to cut the workpiece (sheet metal) stock. Once the sheet is kept in between the die and punch is the replica of shape that is to be produced. It presses the sheet inside the die as compressive load is applied and finally produce the required shape and size of blank (object). Punch is associated with the press ram or slide. Punch is also made out of hardend steel and it comes out in the category of cutting tool.

5.4.1.5 Punch Pad

It helps in holding the punch in proper alignment with respect of die and secured to the punch holder. It is also referred as punch retainer.

5.4.1.6 Pressure Plate

It is usually kept in between the punch and the punch holder. Basics function/objective of pressure plate is to distribute the punch pressure over a wide area on punch holder device.

5.4.1.7 Guide Posts

They are basically used to obtaining the correct alignment of Punch holder with die shoe.

5.4.1.8 Pilot

Correct location of blank is very important if mechanical feeding is adopting. Correct location is ensured by the pilots provided in the punch holder.

5.4.1.9 Stops

They help to restricting the feed of the stock material to a pre-determined length each time so that correct spacing may always occurs. Commonly used stops are **lever stops** and **bottom stops**.

5.4.1.10 Knock out Pins

They assist in knocking away the sheared blank from the face of the punch to make free the latter from former.

5.5 CLASSIFICATION OF DIES

On the basics of operation, dies are classified in two groups:

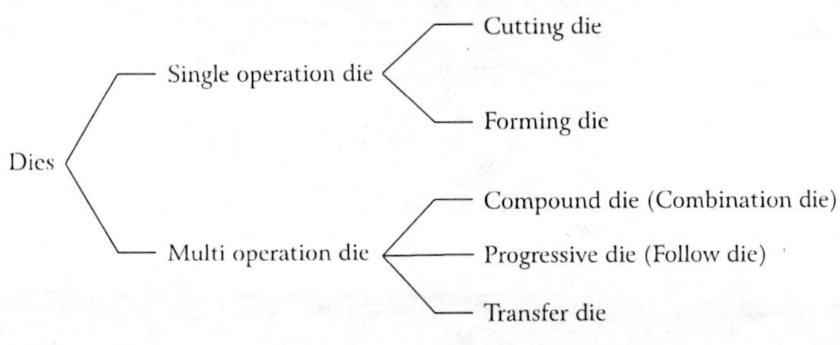

Fig. 5.3 Block diagram of die classification

5.5.1 Single Operation Die or Simple Die

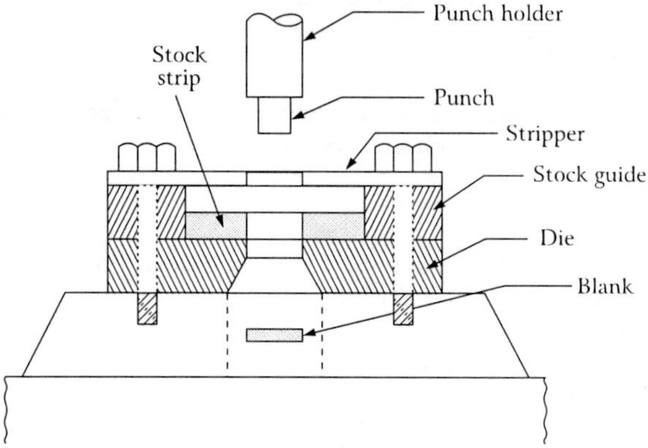

Fig. 5.4 Single operation die

5.5.2 Progressive Die

Progressive die also known as follow die. Progressive die basically used to fulfill the high production rate demand. This progressive die is very suitable for multiple operations such as **blanking, punching, notching**, etc. in progressive die workpiece used in the form of strip.

In this die strip (work piece) travel from one station to another station. Different operation performed separately at each station. **Washer making** is one of the unique example by progressive die. This type of die consist two punches, one for blanking purpose and other for piercing purpose both are mounted on the ram as shown in Fig. 5.5. Ram reciprocate up and down motion, and motion of ram is only possible by hand wheel or power *d* wheel drive.

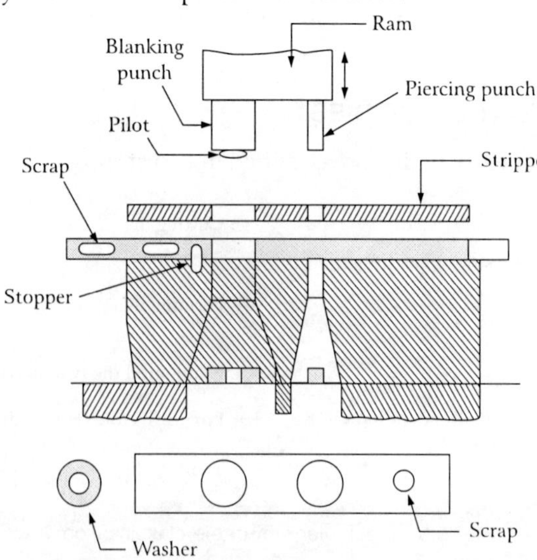

Fig. 5.5 Washer making with progressive die

5.5.3 Compound Die

Compound die is also known as combination die. The unique feature of compound die that it can performed various operation on the workpiece (strip) in single stroke (one up and down of ram) of punch at one station. Basic difference between the compound die and progressive die is that the speed of compound die is slow as compared to progressive die but they are more accurate and economical for mass production as compared to simple die or single die.

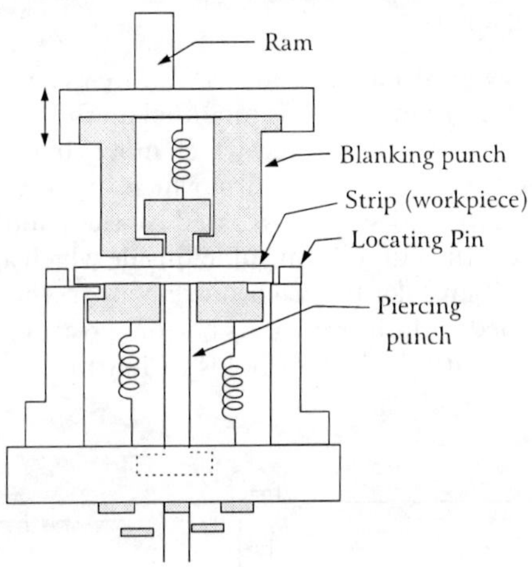

Fig. 5.6 Compound die

5.6 FLY PRESS

It is a manually operated screw press which has a massive screw and flywheel, or ball weighted handle, attached to the top of the screw to provide inertia to drive the screw down, which pushes the ram block/tool holder on to the work. It creates a tremendous force, and allows the operator to bend, forge, pierce, punch, or texture, metals precisely and easily once the necessary tooling has been made. Additionally, it provides tactile feedback to the operator, allowing very delicate work to be done that is pretty much beyond the range of hydraulic presses. Further, it only places the

Fig. 5.7 Fly Press

tool in contact with the work momentarily, preserving the heat in the work much more effectively than hydraulic presses. Besides, it is clean, uses no messy hydraulic oil, has no smell, requires no power supply, and it just works really well.

It is also known as Ball Press and one of the oldest type press with a unique feature that it is very simple press and operated by hand (manually). A single side fly press is shown in figure 5.8.

Fly press are two types:

(i) Single side fly press
(ii) Double side fly press

This type of presses consist a cast iron base. Entire weight of press is beared by base. It consist a cast iron frame which is mounted on the base and the top end of frame there is a provision of vertical screw which moves in the top end of frame in ups and down position. Screw carries an arm which support two cast iron weight (ball) at its two ends arm and screw is rotated by the handle. A portion of the frame is extended below the nut to form guides inside which a ram attached at the bottom of the screw. Ram take the axial movement in the guide and a punch (cutting tool) is attached with the ram at its bottom position where as die is fixed on the based on T-slots with the help of Bolds with firmly.

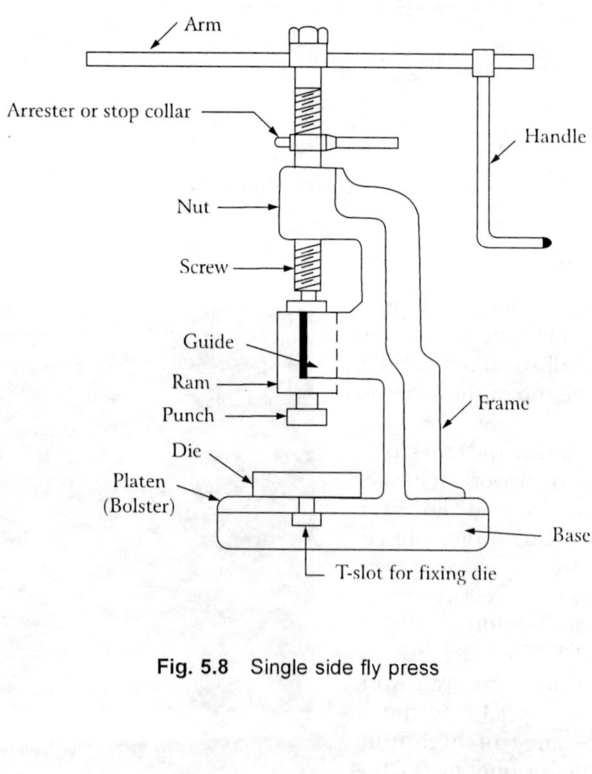

Fig. 5.8 Single side fly press

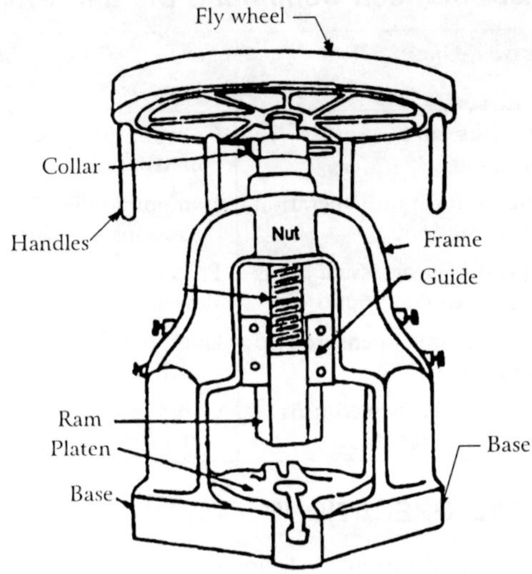

Fig. 5.9 Double side fly press

5.6.1 Working of Press

Sheet is kept at the die and arm given a quick rotation by rotating the handle.

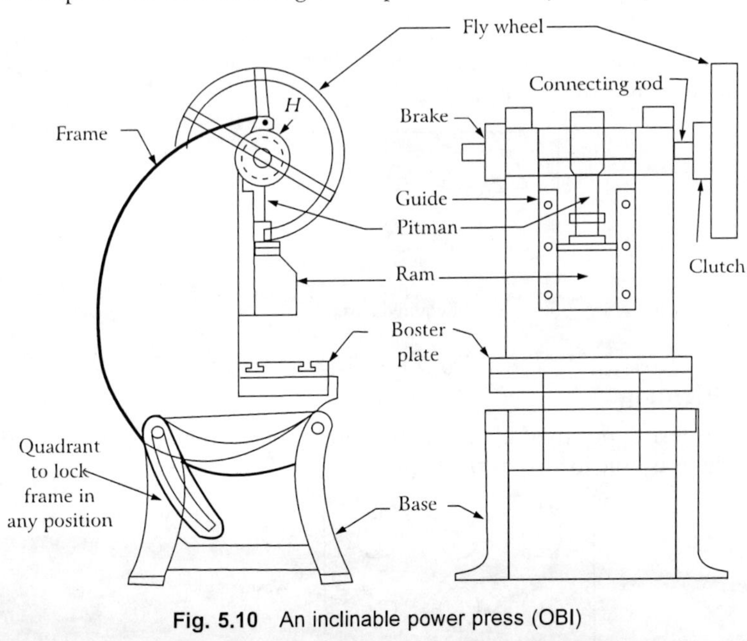

Fig. 5.10 An inclinable power press (OBI)

5.6.2 Differentiate between Compound Die and Progressive Die

Sl. No.	*Progressive Die*	*Compound Die*
1.	Progressive dies are simple in construction.	Compound dies are complicated in construction.
2.	Progressive die perform one operation at a time at one station.	Compound die perform more than one operations at a time at one station.
3.	Production speed of progressive die is high as compared to compound die.	Production speed of compound die is slow as compared to progressive die.
4.	Progressive die may not so ensure grip and position to strip.	Compound die ensure more grip and position to strip.
5.	Progressive die may not requires double and triple action presses.	Compound die usually requires double and triple action presses.

5.7 SHEET METAL OPERATIONS

Common press work operations are as follows:

- Blanking
- Notching
- Drawing
- Trimming
- Slitting
- Piercing
- Lancing
- Parting
- Nibbling

5.7.1 Blanking

Blanking is one of the most important operation performed in sheet metal shop. In blanking operation the cutout part from strip (workpiece) is required shape and size of object while the rest is wastage (scrap).

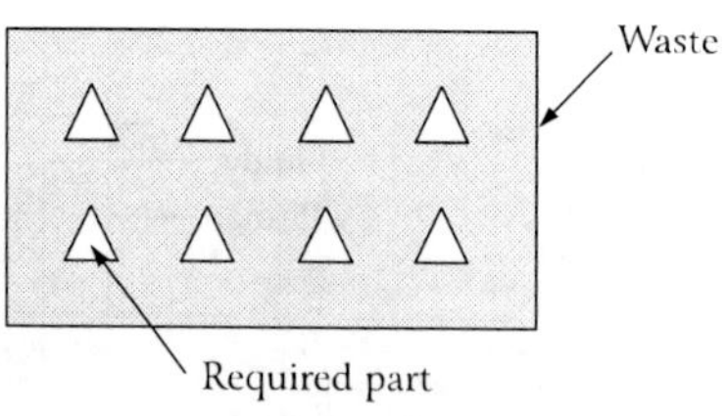

Fig. 5.11 Blanking

5.7.2 Piercing

Piercing is just opposite of blanking. In piercing operation cut part from strip is waste where as hole in strip is the required shape.

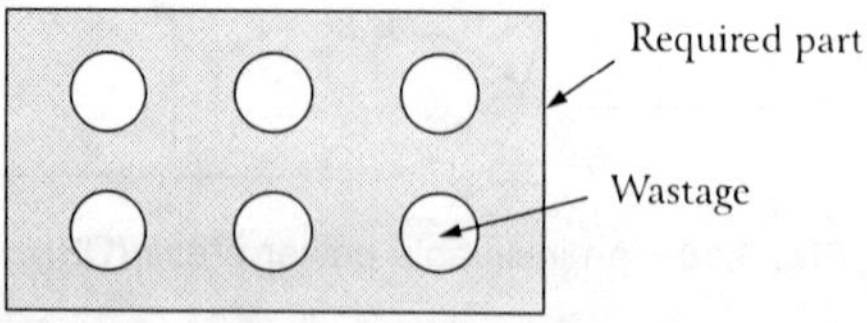

Fig. 5.12 Piercing

5.7.3 Notching

Notching is a cutting operation in which metal is removed (cut) from the edge of strip in small amount.

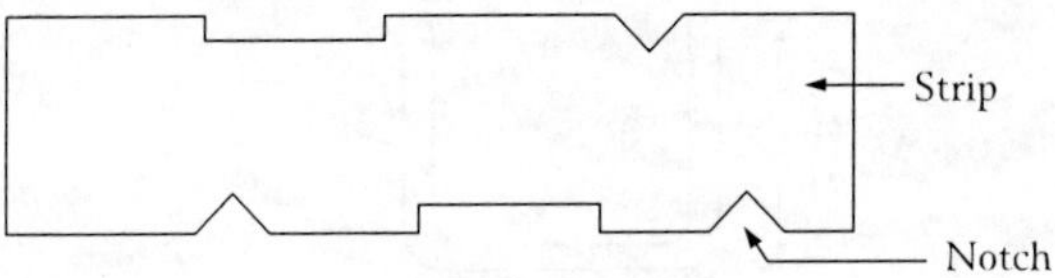

Fig. 5.13 Notching

5.7.4 Lancing

In Lancing operation metal strip is first cut and same is bent at an angle to form a sort of tab or Lau ver, *i.e.*, it is a cutting and bending operation without producing scrap.

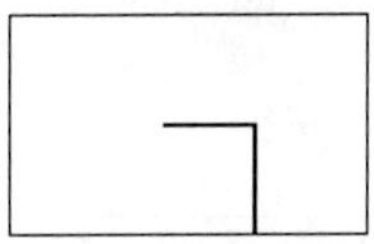

Fig. 5.14 Lancing

5.7.5 Drawing

Drawing is also another press work operation. It is carried out to produce cup shaped component having some depth without producing scrap.

5.7.6 Nibbling

It is a process of cutting out a contour or other shape by punching a series of over lapping round or square holes along edge of part. This operation is a substitute of blanking if numbers of pieces required in small quantity. Many overlapping holes are made and the operation is similar to the making of large slot in a sheet. Intricate slots, holes and notches can be easily produced by this process.

5.8 CALCULATION OF BLANK SIZE

In drawing operation it is the basic step to calculate the blank size which is required for a particular drawn sheet. There are various methods to calculate the blank size. But one of the simplest method for calculating the blank diameter is given by **area method.** In this method it is assumed that surface area of blank size is equal to the surface area of the drawn part and mathematically it is expressed as

$$D = \sqrt{d^2 + 4dh}$$

where D = Blank diameter

d = diameter of finished shell

h = height of finished part

r = corner bottom radius

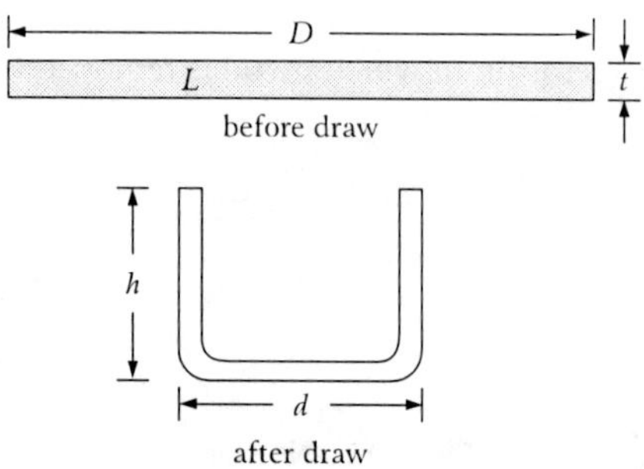

Fig. 5.15

$$\frac{d}{r} = 20$$

or greater

(i) if ratio of $\frac{d}{r}$ is $15 < \frac{d}{r} < 20$ then $D = \sqrt{d^2 + 4dh} - 0.5\,r$

(ii) if $10 < \frac{d}{r} < 5$ then $D = \sqrt{d^2 + 4dh} - r$

(iii) if $\frac{d}{r} < 10$ then $D = \sqrt{(d - 2r)^2 + 4d(h - r) + 2\pi r(d - 0.7)}$

5.9 WORKING PRINCIPLE OF METAL CUTTING IN PRESS WORK

Metal cutting of sheet in a press work by die and punch is known as shearing operation. Shape and size of punch is quite exact as the shape and size of die i.e., die opening having the some clearance for smooth motion of punch inside the die as well as to easy and automatic with draw the object (cutted part). Clearance play an important role for shape and quality of the sheared edge of the slug magnitude of clearance is very much affected by **hardness, thickness and properties** of the **sheet metal**.

In press cutting operation material is subjected to both the **tensile** and **compressive** stresses as shown in the Fig. 5.16 and 5.17. Stress are highest at the edge of die and punch.

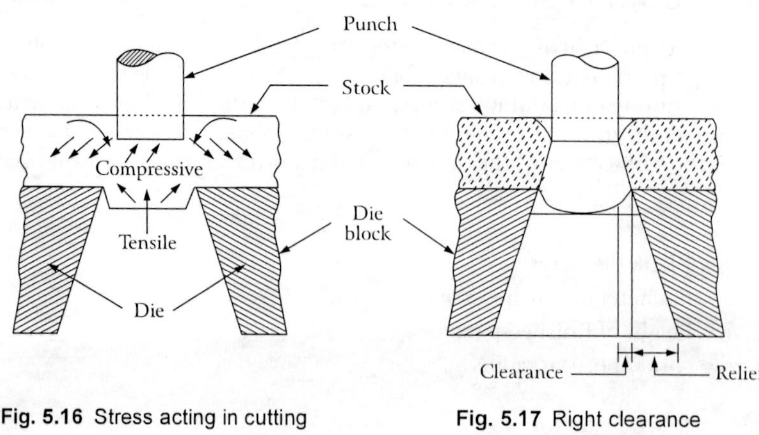

Fig. 5.16 Stress acting in cutting

Fig. 5.17 Right clearance

5.9.1 Clearance between Die and Punch [*UPTU 2003-04, 05-06*]

Clearance: The space between the mating part of die set mean it is the space between die and punch. Proper clearance is required optimum finish of cut edge. **Clearance amount basically depends upon the following factors:**

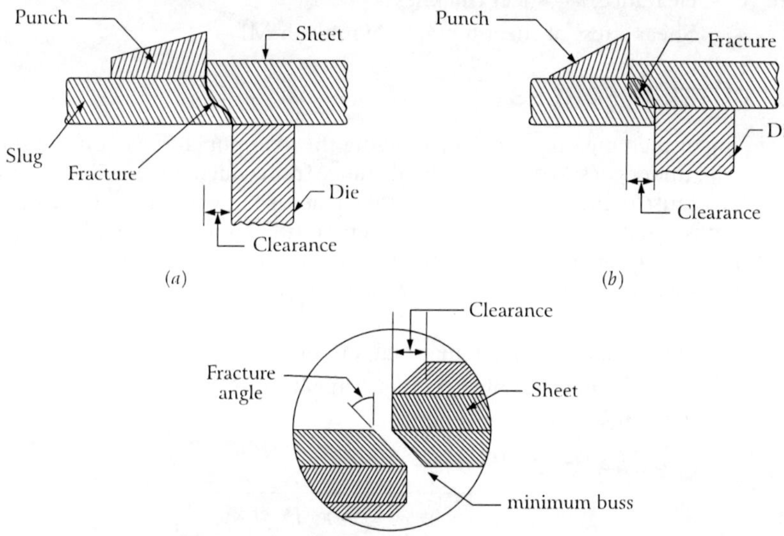

Fig. 5.18 Effects of clearance

(i) Sheet material
(ii) Sheet thickness
(iii) Type of operation performed
(iv) Sheet hardness.

If the clearance is insufficient and its effect is different and if it is excessive then its effect is different. In case of insufficient clearance the fracture lines will start at the cutting edge of the punch and also at the die as shown in Fig. 5.18(*a*).

In excessive clearance, large plastic deformation and large edge radius formed. Edge of sheet to be drawn or pulled in the direction of the working force.

Hence either clearance is insufficient or excessive cracks propagate by the tool edge to avoid this situation proper clearance between die and punch is required.

Harder materials requires more clearance as compared to softer materials. Value of clearance for various materials are given below:

TABLE 5.1 Clearance value for different materials

S.No.	*Material*	*Clearance value*
1.	Brass and soft steel	5% of *t*
2.	Medium steel	6% of *t*
3.	Aluminium	10% of *t*
4.	Hard steel	7% of *t*

where t = thickness of sheet.

Clearance can also evaluated by the following formula

$$C = 0.0032 \cdot t \cdot \sqrt{\tau_s}\text{, in mm}$$

where C = clearance, t = sheet thickness,

τ_s = Shear stress of sheet metal in N/mm^2 or MPa.

5.10 CUTTING FORCES

In blanking or piercing operation on applying the load punch strike on the sheet and material under goes to deformation, distance from which the punch enter into the sheet to cause deformation (rapture) take place is known as **penetration** and penetration is equal to the stock thickness. Penetration basically depends upon the type of sheet being cut and also stock thickness. If hard sheet is being cut by punch then very little penetration is required while if soft material being cuts then large penetration is required.

If D = Diameter of circular blank in mm.

τ_s = Shear stress of sheet metal in MPa or N/mm^2

t = Shear thickness is mm

then cutting force is given by

$$F_{max} = \pi \cdot D \cdot t \cdot \tau_s$$

For a rectangular blank:

If L = length of block in mm.

b = width of block in mm.

then cutting force

$$F_{max} = 2(l + b) \times t \times \tau_s$$

Energy in Press Work

Work done to make a cut or press work energy

$$E = F_{max} \times \text{travel of punch}$$

$$E = F_{max} \times t \times K$$

where K = percentage of penetration required for rapture.

After allowing the energy lost in machine friction and in pushing slugs through the die, above equation can be written as

$$E = F_{max} \times K \times t \times Z$$

Z = constant and for general purpose and

$$Z = 1.16.$$

5.11 SPRING BACK

Spring back is very important phenomenon in bending operation. In bending operation on applying the load (pressure) metal goes to deformation, on releasing the applied load, there is elastic recovery by the material. This known as positive spring back. Even in some certain condition negative spring back is also possible. In bending operation after releasing pressure, plastic stress also release and cause metal movement. Spring back varies 0.5° to 5° for steel and 10°–15° for bronze.

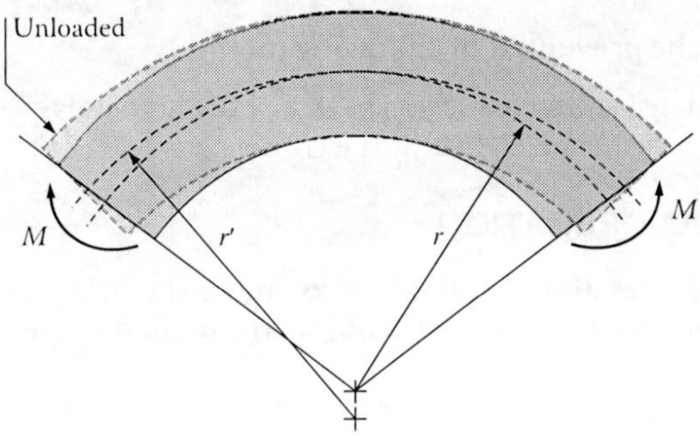

Fig. 5.19(A) Spring back after unloading

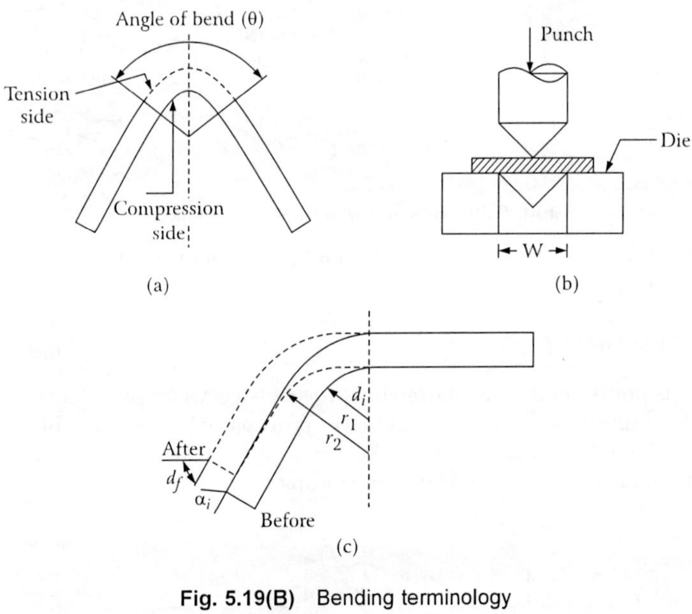

Fig. 5.19(B) Bending terminology

r_1 = bend radius before spring back

r_2 = bend radius after spring back

E = modulas of elasticity of work piece

σ_y = yield stress of work piece material

Spring back basically depends upon the following factors:

(i) Object/material-greater the yield strength, the greater then spring back

(ii) Thickness of sheet-thinner the raw material, the great the spring back

(iii) Hardness of the material (sheet)

(iv) Bend radius

As soon as value of bend radius increases, value of spring back increases.

Spring back can be prevented by fallowing method:

(i) Over bending can reduce spring back in v-die air bends

(ii) By coining the metal slightly at the corner.

5.12 BENDING OPERATION

In metal forming operation bending is also one of the important operation. In bending workpiece is bend at an angle to give the desired shapes such as flanges, seam etc.

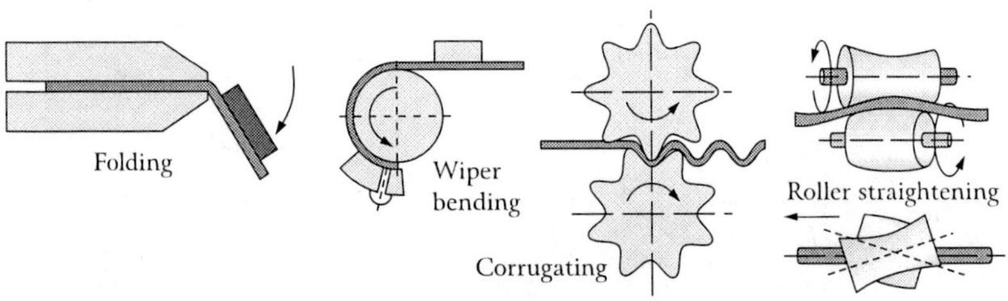

Fig. 5.20 Bending with rotary tool motion

Common bending operations are ***V*-bending, edge bending, roll bending, press brake forming** etc.

5.13 DEEP DRAWING [*UPTU 2010-11*]

For analysis point of view, deep drawing process is very complex. Various type of forces act simultaneously on the workpiece. Workpiece between the blank holder and die subjected to radial drawing. Where as corner of workpiece around the corner of punch and die is subjected to bending.

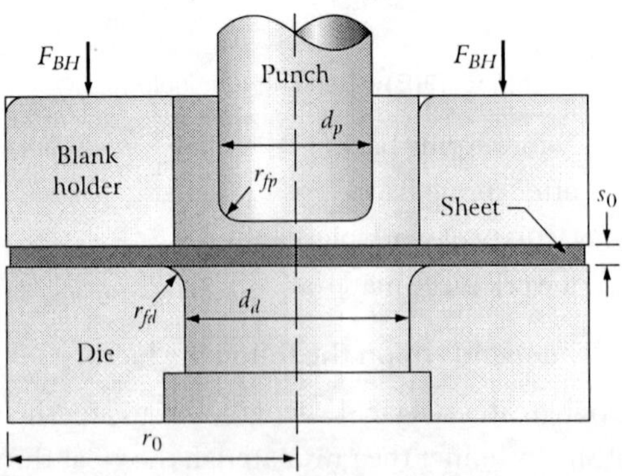

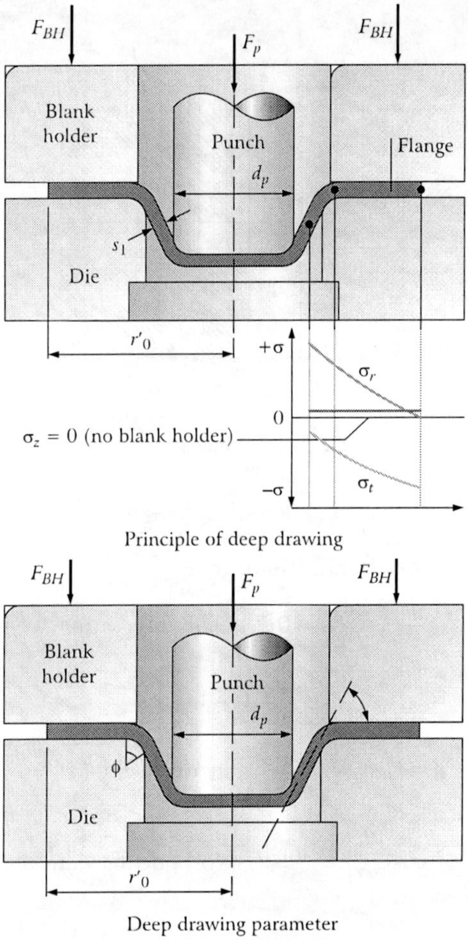

Fig. 5.21 Principle of deep drawing

Figure 5.22 shows a deep drawing operation where r_p, r_J *and* r_d are the radii of punch, job and die respectively. Clearance between the die and punch is $(r_d - r_p)$ is equal to workpiece thickness t, r_{cp} and r_{cd} are the punch and die corners respectively. Clearance C is maintained between the punch and blank holder.

f_n = blank holder force on entire circumstances.

Radial stress due to friction can be represented by an equivalent radial stress $\frac{2\mu f_n}{2\pi r_j t}$ at outer periphery.

Considering the radial equilibrium of element

$$r d\sigma_r + \sigma_r dr - \sigma_\theta dr = 0$$

Using Tresca yield criteria

$$\sigma_1 = \sigma_r \text{ and } \sigma_2 = \sigma_3 = +\sigma_\theta \quad ...(i)$$

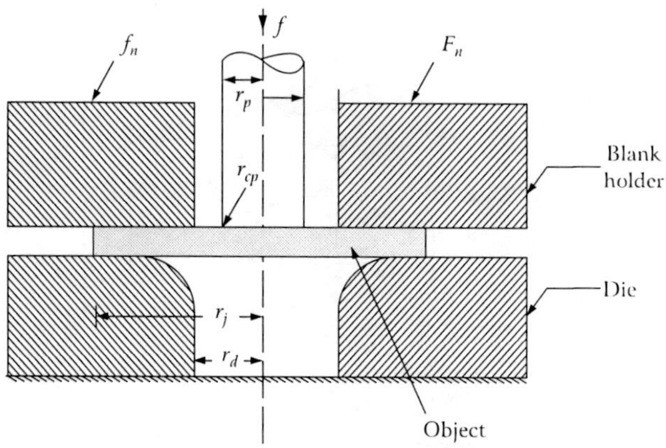

Fig. 5.22 Cup drawing process

$$\frac{\sigma_1 - \sigma_3}{2} = k \quad \text{or} \quad \sigma_r - \sigma_\theta = 2k$$

Putting the value of σ_θ in Eq. (*i*), we get

$$\frac{dr}{r} + \frac{d\sigma_r}{2k} = 0 \qquad \text{on integrating, we have}$$

$$\frac{\sigma_r}{2k} = C - \log_e r \qquad \ldots(ii)$$

Now, $\quad r = r_J,\ \sigma_r = \dfrac{\mu f_n}{\pi r_J t}$ putting in Eq. (*ii*), we get

$$C = \frac{\mu f_n}{2\pi k r_J t} + \log_e r_J$$ Now putting value of C in Eq. (*ii*), we get

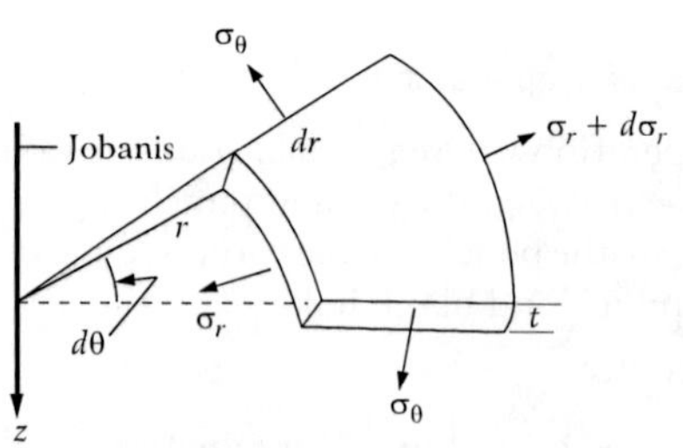

Fig. 5.23 Stress action on element during drawing

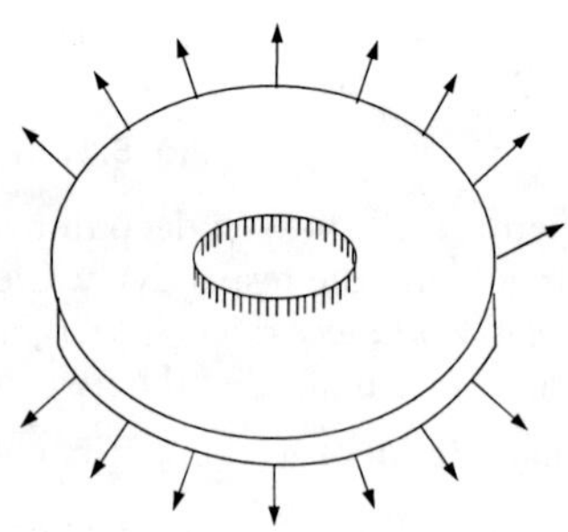

Fig. 5.24 Radial stress due to blank holding pressure

$$\frac{\sigma_r}{2k} = \frac{\mu f_n}{2\pi k r_J t} + \log_e \frac{r_j}{r}$$

$$\left.\frac{\sigma_r}{2k}\right|_{r = r_d} = \frac{\mu f_n}{2\pi \cdot k \cdot r_J \cdot t} + \log_e \frac{r_J}{r_d} \qquad \textbf{Proved.}$$

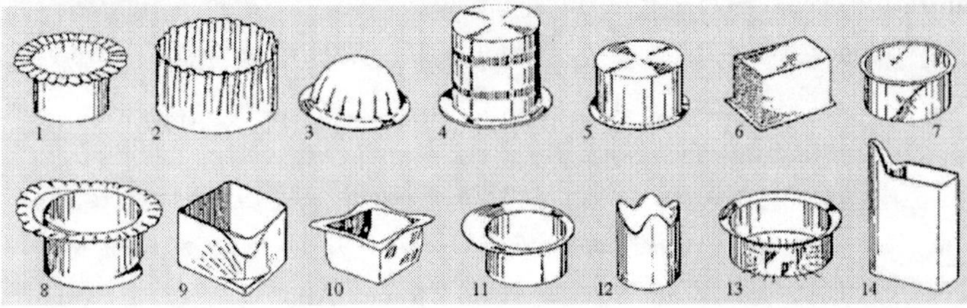

Fig. 5.25 Various failure modes in deep-drawing: 1-flange wrinkling; 2-wall wrinking; 3-part wrinkling; 4-ring prints; 5-traces; 6-orange sking; 7-Lüder's strips; 8-bottom fracture; 9-comer fracture; 10,12-folding; 13,14-corner folding

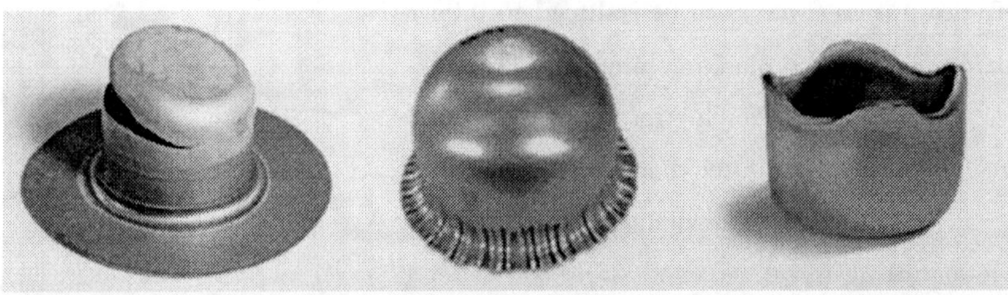

(*a*) (*b*) (*c*)

Fig. 5.26 Typical failure modes in deep drawing: (a) Tearing, (b) wrinkling (courtesy T. Altan, ERC), (c) earing (courtesy Hydro Aluminium Deutschland GmbH)

Example 5.1

Calculate the maximum punch forces necessary to blank a steel washer 45.1 mm outer dia × 25.0 inner dia × 1.58 mm thick, if σ_{max} = 450 N/mm². Estimate the work done if % penetration is 25%.

Solution: Penetration (c) = 25% = 0.25, thickness = 1.58 mm

Work Perimeter $l = \pi(45.1 + 25) = 220.114$ mm.,

$$\sigma_{max} = \frac{F_{max}}{A} = \frac{F_{max}}{t \times l}$$

$$F_{max} = \sigma_{max} \times t \times l = 450 \times 1.58 \times 220.114$$

$$= 156.5 \text{ min}$$

$$\text{Work done} = F_{max} \times (C) \times t = 156.50 \times 0.25 \times 1.58$$

$$= 61.82 \text{ joule} \quad \textbf{Ans.}$$

Example 5.2

Determine the blanking force to cut a blank 30 mm wide and 35 mm long from 1.5 mm thick metal strip. Ultimate shear stress of the metal is 450 N/mm² also determine the work done if percentage penetration is 25.

Solution: Given $C = 25\%$, $\tau_s = 450$ N/mm^2

Work done $\qquad l = \pi\,(30 + 35) = 204.1$ mm

But $\tau_s = \dfrac{f_s}{t \times l}$ $\qquad F_s = \tau_s \times t \times l = 450 \times 1.5 \times 204.1$

$$= 137767.5 \text{ N} \quad \text{or} \quad 137.76 \text{ kN}$$

$$\text{Work done} = F_s \times C \times t = 137.76 \times 25 \times 1.5$$

$$= 51.66 \text{ k Joule} \quad \textbf{Ans.}$$

Example 5.3

A steel washer of 45 mm outer diameter and 25 mm with inner hole diameter and the thickness of washer is 3.2 mm. If the maximum shear stress is 415 N/mm^2 percentage penetration is 24. Then determine the shear to be ground on tool if maximum punch force is to be reduced to 0.05 MN.

Solution: F_1 = Blanking force = $\pi \,.\, D \,.\, t \,.\, \tau_s$,

F_2 = Piercing force = $\pi \,.\, d \,.\, t \,.\, \tau_s$,

where D = Outer diameter

d = Inner diameter

Total cutting force

$$F = F_1 + F_2$$

$$= \pi(D + d)t \times \tau_s$$

$$F = \pi(45 + 25) \times 3.2 \times 415$$

$$= 291.89 \text{ kN}$$

$$\text{Work done} = F \times C \times t = 291.89 \times 0.24 \times 3.2$$

$$= 224.175 \text{ kJ} \quad \textbf{Ans.}$$

$$F_s = \text{max. punch force}$$

$$= 0.05 \text{ Mn} = 50 \text{ kN}$$

$$F_s = \frac{F \times C \times t}{S}$$

$$50 = \frac{224.175 \times .24 \times 3.2}{S}$$

$$S = 3.44 \text{ mm} \quad \textbf{Ans.}$$

Example 5.4

Calculate the blank diameter to make a steel shell shown in Fig. 5.27.

Solution: t = thickness steel plate = 3 mm

d = mean diameter

$= 66 - 3 = 63$ mm

h = Mean height

$$h = 40 - \frac{t}{2} = 40 - \frac{3}{2} = 38.5 \text{ mm}$$

D = Blank diameter

$$D = \sqrt{d^2 + 4dh}$$

$$= \sqrt{63^2 + 4 \times 63 \times 38.5}$$

$$= \sqrt{13671} = 116.9 = 117 \text{ mm} \quad \textbf{Ans.}$$

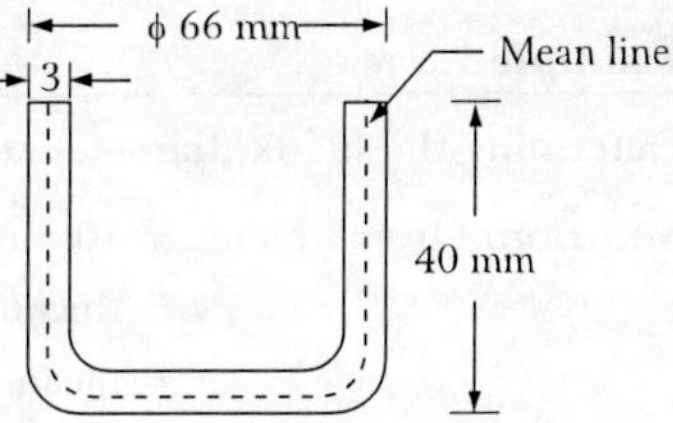

Fig. 5.27

Example 5.5

Figure 5.28 shown a cup is to be made from cold rolled steel of 1.0 mm thickness design die of drawing operation.

Solution: Shell diameter

d = 60 mm

h = 50 mm

r = 1.5 mm

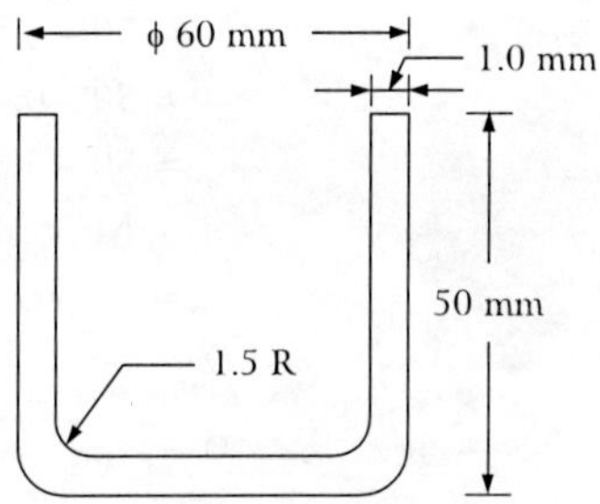

Fig. 5.28

(i) Size of blank $\frac{d}{r} = \frac{60}{1.5} = 40$

$$D = \sqrt{d^2 + 4dh}$$

$$= \sqrt{60^2 + 4 \times 60 \times 50}$$

$$= \sqrt{15600} = 124.89 \simeq 125 \text{ mm}$$

This is a theoretical size. Trimming is required for smooth edge according to Themp rule add 3.2 mm in blank diameter for size 25 mm since cup diameter is 60 mm, 7.7 mm for trimming.

$D = 124.89 + 7.68 = 135.6$

(ii) Percentage reduction $= 100\left(1 - \frac{d}{D}\right)$

$$= 100\left(1 - \frac{60}{125}\right) = 52\%$$

as per practice, a reduction of about 45 to 50% is permissible for Ist draw, hence, it is clear that above cup can not be drawn in one draw.

(iii) No. of draw $\frac{h}{d} = \frac{50}{60} = 0.833 \simeq 1.0$

No. of draw from table = 2

d_1 at Ist draw = 124.84 – 0.45 × 124.86 = 68.86 mm

Diameter at the end of second draw has to be 60 mm

Reduction for second draw $= 110\left(1 - \frac{60}{68.60}\right) = 12.65$

which is less than 30% the permissible reduction for the second draw.

Example 5.6

Determine the blank diameter to draw a shell as shown in Fig. 5.29.

Solution: Inside radius = 10 mm

t = thickness of shell = 4 mm

r = mean radius

$$= 10 + \frac{t}{2}$$

$$= 10 + \frac{4}{2} = 12 \text{ mm}$$

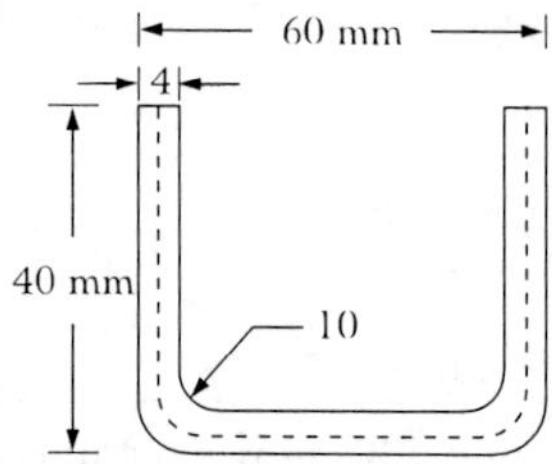

Fig. 5.29

To convert the shell to a square corner shell

h = Mean height

$$= 50 - 0.43 \times r$$

$$= 50 - 0.43 \times 12$$

$$= 50 - 5.16 = 44.84 \text{ mm}$$

d = Mean diameter

$$= 60 - t = 60 - 4 = 56 \text{ mm}$$

D = blank diameter

$$D = \sqrt{d^2 + 4dh}$$

$$= \sqrt{56^2 + 4 \times 56 \times 44.84}$$

$$= \sqrt{13180.16}$$

$$= 114.8 = 115 \text{ mm} \quad \textbf{Ans.}$$

Example 5.7

A shell shown in Fig. 5.30 has a height of 48 mm and a diameter of 50 mm. The corner radius is 4 mm and workpiece material is medium carbon steel, having 1.5 mm thickness. Design die for drawing operation.

Solution:

h = height of shell = 48 mm

d = diameter of shell = 50 mm

r = corner radius = 4 mm

t = material thickness = 1.5 mm

$$\frac{d}{r} = \frac{50}{4} = 25$$

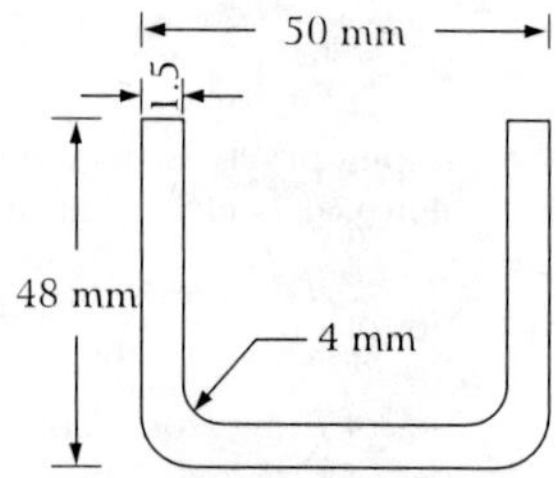

Fig. 5.30 Drawn shell

Blank diameter, $D = \sqrt{d^2 + 4dh}$

$$= \sqrt{50^2 + 4 \times 50 \times 48} = \sqrt{12100} = 110 \text{ mm}$$

This is a theoretical value of blank size for extra trimming metal 3 mm is to be added to blank diameter for each 25 mm of shell diameter.

$$\text{Blank diameter} = D + \frac{3}{25} \times 50$$

$$= 110 + 6 = 116 \text{ mm}$$

(*b*) To find out percentage reduction,

$$P = \text{Percentage reduction}$$

$$= 100\left(1 - \frac{d}{D}\right) = 100\left(1 - \frac{50}{116}\right)$$

$$= 100 \times 0.5689 = 56.896$$

$$P = 56.89\% \quad \textbf{Ans.}$$

Example 5.8

A steel washer of 50 mm outer diameter and 25 mm inner diameter is to be made from 1.5 mm thick sheet in one operation. If percentage penetration is 35% and necessary shear stress is 400 N/mm² then find out max. Punch force if both punches operate at same time and also find out reduction in punch force if 0.6 mm double shear ground on tool.

Given: $D = 50$ mm, $d = 25$ mm, $t = 1.5$ mm, $\tau_s = 400$ N/mm^2

Solution:

$$F = \pi(D + d) \times t \times \tau_s$$

$$= \pi(50 + 25) \times 1.5 \times 400 = 141.3 \text{ kN} \quad \textbf{Ans.}$$

$$\Rightarrow \quad \text{Reduction in punch } f_s = \frac{f \times p \times t}{S}$$

$$\Rightarrow \quad = \frac{141.3 \times 0.35 \times 1.5}{0.6} = 123.6 \text{ kN} \quad \textbf{Ans.}$$

Example 5.9

Two holes, one 8 cm² and another 10 cm² diameter to be produced in a sheet, having the 2.5 mm thickness. Determine the cutting force and strip force if the strength of material is 2000 kg/cm².

Solution:

$$L = \pi d + l \times b$$

$$= \pi \times 10 + 8 \times 5$$

$$= 71.4 \text{ cm}$$

$$\Rightarrow \quad \text{Cutting force } f = L \times t \times \tau_s$$

$$= 71.4 \times 2.5 \times 2000$$

$$= 35.7 \text{ tonnes} \quad \textbf{Ans.}$$

$$\Rightarrow \quad \text{Strip force } F_{SF} = 20\% \, F$$

$$= \frac{20}{100} \times 35.7$$

$$= 7.14 \text{ tonnes} \quad \textbf{Ans.}$$

Example 5.10

A washer of 12.5 mm internal diameter and 25.0 mm outer diameter is to be made from 1.5 mm thick strip of 0.3% of carbon steel. Find out

(i) clearnace,
(ii) blanking die opening size,
(iii) blanking punch, size
(iv) piercing punch size,
(v) piercing die opening size.

Solution:

(i) Clearance between die and punch = 5% of t (standard)

$$C = \frac{5}{100} \times 1.5 = 0.075 \text{ mm} \quad \textbf{Ans.}$$

(ii) Blanking die opening = 25.0 – 0.05 = 24.95 mm **Ans.**

(iii) Blanking punch size = blank die opening – 2C

= 24.45 – 2 × 0.075 = 24.8 mm **Ans.**

(iv) Piercing punch size = internal diameter of washer + 0.05

= 12.5 + 0.05 = 12.55 mm **Ans.**

(v) Piercing die size = Piercing size + 2C

= 12.55 + 2 × 0.075

= 12.7 mm **Ans.**

Example 5.11

If ultimate shear stress for washer material is 300 N/mm^2 in Example 5.10 then find:

(i) total cutting force if both punches work at same time,
(ii) cutting force if punches are straggered, only one punch work at a time,
(iii) 60% penetration and shear on punch 1 mm, what will be the cutting forces if both punches work together.

Solution:

(*i*)
$$D = 25 \text{ mm}$$
$$d = 12.5 \text{ mm}$$
$$t = 1.5 \text{ mm}$$
$$\tau_s = 300 \text{ N/mm}^2$$
$$F = \pi (D + d) \times t \times \tau_s$$
$$= \pi (25 + 12.5) \times 15 \times 300$$
$$= 52.98 \text{ kN} \quad \textbf{Ans.}$$

(*ii*)
$$F = \pi D t\, \tau_s = \pi \times 25 \times 1.5 \times 300$$
$$= 35325 \text{ N} \quad \text{or} \quad 35.32 \text{ kN} \quad \textbf{Ans.}$$

(*iii*)
$$F = \frac{t \times k \times F_{\max}}{(k \times t + 1)} = \frac{1.5 \times 0.60 \times 52.98}{(0.6 \times 1.5 + 1)}$$
$$= \frac{47.682}{1.9} = 25.09 \text{ kN} \quad \textbf{Ans.}$$

Example 5.12

A hole of 100 mm is to be punched in a cold rolled medium carbon steel plate of 5.6 mm thickness. Ultimate sear stress of plate material is 550 MPa with normal clearance on the press tool, cutting is completed at 40% penetration of the punch. Determine the diameter for the punch and die required for the purpose. If the press capacity is 30 tonnes. Calculate the shear angle to be provided on the punch in order to bring the work with in capacity of the existing press.

[UPTU 2003-04, 2009-10]

Solution: Clearance = 5% (assume)

$$C = \frac{5}{100} \times 5.6 = 0.28 \text{ mm}$$

Punch diameter = hole diameter

= 100 mm **Ans.**

Diameter = hole diameter + 2*C*

= 100 + 2 × 0.28 = 100.56 mm **Ans.**

Without any shear $F_{max} = \pi \cdot D \cdot t \cdot \tau_s$

$= \pi \times 100 \times 5.6 \times 550 = 967.6$ kN

Work done in punching $= f \times p \times t$

$= 967.6 \times 10^3 \times 0.4 \times 5.6 \times 10^{-3}$

= 2167.45 Nm

Work done with shearing and 30 tonnes maximum load

$= 30 \times 10^3 \times 9.81\ (0.4 \times 5.6 \times 10^{-3} + S)$,

S = depth of shear on punch

$$S = \frac{2167.45}{20 \times 1000 \times 9.81} - 0.4 \times 0.0056$$

= 5.12 mm

$$\tan\theta = \left(\frac{S}{D/2}\right)$$

$$\theta = \tan^{-1}\left(\frac{5.12}{100/2}\right) = 5.85°$$ **Ans.**

Example 5.13

A hole of 100 mm diameter is to be punched in a steel plate having 6 mm thickness. Material is cold rolled carbon steel for which shear stress is 400 N/mm^2. Normal radial clearance is 10% of strip thickness percentage penetration is 50%. Determine:

(i) punch and die diameter,

(ii) press capacity (in tonnes) *[UPTU 2001-02]*

Solution:

(i) Clearance *C* = 10% of strip (given)

$$= \frac{10}{100} \times 6 = 0.6 \text{ mm}$$

Punch diameter = hole diameter

= 100 mm **Ans.**

Die diameter = Punch diameter + 2*C*

= 100 + 2 × 0.6

= 101.2 mm **Ans.**

(ii)

$$F = \pi \times D \times \tau_s \times t$$

$$F = \pi \times 100 \times 6 \times 400 = 753982.2 \text{ N}$$

$$\text{Press capacity} = \frac{753982.2 \times 10^{-3}}{9.81} = 76.86 \text{ tonnes} \quad \textbf{Ans.}$$

Example 5.14

A shell of diameter 75 mm and height 200 mm is to be drawn. How many drawing operation would be necessary. If there is no annealing operations intervening. Assume reductions of 50%, 40% and 30% etc. for each drawing without annealing.

Solution: D = blank diameter,

d = shell diameter = 75 mm,

h = height of shell = 200 mm

$$D = \sqrt{d^2 + 4dh} = \sqrt{75^2 + 4 \times 75 \times 200}$$

$$= \sqrt{65625} = 256.2 \text{ mm}$$

d_1 = diameter after 1st draw,

P_1 = percentage reduction for 1st draw = 50%

$$P_1 = 100\left(1 - \frac{d_1}{D}\right), \quad 50 = 100\left(1 - \frac{d_1}{256}\right), \quad d_1 = 128 \text{ mm}$$

d_2 = diameter after IInd draw, P_2 = percentage reduction for IInd draw = 40%

$$P_2 = 100\left(1 - \frac{d_2}{D}\right), \quad 40 = 100\left(1 - \frac{d_2}{128}\right) \quad \text{or} \quad d_2 = 77 \text{ mm}$$

Third draw may be necessary to bring the shell diameter to 75 mm as the area of cylinder is the circumference times the height, therefore the height may be found by dividing the net area by circumference.

$$A = \text{Area of blank} = \frac{\pi}{4}D^2 = \frac{\pi}{4} \times 256^2$$

$$= 51445 \text{ mm}^2$$

A_1 = Area of base at Ist stage

$$= \frac{\pi}{4}d_1^2 = \frac{\pi}{4} \times 128^2 = 12861 \text{ mm}^2$$

AV_1 = Area of vertical of cylinder = $A - A_1$

$$= 51445 - 12861 = 38584 \text{ mm}^2$$

h_1 = height of Ist draw

$$= \frac{AV_1}{\pi d_1} = \frac{38584}{\pi \times 128} = 96 \text{ mm}$$

For IInd draw:

$$A_2 = \text{Area of base} = \frac{\pi}{4}d_2^2 = \frac{\pi}{4}\times 77^2 = 4654 \text{ mm}^2$$

$$AV_2 = \text{Area of vertical cylinder} = A - A_2 = 51445 - 4654$$

$$= 46791 \text{ mm}^2$$

$$h_2 = \text{height of II}^{nd} \text{ draw}$$

$$= \frac{AV_2}{\pi d_2} = \frac{46791}{\pi\times 77} = 193.5 \text{ mm}$$

For IIIrd draw:

$$A_3 = \text{Area of base} = \frac{\pi}{4}\times 75^2 = 4415 \text{ mm}^2$$

$$AV_3 = \text{Area of cylinder} = A - A_3 = 51445 - 4415$$

$$= 47030 \text{ mm}^2$$

$$h_3 = \text{Height of third draw}$$

$$= \frac{AV_3}{\pi d} = \frac{47030}{\pi\times 75} = 200 \text{ mm}$$

Thus three drawing operation are needed to draw the shell of 200 mm height.

OBJECTIVE QUESTIONS

1. In a shear cutting operation, a sheet of 5 mm thickness is cut along a length of 200 mm. The cutting blade is 400 mm long (see Fig. 5.31) and zero-shear (S = 0) is provided on the edge. The ultimate shear stress of sheet is 100 MPa and penetration to thickness ratio is 0.2. Neglect friction

   (*a*) assuming force vs displacement curve to be rectangular the work done (in jule) is:

   (*a*) 100 (*b*) 200
   (*c*) 250 (*d*) 300

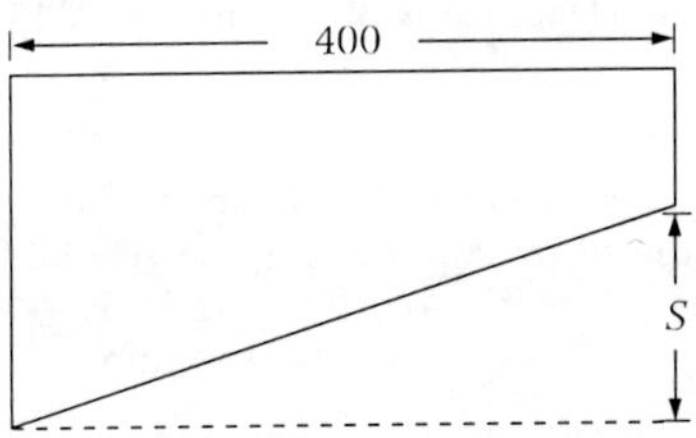

Fig. 5.31

   (*b*) A shear of 200 mm (S = 20 mm) is now provided on the blade assuming force vs displacement curve to be trapezoidal, the maximum force (in kN) exerted is:

   (*a*) 5 (*b*) 10
   (*c*) 20 (*d*) 40 [*GATE 2010*]

2. In the deep drawing of cups, blanks shows a tendency to wrinkle up around the periphery (flange), the most likely cause and remedy of the phenomenon are respectively.

(a) Buckling due to circumferential compression, increase blank holder pressure
(b) High blank holder pressure and high friction reduce blank holder pressure and apply lubricant
(c) High temperature causing increases in circumferential length apply coolant to be blank
(d) Buckling due to circumferential compression, decreases blank holder pressure.

[GATE 2008]

3. Match the correct combination for following metal working process:

	Processes	*Associated state of shear*
P	Blanking	1. Tension
Q	Stretch forming	2. Compression
R	Coining	3. Shear
S	Deep Drawing	4. Tension and compression
		5. Tension and shear

(a) P-2, Q-1, R-3, S-4 (b) P-3, Q-4, R-1, S-5
(c) P-5, Q-3, R-3, S-1 (d) P-3, Q-1, R-2, S-4 [GATE 2007]

4. A 2 mm thickness metal sheet is to be bent at an angle of one radian with a bend radius of 100 mm. If the stretch factor is 0.5 the bend allowance is

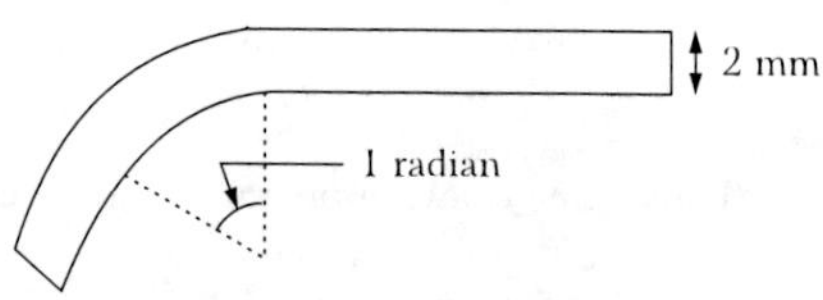

Fig. 5.32

(a) 99 mm
(b) 100 mm
(c) 101 mm
(d) 102 mm

[GATE 2005]

5. 10 mm diameter holes are to be punched in a steel sheet of 3 mm thickness. Shear strength of the material is 400 N/mm^2 and percentage penetration is 40. Shear provided on the punch is 2 mm. Blanking force during the operation will be
(a) 22.6 kN (b) 37.7 kN
(c) 61.6 kN (d) 94.3 kN [GATE 2004]

6. A metal disc of 20 mm diameter is to be punched from a sheet of 2 mm thickness. The punch and the die clearance is 3% then required punch diameter is,
(a) 19.88 mm (b) 19.84 mm
(c) 20.06 mm (d) 21.12 mm [GATE 2003]

7. A sheet of 100 mm diameter and 100 mm height with the corner radius of 0.4 mm is to be produced by cup drawing the required blank diameter is
(a) 118 mm (b) 161 mm
(c) 224 mm (d) 312 mm [UPTU 2003]

8. Force required in a blanking operation of low carbon steel sheet is 5.0 kN. Thickness of the sheet is '*t*' and diameter of blanked part is '*d*'. For the same work material if the diameter of the blanked part is increased to 1.5*d* and thickness is reduced to 4*t*. The new blanking force is kN is:
(a) 3.0 (b) 4.5
(c) 290.7 (d) 8.0 [GATE 2007]

9. Calculate the Punch size in mm for a circular blanking operation for which detail are given below.
Size of blank – 25 mm, Thikness of sheet – 2 mm, Radial clearance between Punch and die – 0.06 mm, Die allowance – 0.05 mm
(a) 24.83 (b) 24.89
(c) 25.01 (d) 25.17 [GATE 2012]

10. Production of different shapes and figure on the surface of a blank using a punch and die is known as
   (*a*) Forming (*b*) Piercing
   (*c*) Embossing (*d*) Punching

11. Cold working process used for making small bolts, pins, rivets etc. is known as
   (*a*) Punching (*b*) Trimming
   (*c*) Cold heading (*d*) None of these

12. Circular holes in a sheet is produced by
   (*a*) Piercing (*b*) Punching
   (*c*) Blanking (*d*) Shearing

13. When a number of holes are produced in sheet and the diameter of punch point is less than or equal to the thickness of the sheet, opertion is known as
   (*a*) Blanking (*b*) Shearing
   (*c*) Piercing (*d*) Punching

ANSWERS

1. (*a*)–(*a*); (*b*)–(*b*)	2. (*a*)	3. (*d*)	4. (*c*)
5. (*a*)	6. (*a*)	7. (*c*)	8. (*a*)
9. (***)	10. (*c*)	11. (*c*)	12. (*b*)
13. (*c*)			

HINTS

1. (*a*) $f = L(t \times p)\tau$

where
L = Perimeter of cut,
t = Thickness of cut
P = Penetration

$F = 200 \times (5 \times 0.2) \times 100 = 20$ kN **Ans.**

(*b*) Workdone $= f \times$ displacement
$= 20000 \times 5 \times 10^{-3} = 100$ Nm
$= 100$ watt **Ans.**

4. We know that

Bend allowance $= \alpha(R + kt)$

where
R = inside radius of bend = 100 mm
α = bend angle = 1 radian
t = thickness of sheet
k = location of neutral axis from bottom
$= 0.33\text{ R} < 2t$
$= 0.50\text{ R} > 2t$

Bend allowance $= 1 \times (100 + 0.5 \times 2) = 101$ mm **Ans.**

5. $\tau = 400$ N/mm^2
$t = 3$ mm
$d = 10$ mm

Shear on punch

$$t_1 = 2 \text{ mm}$$
$$p = 0.4$$

Blanking force when shear is applied on the punch $= \pi \cdot d \cdot t\left(\frac{t \cdot p}{t_1}\right)\tau_s$

$$= \pi \times 10 \times 3 \times 400\left(\frac{3 \times 0.4}{2}\right) = 22.6 \text{ kN} \quad \textbf{Ans.}$$

6. Given punch diameter $d = 20$ mm

$$t = 2 \text{ mm (sheet thickness)}$$
$$\text{Clearance} = 3\% \text{ of } t$$

Punch diameter $D = d - 2c = 20 - 2 \times \frac{3}{100} \times 2 = 20 - 0.12 = 19.88$ mm **Ans.**

7. Given,

$$d = 100 \text{ mm (diameter)}$$
$$h = 100 \text{ mm (height)}$$
$$r = 0.4 \text{ mm}$$
$$d > 20r$$

Blank diameter

$$D = \sqrt{d^2 + 4dh} = \sqrt{100^2 + 4 \times 100 \times 100} = 224 \text{ mm} \quad \textbf{Ans.}$$

8. We know that blanking force $F_b \alpha\, d \cdot t$

$$\frac{F_{b2}}{F_{b1}} = \frac{1.5d \times 0.4t}{d \times t} = 0.6$$

or $\quad F_{b2} = 0.6, \; F_{b1} = 0.6 \times 5 = 3$ kN **Ans.**

REVIEW QUESTIONS

1. What is a press. Explain the presses in detail?
2. Give the classification of presses.
3. Define the following press operations
   (*a*) Blanking (*b*) Piercing
   (*c*) Trimming (*d*) Nibbling
   (*e*) Notching (*f*) Lancing
   (*g*) Parting.
4. Describe the fly press with the help of a neat sketch.
5. Describe the inclinable power press. Giving its main application.
6. Describe what happens when sheet metal is sheared.
7. Compare open and close frame press.
8. What are single action, double action and triple action presses.
9. Compare mechanical presses with hydraulic presses.
10. Draw a Punch and Die assembly, indicating its main parts and their functions.
11. Define in detail:
    (*a*) Die set (*b*) Die material
    (*c*) Method of die support (*d*) Cutting action of die.
12. Explain the **Die block** design and **Punch design.**
13. Explain the clearance between die and punch.

14. Define the methods for reducing the cutting forces.
15. Describe the method for calculating the cutting forces in blanking and punching.
16. Differentiate between the blanking and piercing operation.
17. Why angular clearance is provided on die?
18. Prove through analysis of deep drawing of cup, the radial stress σ_r, at radius r is given by

$$\frac{\sigma_r}{2k} = \frac{\mu F_n}{2\pi k r_j} + \log_e \frac{r_j}{r}$$

also show that the limiting case for fracture at bottom is given by

$$e^{-\frac{\mu\pi}{2}} = \frac{\mu F_n}{2\pi k r_j t} + \log_e \frac{r_j}{r_d}$$

where F_n = blank holding force,
r_j = initial blank radius,
r_d = die radius,
k = yield shear stress,
t = plate thickness and
μ = coefficient of friction with occurrence at the edge between blank and holder and between blank and die corner.

[UPTU 2009-10, 2010-11]

19. How does compound die differ from progressive die? Giving a neat sketch, describe constructional feature and working of a compound die. *[UPTU 2001-02, 2010]*
20. Describe the process for washer making with the help of progressive die. How does compound die differ from progressive die. *[CCSU 2010, UPTU 2002-03, 2004-05]*
21. Why maximum cutting forces required for blanking and Punching are reduced? Explain all the methods of reducing cutting forces. *[CCSU 2010]*
22. What is deep drawing? Discuss deep drawing operation also explain various defects in deep drawing operation. *[CCSU 2010]*
23. Sketch **V-bending** die and **U-bending** die. Briefly Explain **spring back** in bending and various techniques for compensation of spring back. *[CCSU 2010]*
24. What is bending operation? Explain.
25. Describe following types of bending:
    (*a*) V-bending
    (*b*) Edge bending
    (*c*) U-bending.
26. What is spring back phenomenon in bending? *[UPTU 2004-05]*
27. On what factors the selection of press for sheet metal forming operation depends. *[UPTU 2006-07]*
28. Explain the difference between die and punch. *[UPTU 2006-07]*
29. Draw a neat sketch of die-punch assembly with accessory parts/components also explain how cutting operation such as blanking or piercing differ from deep drawing operation. *[UPTU 2002-03]*
30. What is the rate of clearance in blanking and piercing operations? How are clearances provided in punch and die. On what factors die and punch clearance depends. Discuss them in detail. *[UPTU 2003-04, 2005-06]*
31. How can the cutting forces be reduced in blanking operation? Estimate the blanking force to cut a blank of 25 mm wide and 30 mm long a 1.5 mm thick metal strip.

If the ultimate shear stress is 450 N/mm^2. Also determine the workdown. If percentage penetration is 25% of material thickness.
(**Ans.** 74.25 kN, 27.84 kJ) [*UPTU 2008-09*]

32. A hole of 100 mm is to be punched in a cold rolled medium carbon steel plate of 5.6 mm thickness. Ultimate shear strength of plate material is 550 MPa. with normal clearance on the press tool, cutting is completed at 40% penetration of the punch. Calculate the diameter for the punch and die required for purpose. If the shop has press of 30 tonnes capacity, calculate the shear angle to be provided on the punch in order to bring the work within the capacity of the existing press. [*UPTU 2009-10*]

33. Determine the punch forces to blank a steel washer 40 mm outer diameter and 25 mm inner diameter and 1.5 mm thick. If ultimate shear stress is 400 N/mm^2 for steel strip also determine the work done if percentage penetration is 35%.

34. Calculate the maximum possible axial force that may be required to punch a hole of 8 mm diameter in 1.5 mm thick sheet. If τ_s = 200 N/mm^2. Also calculate the diameter of the smallest hole that can be safety punched in the above steel with a factor of safety 2. Compressive strength is 800 N/mm^2.

35. A steel cup having 25 mm height and 40 mm diameter with a large width of 8 mm drawn from 1.2 mm thick sheet. Find out the blank diameter and drawing force what is drawn ratio? Can cup be drawn in a single pass.

36. A aluminium cup having 15 mm diameter and 40 mm height out of 2.3 mm sheet reduction ratio in first draw and in subsequent draws may not exceed 35% and 12% respectively. Find the blank size and number of draws necessary.

37. Calculate the blank diameters to make the shell as shown in Fig. 5.33.

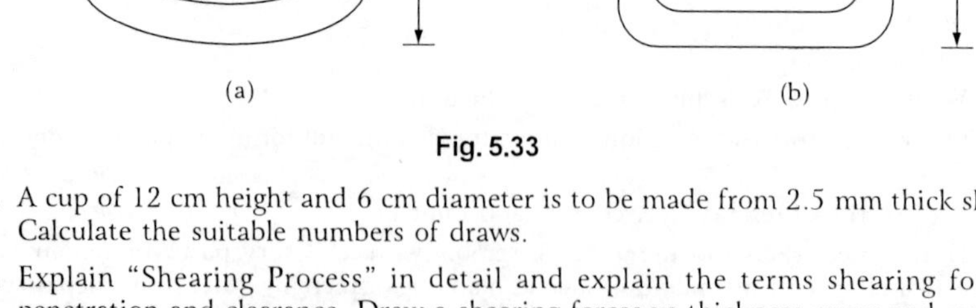

Fig. 5.33

38. A cup of 12 cm height and 6 cm diameter is to be made from 2.5 mm thick sheet. Calculate the suitable numbers of draws.

39. Explain "Shearing Process" in detail and explain the terms shearing forces, penetration and clearance. Draw a shearing forces vs thickness curve and explain how this can be used to determine the energy required in shearing. How can the maximum shearing forces be reduced. [*UPTU 2010-11*]

❑❑❑

UNIT-IV

CASTING
(FOUNDRY)

6

CHAPTER

CASTING AND PATTERNS

6.1 BASIC PRINCIPLE AND SURVEY OF CASTING PROCESS

Casting is one of the oldest method of producing the object. It is an oldest as Egyptian culture. Foundry is an art, which is made by Latin word 'Fundere' means melting and pouring. It is still the most popular method of producing machine components and parts of varieties of different metals and alloys. Casting basically recognized for producing heavy component on one side and highly complicated light component on other side.

Casting is defines as a process involving pouring of molten metal into a mould cavity which is formed in some moulding material such as sand. Mould cavity is the true shape or replica in size and shape that to be casted.

In casting process, molten metal is poured in the mould cavity and allowed to solidified. Important factor in casting that process is basically based on the flowability property of the molten metal by virtue of which it flows from different parts of a Gating System.

In casting process, the Ist step is to prepared the mould cavity which is prepared by the some suitable moulding material such as sand with some organic components, binders, etc. mostly all types of material can be casted by one method or the other, some common materials that can be casted are all types of **cast iron, steels, stainless steel, aluminium, copper, brass, bronze, magnesium, bearing metals**, etc.

6.2 ADVANTAGES OF CASTING PROCESS

Casting have the following advantages:

1. Very heavy and bulky components can be easily produce.
2. Parts can be produced in a single piece or multipiece in a single stroke.
3. Casting produce the strong (rigid) component.
4. High strength of part/component can be produce by casting process.
5. Intricated shapes can be produce by very easily.
6. Process is very suitable for mass production.
7. A large variety of alloys can be casted very easily.

6.3 STEPS INVOLVE IN THE CASTING PROCESS

Following steps are involve in the casting (sand) process:

- Pattern making
- Core making
- Moulding
- Melting and pouring
- Cooling and solidification
- Cleaning

1. **Pattern Making:** First step in the sand casting is making of a pattern and finally to produce a mould cavity by pattern which would be filled by molten metal. Pattern is the replica of shape that to be casted. Pattern can be formed out of **wood, metal (ferrous and non-ferrous metal), wax, P.O.P., plastic,** etc. one of the unique feature of pattern is that patterns are always **larger in size** as compared to actual product to be casted.

   If the casting is hollow like pipe fitting (T, elbow, etc.) additional pattern that is referred as core used to produce the hollow surface.
2. **Core Making:** If casting containing any hole either through or blind, an additional pattern is required this additional pattern is referred as core and normally made by sand. Core mixture usually comprises some refractory sand with large grain, binds for strength to core after baking. Cores are placed in the cavity to produce hole in the casting and they are different types.
3. **Moulding:** Mould is a void (hole/cavity) created in compact sand mass with the help of a pattern which is the true copy or replica of shape that to be casted. This cavity is filled by the molten metal. Process of creating or producing of mould (in the sand) is termed 'moulding'. Moulding related activity includes **pattern making, preparing moulding sand mixture, core making, baking of mould and core preserving mould.**
4. **Pouring and Melting:** After preparing the mould (cavity for casting purpose) the next and important step is melting of metal in a furnace and finally poured the same in the mould cavity, melting is done some specified area of the foundry, where as pouring is done in some other area. In pouring process, molten metal (liquid metal) is poured in the mould cavity through the gating systems.
5. **Cooling and Solidification:** This is also very important step in the casting (sand) process as every metals shrink after solidified. Once the pouring of molten metal is completed, the next step is to allow to cool the molten metal. During solidification, cast form develops cohesion and acquires structural characteristics. Mode of solidification affect the properties of the casting, as casting acquires a metallographic structure which is evaluated during solidification. Metallographic structure consists:
   (*a*) Grain size, shape and orientation
   (*b*) Distribution of alloying elements
   (*c*) Crystal structure and imperfections
6. **Cleaning:** Once the molten metal poured in the mould cavity and it is allowed to cool and solidify in mould cavity for a long time. Now casting required to removed from the moulding box, removal of casting from moulding

box is referred as shakeout. In casting process, clearing involve the all operations such as removal of sand, scale and addition metal (in terms of runner and riser) from casting. Outside surface and inside surface of casting are cleared of adhering refractory particles and oxides scale and they look smooth and pleasing.

Cleaning surface of casting may be completed by following ways–

(i) **Hand method:** such as wire brush, file, pick.

(ii) **Mechanical method**: such as tumbling, shot blasting, hydroblasting, chemical cleaning, etc.

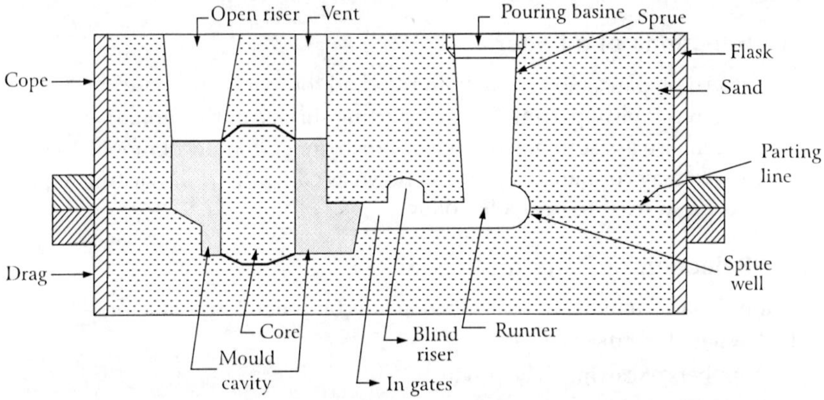

Fig. 6.1(A) Illustrating Salient Feature of Sand Casting

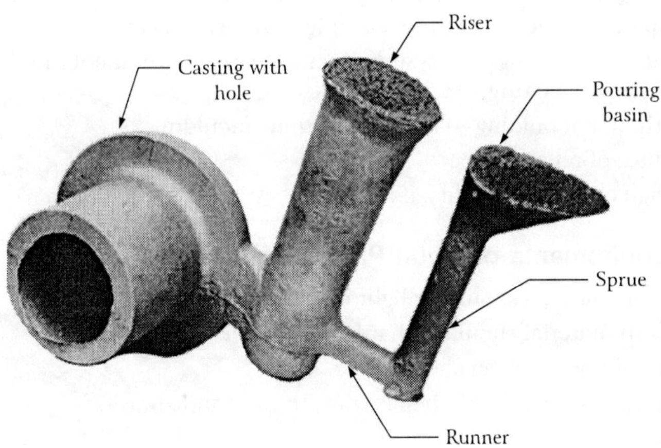

Fig. 6.1(B) Casting with riser, pouring basin etc.

6.4 PATTERN

Pattern is defined as the replica or true copy of shape that to be casted. Pattern is used for moulding a cavity in the moulding sand mixture, such formed cavity is similar to the shape of the casted part. Pattern is always **larger** in the size as compared of part that to be casted.

Pattern is the principal tool used in the casting process. It is the replica of the object to be made by the casting process, with some modifications. The main modifications are the addition of pattern allowances, and the provision of core prints. If the casting is to be hollow, additional patterns called **cores** are used to create these cavities in the finished product. The quality of the casting produced, depends upon the material of the pattern, its design, and construction. Cost of the pattern and the related equipments are reflected in the cost of the casting. The use of an expensive pattern is justified when the quantity of castings required is substantial.

6.4.1 Functions of a Pattern

(*a*) Pattern prepares mould cavity in the sand for casting purpose.
(*b*) Pattern establish parting line and parting surface in the mould.
(*c*) Pattern also assist to positioning a core, before ramming of moulding sand.
(*d*) Pattern may contain projections known as core print if the casting requires a core and need to be made hollow.

6.4.2 Selection of Pattern Material

Selection of pattern material basically depends on the following factors:

1. Design of casting.
2. Numbers of casting to be produce.
3. Types of moulding process.
4. Casting quality *i.e.*, **rough, medium, finish.**
5. Shape of casting *i.e.*, **simple, complex, intricated** etc.
6. Nature of moulding process such as **sand casting, permanent mould casting, investment casting, etc.**
7. Method of moulding — hand or machine moulding,
8. Chance of repeat order.
9. Design change possibility, if any, etc.

6.4.3 Requirements of Good Pattern Material

1. Pattern material should be light in weight.
2. Pattern material should be hard and durable.
3. It should be easily available.
4. Pattern material should be resistant to wear and abrasion,
5. It should also corrosion resistant.
6. It should be low in cost (economical).
7. Machining process should be perform very easily.
8. Surface finish should be good.

6.4.4 Pattern Material

Usually pattern materials are **wood, metal,** and **plastics**. The most commonly used pattern material is wood, since it is readily available and its low weight. Also, it can be easily shaped and relatively cheaper. The main disadvantage of wood is its absorption of moisture, which can cause distortion and dimensional changes.

Hence, proper seasoning and upkeep of wood is almost a prerequisite for large-scale use of wood as a pattern material.

Patterns may be constructed out of various materials. Each material has its own **advantages, limitations,** and field of **applications**.

Some materials used for making patterns are: **wood, metals** (ferrous and non-ferrous) and **alloys, plastic, plaster of paris, rubbers, wax** and **resins**.

To be suitable for use, the pattern material should be:

1. Easily worked, shaped and joined.
2. Light in weight.
3. Strong, hard and durable.
4. Resistant to wear and abrasion.
5. Resistant to corrosion, and to chemical reactions.
6. Dimensionally stable and unaffected by variations in temperature and humidity.
7. Available at low cost.

6.4.4.1 Wood

Wood is one of the most common material used for pattern material because it fulfill the various desired requirements. Wood use as a pattern material should be properly **dried**, free from **knots** and free from **excessive sap.**

6.4.4.1.1 Advantages of woods using as a pattern material

- Light in weight.
- Pattern can be made in a single piece or in a multipiece.
- Surface finish is good.
- Easily available.
- Low in cost.
- Wood can be easily joint.
- Pattern can be easily repaired.

6.4.4.1.2 Limitations of wood using as a pattern material

- More chance of moisture attack.
- Required proper handling.
- Required proper storage.
- Chance to warpage, wears out quickly due to sand abrasion.

  Wood such as

  ⇒ **shisham,**

  ⇒ **teak,**

  ⇒ **mahogany,**

  ⇒ **sal,**

  ⇒ **deodar**, etc. are the common in used for a pattern materials.

6.4.4.2 Metals

Metals are used as a pattern material since very old time as they have some their very common advantages over wood pattern material and they are used where repetitive production of casting is required in large numbers.

Various metal such as **aluminium, cast iron, copper, brass, white metal**, etc. are used for making patterns.

6.4.4.2.1 Aluminium and Its Alloys

Aluminium is also another material which is commonly used as a pattern material in foundry industries. Aluminium having some special properties which differ it with other materials such as it is a ductile material and not found in metallic state. One of the main reason of Aluminium using as a pattern material its weight. It is light in weight and can be shaped easily.

Aluminium may be classified into three categories:

(i) Pure aluminium.
(ii) Wrought aluminium.
(iii) Casting aluminium.

Aluminium Alloys	*Number*
Aluminium 99%	1XXX
Copper	2XXX
Manganese	3XXX
Silicon	4XXX
Magnesium	5XXX
Magnesium + Silicon	6XXX
Zinc	7XXX

Aluminium series

Advantages

(i) Aluminum is light in weight as compared to cast iron and steel. Surface finish of Al is very high degree.
(ii) Aluminium is very easy to cast in any shape.
(iii) Machining process can be easily performed on aluminium.
(iv) Anti-corrosion resistant.
(v) Aluminium alloys do not rust.
(vi) Aluminium is not so costly.

Limitations

(i) Aluminium patterns requires great care while handling or using.
(ii) Aluminium patterns cannot easily repared.
(iii) Aluminium patterns are subjected to wear by abrasive action of moulding.

6.4.4.2.2 Cast Iron

Cast iron also frequently used as a pattern material in casting industries due to the following advantages.

Advantages

⇒ Cast iron can be casted into any shape due to its good casting (ability) properties.

⇒ Cast iron is very cheaper and, easily available every where.

⇒ Cast iron is more durable than other metals.

⇒ Machining process can be easily performed on a cast iron.

⇒ It can withstand the abrasive action of the moulding sand.

Limitations

⇒ Cast iron is bulky in weight.

⇒ Due to brittle material it can easily break.

⇒ More chance of rusting and reparing of cast iron pattern is not so easy.

6.4.4.2.3 Brass

Brass is frequently used in casting process to produce those parts which requires very good surface finish. Main drawback of brass pattern that it is used to produced only small size parts.

Advantages

⇒ Less chance of rusting of brass.

⇒ Machining can be easily performed on a brass metal.

⇒ Brass is a strong and tough.

⇒ Brass can be easily adhere by soldering and brazing.

⇒ Good surface finish can be obtained.

⇒ No effect of abrasive of sand on brass.

⇒ Small pattern can be easily rectified and fitted.

Limitations

⇒ Brass is very costly as compared to cast iron and other metals.

⇒ Heavier than cast iron.

⇒ Very big size parts can not be produced by brass patterns.

6.4.4.2.4 White Metal

White metal is also used as a pattern material and having different advantages and Limitations.

Advantages

⇒ It possesses low melting point, i.e., 260°C.

⇒ White metal have very good castability properties.

⇒ White metal possesses light weight, can be casted into narrow cavity.

⇒ They can be easily join by soldering, white metal is die casting alloy.

Limitations

⇒ Very soft metal.

⇒ Pattern can easily wear out due to abrasion action of sand.

⇒ Proper care during transportation and storing is required.

6.4.4.3 Plastic

Thermoplastic and thermosetting plastic can be used as a pattern materials. Usually thermoserting epoxy and polyster resins are used. Plastics having more advantages over any other pattern materials.

Advantages

(i) Plastics are light in weight.
(ii) No chance of corrosion.
(iii) Less costly as compared to brass, aluminium, and white metal.
(iv) Easily available globally.
(v) Can be casted in any shape very easily.
(vi) It possesses low solid shrinkage.
(vii) It possesses good surface finish.
(viii) It possesses good tensile strength.
(ix) It is hard and very durable.
(x) It possesses good compressive and flexural strength.
(xi) It possesses good impact strength.
(xii) Less chance of chemical reaction.

Limitations

(i) Plastic patterns are fragile and thus light sections and may need metal reinforcement.
(ii) Plastics are very soft hence can be deformed under applied load easily.
(iii) They have poor ductility and low heat resistant.
(iv) Plastic is a non-renewable resources.

6.4.4.4 Plaster of Paris (PoP)

Advantages

(i) It can be easily worked by using wood working tools.
(ii) It is low in cost and easily available.
(iii) It is light in weight.
(iv) It possesses very good compressive strength.
(v) It can be casted in any shape.

Limitations

(i) Expands on solidification.
(ii) It do not possesses a very good tensile strength.

6.5 TYPES OF PATTERNS *[UPTU 2010, C.C.S. University 2010]*

Following types of patterns are very common in use:

(a) Single piece pattern.
(b) Split pattern or two piece pattern.
(c) Multipiece pattern.
(d) Cope and drag pattern.
(e) Segmental pattern.
(f) Gated pattern.
(g) Match plate pattern.

(h) Skeleton pattern.
(i) Loose piece pattern.
(j) Follow board pattern.

6.5.1 Single Piece Pattern

Single piece pattern is also known as *solid pattern.* As the name depict this type of pattern is made out of one piece and they are very simple in construction.

Single piece pattern may be out of wood or a metal piece. For producing the mould cavity, single piece pattern is accommodated either in cope or drage. **Stuffing box** of the steam engine is casted by the single piece pattern.

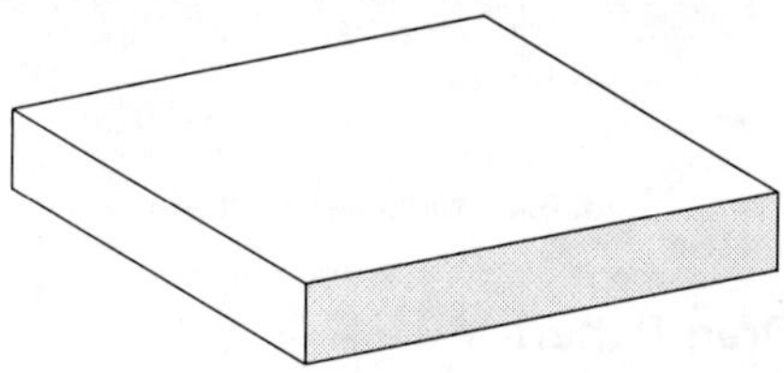

Fig. 6.2 Solid pattern/single Piece Pattern

6.5.2 Split or Double Piece Pattern

Figure 6.3 shows the split pattern. As the name suggested this type of patterns are made by two pieces known as upper and lower part. Both upper and lower parts are fastened by a dowel pins. Surface which informed at the line of separation of the two parts, usually at the centre line of the pattern is called *parting surface or parting line.* Casting such as **boll valve, tapes** and **water stop-cocks** are produced by the split pattern.

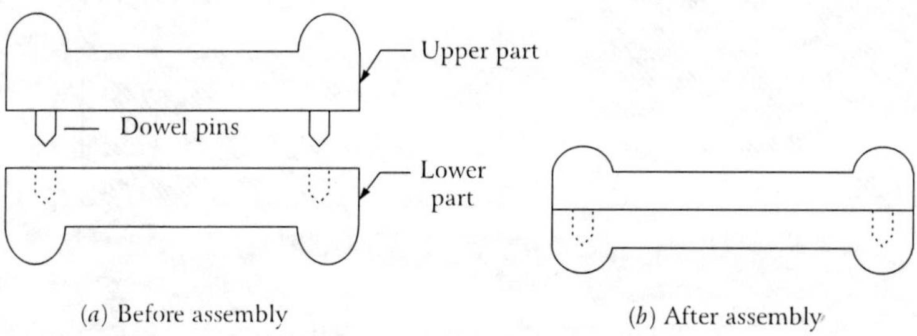

Fig. 6.3 Split pattern

6.5.3 Multipiece Pattern

Sometime casting is larger in size and have a complicated structure, then producing this type of object (casting) multipiece pattern are required. This type of pattern are made more than two parts. They may be 3, 4 or more number of parts depending upon design of product. In multipiece patterns, all patterns are fastened with each other by the means of dowel pins. If pattern is made out of three parts then top part is known as **cope**, middle as **check** and bottom as **drag.**

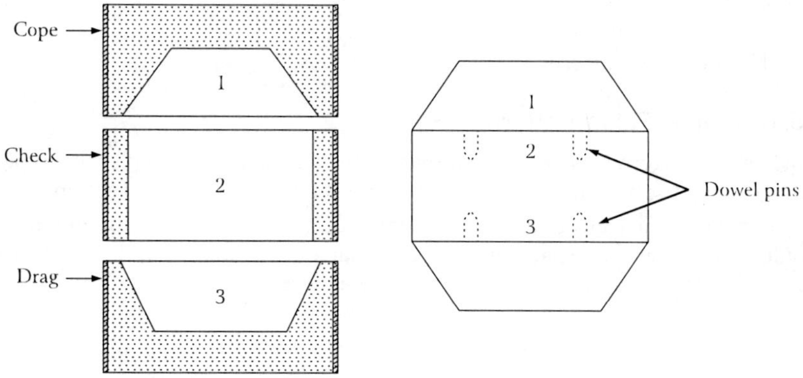

Fig. 6.4 Multipiece pattern

6.5.4 Cope and Drag Pattern

It is also another type of split or two piece pattern but very large in size. Each half of pattern is fastened to a separate wood/metal plate. Every metal plate have the proper provision of runners and gates. This type of patterns are basically used for producing a big casting which is not possible by one piece.

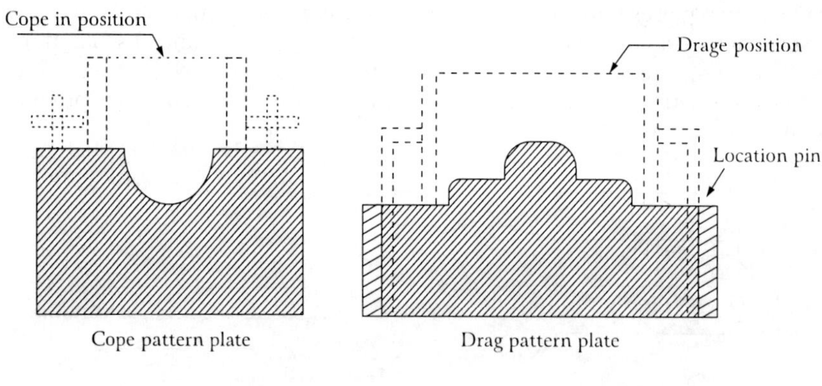

Fig. 6.5 Cope and Drag Pattern

6.5.5 Segmental Pattern

This type of patterns are usually made in the segments and patterns are used to produce a casting which have a circular ring shape of large section. Segmental patterns are the sections of a full pattern. Moulding after one location (place) now pattern is rotated for next location and moulding is done. This type of pattern is differ with sweap pattern in manner of that it do not revolve continuously about the post to make the mould.

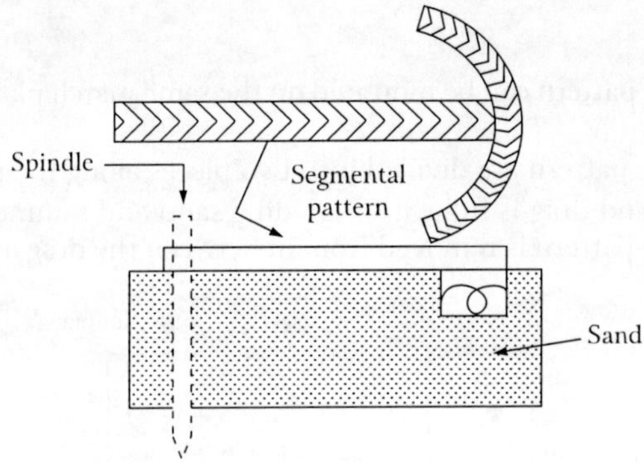

Fig. 6.6 Segmental pattern

6.5.6 Gated Pattern

If a large number of casting (object) to be produced in a single stroke then gated type patterns are used. This type of patterns are made out of metal to increase the strength. In this type of casting, mould cavity is connected to each other by means of gates, formers which provide channels or gates for feeding the molten metal. This type of pattern are used to produce small parts usually same shape and size.

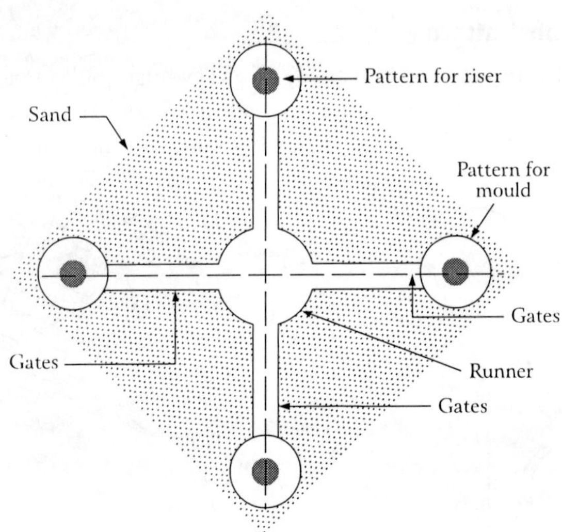

Fig. 6.7 Gated pattern

6.5.7 Match Plate Pattern

Match plate pattern is mounted on the metal plate/wood plate with gates and runner which are the integral part of the pattern.

Features:

- Numbers of pattern can be mounted on the same match plate in match plate pattern.
- Match plate pattern are divided into two pieces along the parting line.
- Once cope and drag is full with moulding sand and rammed with properly, match plate pattern is removed from in between the drag and cope.

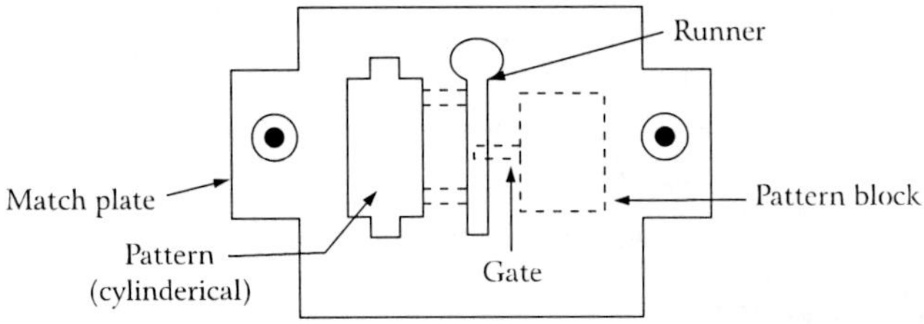

Fig. 6.8 Match plate pattern

- Now cope and drag finally assemble for casting purpose.
- Match plate pattern is used for small parts for mass production.
- Castings are very accurate in size.
- I.C. Engine piston rings are produce by the match plate pattern.

6.5.8 Skeleton Pattern

This type of patterns are used for casting of a very huge (bulky) part.

- To making solid frame, the pattern is made of wooden frame having the strips.
- Skeleton pattern is filled with moulding sand and then rammed.

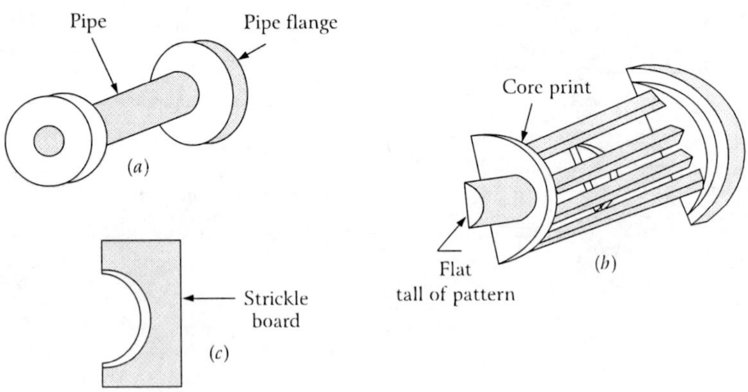

Fig.6.9 Skeleton pattern

- Proper countour is made by using a template or strickle which removes excess sand out of the space round.
- Skeleton pattern are basically made into two parts, upper and lower.
- Skeleton pattern is less costly as compared to a solid pattern.

Turbine casting water pipe, L-bend and many more items are casted by this pattern.

6.5.9 Loose Piece Pattern

In casting process, there are some pattern which cannot be withdraw from mould cavity as they are embedded in the moulding sands.

- Loose piece pattern are made by more than one loose piece.
- This type of pattern are basically used to produced a part which having some projecting part as above or below of the main parts.
- This type of projections are very difficult to lift the pattern from mould cavity.
- Therefore, it is must to make a pattern in such a way that the projecting portions or part should be loose piece manner with main pattern (Fig. 6.10) by the help of dowel pins.
- Main body of pattern first drawn from the mould cavity and then loose piece. Loose piece pattern consume more labour and consume more timing in casting.

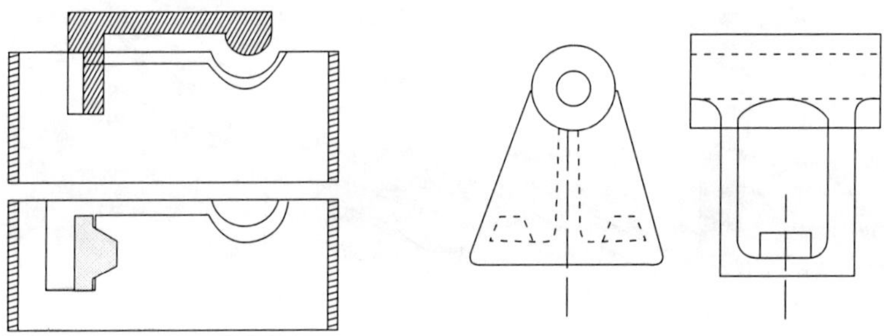

Fig. 6.10 Loose piece pattern

6.5.10 Follow Board Pattern

This type of patterns are basically used to produce a casting whose some parts are structurally weak and if supported not properly, chance of break under the force of ramming. Weaker pattern is supported by follow board finally drag is rammed and then follow board is withdrawn. Rammed drag is invested, cap is mounted on it and rammed.

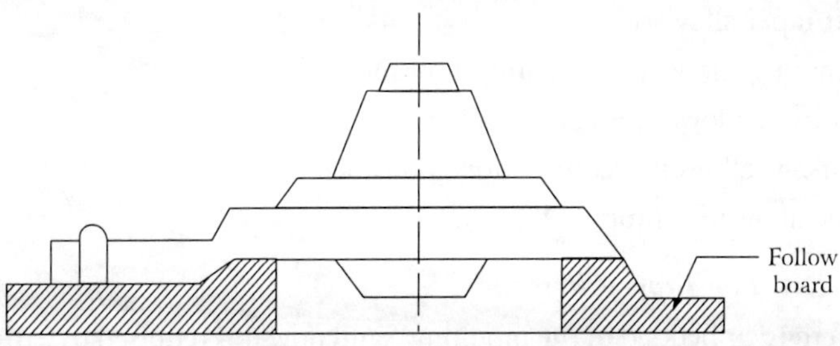

Fig. 6.11 Follow board pattern

6.6 PATTERN ALLOWANCES

[UPSC 2010; UPTU 2010; C.C.S. University 2011]

Pattern allowances are vital feature as it affects the dimensional characteristics of the casting. Thus, when the pattern is produced, certain allowances must be given on the sizes specified in the finished component drawing, so that a casting with the particular specification can be made. The selection of correct allowances greatly helps to reduce machining costs and avoid rejections.

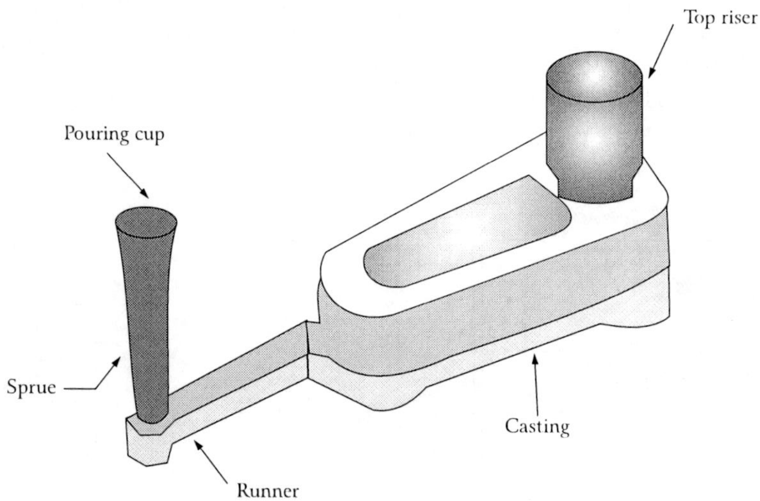

Fig. 6.12 A typical pattern attached with gating and risering system

Pattern of any shape (casting) is always large in the size as compared to the casting. It is due to mechanical reasons as well as metallurgical reasons like, every material gets shrinkage on solidification, machining is required, etc.

6.6.1 Type of Allowances

Following are the type of allowances:

- Draft/taper allowances,
- Machining allowance/finishing allowance,
- Distortion allowances/camber allowance,
- Shrinkage allowance/contraction allowances,
- Shape allowance/rapping allowance.

6.6.1.1 Draft or Taper Allowance

Once pattern is embedded in the moulding sand now it is required to withdrawing the pattern from moulding sand, for this purpose a vertical force is required to lift the pattern. For easy removal of pattern from mould cavity there should be some degree or mm taper in the pattern.

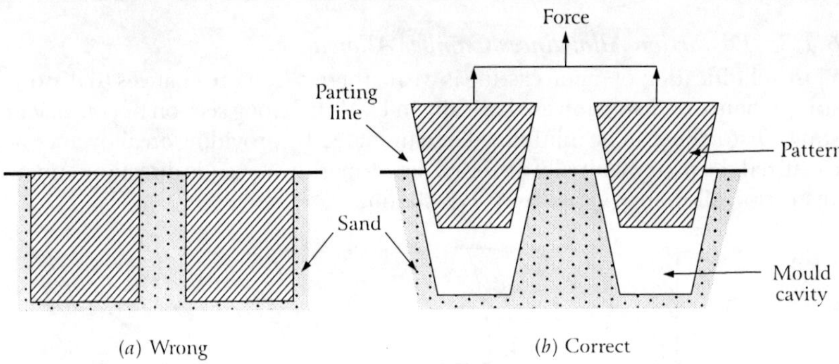

Fig. 6.13 Draft Allowance

If there is no taper in the pattern then more chance to break the mould cavity wall during withdrawing the pattern. Draft allowance varies with the complexity of job.

TABLE 6.1 Draft Allowances

Pattern material	*Surface height*	*Draft allowance in degree*	
		External surface	*Internal surface*
Wood	up to 20	3.0	3.0
	21-50	1.5	2.5
	51-100	1.0	1.5
Metal	101-800	0.75	0.50

6.6.1.2 Machining or Finishing Allowances

In sand casting operation machining/finishing allowances is given in the pattern as on solidification every metal get shrinked up to some extend and the outer surface of object (casting) is rough, therefore, to smoothing the outer surface of casting machining allowance is must.

- Ferrous metals have scales on the skin which are to be removed by cleaning.
- Therefore, extra material is to be provided which would be subsequently turned by machining.
- Range of the machining allowance is from 2-20 mm.

Machining allowance basically depends upon the following. Such as

- **type of casting,**
- **type of material finish required,**
- **type of machining operation required,**
- **degree of finish required,**
- **moulding process required, etc.**

TABLE 6.2 Machining allowances

Metal/Alloys	*Machining allowance (mm)*
(i) Cast iron	3 mm-10 mm
(ii) Cast steel	4.5 mm-12 mm
(iii) Aluminium, brass, bronze, etc.	1.5 mm-5.0 mm

6.6.1.3 Distortion Allowances/Camber Allowance

On just solidification of metal casting is weak, therefore, more chances to distorted casting. Chance of distortion are basically in U, V, then long section or complicated casting. Distortion can be minimized or eliminated by providing an allowance and constructed the pattern initially distorted, i.e., opposite in outside direction. Amount of distortion allowance varies from 1.5-20 mm.

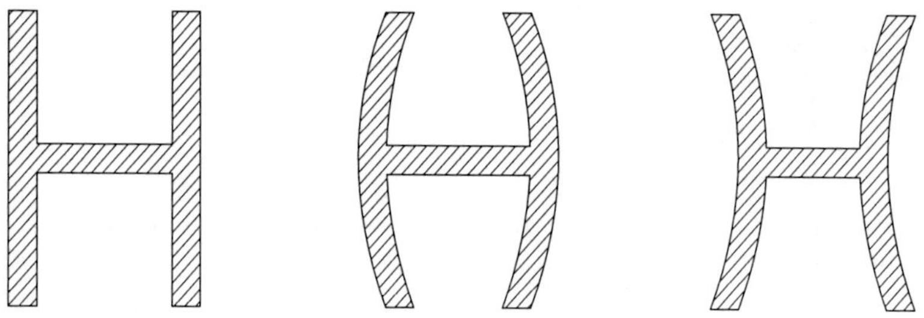

Fig. 6.14 Distortion allowance

6.6.1.4 Shrinkage/Contraction Allowances

It is the tendency of all metals that they shrink after cooling except bismuth. Shrinkage is only due to inter atomic vibration which are implified by increasing in temperature. Metal shrink or contract in three ways:

- Liquid Contraction
- Solidifying Contraction
- Solid Contraction.

(*i*) **Liquid Contraction**

- Liquid contraction means reduction in volume as metal convert from liquid to solid state.
- This type of shrinkage can be compensated by the feeding of metal from riser where metal remains in molten state till object (casting) solidified.

(*ii*) **Solid Contraction:** It referred as reduction in volume due to metal losses temperature in solid state. Rate of shrinkage basically depends upon the type of material. For example, steel shrink at a high rate as compared to aluminium. Shrinkage also depends upon the metallurgical change in metals during solidification.

Different metals have different contraction allowance: Shrinkage basically depends upon the following such as:

- types of material,
- moulding material,
- mould design,
- size of casting and type of moulding procedure.

TABLE 6.3 Shrinkage Allowances for different metals

Material	*Shrinkage allowance (mm/m)*
Cast iron	6.6–10.5 mm/m
Steel	20.8
Aluminium	15.0
Aluminium alloys	12–15
Copper	16.0
Zinc	10–15
Magnesium	13.0

Example 6.1

A casting shown in Fig 6.15 is to be made in steel using a wooden pattern. Assuming only shrinkage allowances, calculate the dimensions of the pattern.

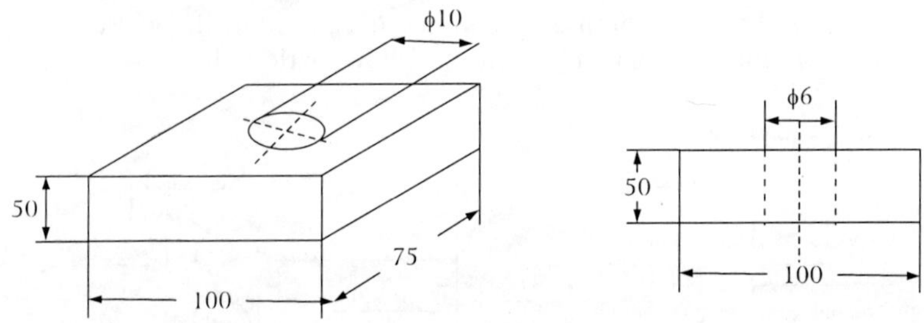

Fig. 6.15

Solution: From table shrinkage allowance is 21 mm/m for steel.

For dimension 100 mm, allowance is 100 × 21/100 = 2.1

For dimension 75 mm, allowance is 75 × 21/1000 = 1.575 = 1.6

For dimension 50 mm, allowance is 50 × 21/1000 = 1.05

For dimension 10 mm, allowance is 10 × 21/1000 = 2.1

Pattern drawing with required dimension taking shrinkage into account is shown in Fig. 6.16 double shrinkage allowance is to be provided on the pattern, if it is to be used for casting the metallic pattern which would ultimately be used for moulding to take cure of the shrinkage of the actual metal cast as sell as the shrinkage of the pattern metal.

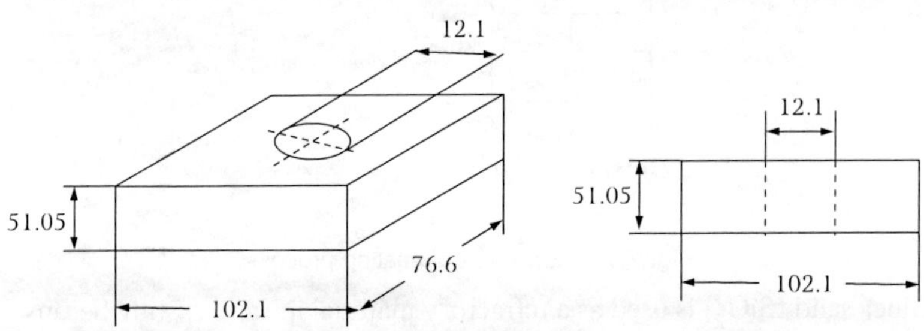

Fig. 6.16 Provision of allowance of pattern is sand casting.

7

CHAPTER

SAND CASTING PROCESS

7.1 INTRODUCTION

Sand casting is the most common technique used around the world. A typical process flow of sand casting is shown in the following flow chart.

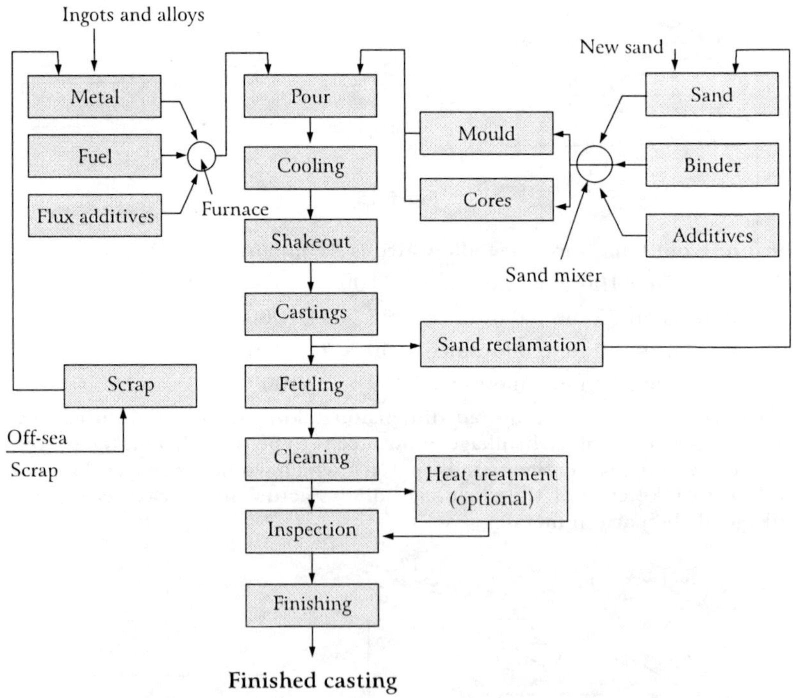

Fig. 7.1 Flow chart of casting process

Silica sand (SiO_2) is used as a refractory material in sand casting because it is relatively cheap. In sand casting process a binder (usually clay) maintains the shape of the mould while pouring molten metal into mould cavity. There are a wide range of sand binder systems that are used in sand casting process. **Bentonite,**

clay is used upto 4-10% of the sand mixture in green sand systems, which is the most common sand casting system. Water, which makes up around 2-4% of the sand mixture, activates the binder. **Carbonaceous** material such as charcoal [2-10% of total volume] is also added to the mixture to provide a reducing environment. It helps in preventing the metal from oxidizing while pouring. The remaining 85-95% of the total mixture contains sand.

A range of chemical binders is used by other sand moulding processes: Oil binders are mixture of **animal oils**, **petrochemicals** and **vegetables**.

Some of the popular synthetic resin binders are:

⇒ urea formaldehyde,

⇒ phenolics, phenol formaldehyde,

⇒ urea-formaldehyde/furfuryl alcohol,

⇒ alkyl isocyanate and

⇒ phenolic isocyanate.

Chemical resin binders are frequently used for foundry cores and less extensively for foundry moulds.

7.2 HOW PATTERNS ARE MADE

The first stage for developing a new casting is pattern making. Pattern is the replica of the finished product. Generally, **Sand Casting Process Patterns** are made of **wood** but **metal, plastic, plaster of paris** can also be used. These patterns are permanent so can be used to form a number of moulds. Pattern making is a highly skilled and precise process *i.e.*, critical to the quality of the final product. Many modern pattern shops make use of Computer-Aided Design (CAD) to design patterns. These systems can also be integrated with automated cutting tools that are controlled with Computer-Aided Manufacturing (CAM) tools. Cores are produced in conjunction with the pattern to form the interior surfaces of the casting. These are produced in a core box, which is essentially a permanent mould that is developed.

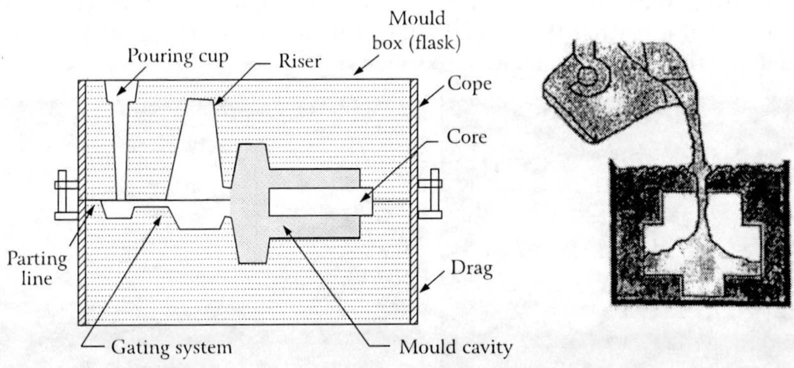

Fig. 7.2 Principle of sand casting

The mould is formed in a mould box with two halves that helps in removing the pattern. As sand moulds, are temporary in nature, a new mould has to be formed each time for individual casting. The above Fig. 7.2 describes a typical two part sand mould.

When the core is inserted on the top of the furnace, its burner starts the melting process immediately. Drag, the bottom half of the mould is made on a moulding board. Cores require greater strength to hold their form during pouring. Dimensional precision also needs to be greater because interior surface are more difficult to machine making errors costly to fix. One of the chemical binding systems is used in forming the cores. Once the core is inserted, the top half of the mould or the cope is placed on top. The interface between the two mould halves is called *parting line.* Sometime, weights are placed on the cope, which helps in securing the two halves together.

Mould designs include a gating system which is designed to carry molten metal smoothly to all parts of the mould. The gating system typically includes a **sprue, gates, runners** and **risers**. The sprue is, where the metal is poured. Gates allow the molten metal to enter the running system. Runner carry the molten metal towards the mould cavity.

Risers may have several functions including vents to allow gases to be released, reservoirs prior to the casting cavity to aid progressive solidification, and waste cavities to allow metal to rise from the casting cavity to ensure it is filled and to remove the first poured metal from the casting cavity, thus avoiding solidification problems.

7.2.1 Melting and Pouring

Many ferrous foundries use a high proportion of scrap metal to make up a charge. As such, foundries play an important role in the metal recycling industry. Internally generated scrap from runners and risers, as well as rejected product, is also recycled. The charge is weighed and introduced to the furnace. Alloys and other materials are added to the charge to produce the desired melt. In some operations the charge may be preheated, often using waste heat. In traditional processes, metal is superheated in the furnace. Molten metal is transferred from the furnace to **ladle** and held until it reaches the desired pouring temperature. The molten metal is poured into the mould and allowed to solidify.

(*a*) (*b*)

Fig. 7.3 Pouring of molten metal into mould cavity

7.2.2 Cooling and Shakeout

The mould is transported to a cooling area, immediately after the molten metal has been poured in, the casting needs to cool for a long time, often overnight, before it can be removed from the mould. Castings may be removed manually or using vibratory tables that shake the refractory material away from the casting. For rapid cooling of castings, many foundries also use quenching baths. This speeds up the process and also helps achieve certain metallurgical properties. To prevent oxidation, the quench bath may contain chemical additives.

7.2.3 Sand Reclamation

Foundries recover a significant proportion of the waste sand for internal reuse. It significantly reduces the quantity of sand that must be purchased and disposed off. Generally, sand casting is reclaimed mechanically. Cores and large metal lumps are removed by vibrating screens and the binders are removed by attrition in which the sand particles rub together.

Fine sand casting and binders are removed by extraction and collected in a baghouse. In some systems, metals are removed using magnets or other separation techniques. For operations using mechanical reclamation, the recycle rate is often limited to around 70%.

This is due to the need to maintain a minimum sand casting quality. For large iron foundries, where sand casting quality requirements are less stringent, over 90% reclamation can be achieved by mechanical means. For many processes, mechanically reclaimed sand is not of sufficiently high quality to be used for core production. Thermal reclamation is becoming more widely used in Queensland, sand casting process heat the sand to the point where organic materials, including the binders, are driven off. Sand casting process can return the sand to an 'as new' **slate**, allowing it to be used for core making. Thermal **reclamation** is more expensive than mechanical systems.

Sand casting process can also be reclaimed using wet washing and scrubbing techniques. These methods produce sand of a high quality but are not commonly used because they generate a significant liquid waste stream and require additional energy input for sand drying. The amount of internal reuse depends on the type of technology used and the quality requirements of the casting process. Reclamation

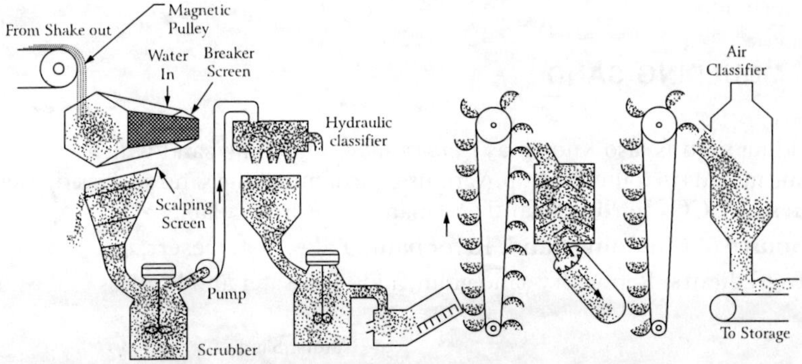

Fig. 7.4 Combined wet and thermal sand reclamation

processes, particularly mechanical ones, break down the sand particles and this can affect the quality of some metals. Also, for mechanical reclamation techniques, impurities may build up in the sand over time, requiring a proportion of the material to be wasted. Large iron foundries do not require a high sand quality so typically achieve the highest rate of reuse in the industry. Often sand cycles through the operation until it is ground down to a fine dust and removed by baghouse.

7.2.4 Fettling, Cleaning and Finishing

The gating system is removed, after the casting has cooled, using abrasive cut-off wheels, Handsaws, or electrical cut-off devices A 'parting line flash' is typically formed on the casting and must be removed by grinding or with chipping hammers. Castings may also need to be repaired by welding, brazing or soldering to eliminate defects.

The casting may undergo additional grinding and polishing operations to achieve the desired surface quality. The casting may then be cooled using either a paint or a metal finishing operation such as galvanizing, powder coating or electroplating.

7.3 ADVANTAGES OF SAND CASTING

- Low tooling cost.
- Largest casting sizes achievable.
- Less expensive than other casting techniques.
- Capable of holding detail and resist deformation when heated.
- Process is suitable for both ferrous and non-ferrous metal castings.
- Handles a more diverse range of products than any other casting method.
- Produces both small precision castings and large castings of up to 1 tone.
- Can achieve very close tolerances if uniform compaction is achieved.
- Mould preparation time is relatively short in comparison to many other processes.
- The relative simplicity of the process makes it ideally suited to mechanization.
- High levels of sand reuse are achievable. It produces less wastge than other techniques.

7.4 MOULDING SAND

[U.P.S.C.-2010]

Moulding sand is also known as *foundry sand,* moulding sand which is used for making moulds in foundry shop. Various type of moulding sand such as **dry sand, green sand, CO_2 hardend sand** and many more are available.

Source of Moulding Sand: River bank, Lake, Sea, Desert.

Ingredients: Refractory sand grain, additive, water and Bindess

Moulding Sand
- Natural Sand
- Synthetic Sand

7.4.1 Natural Sand

Natural sand can be directly used for mould making as received from source. Natural sand basically consist some binders.

Natural sand is free of cost available from the source. Natural sand have moisture contents for a long time. Natural sand requires only upto 10% water for mixing purpose.

7.4.2 Synthetic Sand

Synthetic sand have the sufficient amount of natural sand with or without clay, with having binders and moisture too. As the name depict they are made by artificially by mixing natural sand with some other binding materials. Synthetic sands are made to achieve the desire properties which is not present in natural sand. Synthetic sand basically used for casting of ferrous and non-ferrous materials.

Grain Size of Sand: Grain size of the sand is very important factor for influencing the sand properties such as **refractoryness**, **flowability**, **strength**, **permeability**, **etc.**

Grain size of sand are basically two types:

(*a*) Fine grain size.

(*b*) Coarse grain size.

Fine grain size of sand is responsible for surface finess as well as responsible for lowering the permeability. On the other hand **coarse** grain size is responsible for good permeability, high refractoryness and high flowability. Usually the grain size of moulding sand varies from **0.1** to **2.0 mm.** Fine grain size sand is frequently used for producing the ornaments as well as small size casting where as coarse grain size sand is used for producing large casting.

Grain Shape: Shape of the grain also influence the sand properties like strength (dry/green), flowability, etc. Sand grain shape may be the following type:

(i) Round grain shape

(ii) Angular grain shape

(iii) Sub-angular shape

(iv) Compound shape

Angular Sub-angular Angular Compound

Fig. 7.5 Sand grain shapes

7.4.3 Properties of Moulding Sand [*C.C.S.U. 2010*]

A good moulding sand should have possesses the following properties:

(i) Refractoryness: Refractoryness is the property of sand by vartue of it, sand can withstand at high temperature without fusion, cracking, buckling or any other physical change.

(ii) **Permeability or Porosity:** It is the property of sand which allow the gases/ steam to escape in the atmosphere from the mould sand. During the casting process as moulding sand have the organic binders, moisture content, etc. as this moulding sand come to contact with molten metal (liquid material) steam and gases produce in the moulding sand and escaping of these **steam/ gases** are very essential otherwise more chance to defective casting. Therefore a passage provided for escaping these steam/gases moulding sand should be porous. Soft ramming and adding some clay into sand improves the permeability property. Coarse grain size sand particles exhibit more permeability.

(iii) **Green Strength:** The strength of the sand in the moisture state means when there are some moisture contents in the sand. Green strength possesses in preparing the sand and handling the sand (mould).

(iv) **Dry Strength:** The strength of sand when it do not contain any moisture particles i.e., strength of the sand at dry position is referred as dry strength of sand. Dry strength basically related with grain size, binders. Mould should have sufficient dry strength to withstand the pressure of molten metal and erosive force.

(v) **Hot Strength:** The strength of the sand (mould) when mould is in hot position due to the hot strength property sand hold its proper shape in mould cavity. If there is no hot strength properties in sand then mould may cracked, enlarge, erode or break.

(vi) **Adhesiveness:** Adhesiveness is the property of moulding sand by virtue of it adhere with the other materials (moulding box walls). Only by this properties sand adhere with the moulding box walls.

(vii) **Cohesiveness:** It is the properties of sand by virtue of it (sand) adhere or join with the same particles. During the ramming operation due to cohesiveness sand particles bind together firmly and thus give green strength to the moist moulding sand in maintaining the shape of mould cavity. Cohesiveness basically depends upon the **shape, size** and **sand grain distributions, type of bonding material used** and **moisture contents.**

(viii) **Flowability:** Flowability is the property of sand due to which the sand flows under the effect of ramming to all area of moulding sand box and specially pack around the pattern while distributing the ramming pressure evenly on the sand in all part of the moulding box, flowability improves on increasing clay and water contents.

(ix) **Collapsibility:** It is the property of sand due to which sand mould collapses automatically after solidification of casting. Thus avoid hot tear and cracks in the casting.

(x) **Bench Life:**Bench life is the properties of casting sand by virtue of it (sand) retain its properties during storage or in standing condition.

(xi) **Chemical Resistivity:** According to this properties, sand should not react with the molten metal.

(xii) **Durability:** Moulding sand should be used more than one times.

7.5 MOULD

When moulding box is completely filled with sand and pattern is kept into the sand and properly rammed and pattern is withdrawn from the moulding box, a cavity (shape of pattern) is formed in sand is known as *mould* or *mould cavity.*

7.5.1 Types of Mould

Various types of mould are available, three main types of moulds are given below:

- Expendable mould
- Composite mould
- Permanent mould

7.5.1.1 Expendable Mould

Expendable mould is prepared out of refractory items which are capable to withstand at a high temperature of molten metal like ceramics, silica, gypsum, etc. this type of the mould is used as after casting, is solidified and after solidification mould has to be broken up to remove casting. Expendable moulds are different type such as

- **green sand mould,**
- **dry sand mould,**
- **CO_2 dried sand mould,**
- **steel mould,**
- **investment casting mould, etc.**

7.5.1.2 Permanent Mould

As the name depicts it is a permanent type mould, made by some metal. One of the main characteristic of this type of mould that it can retain their strength at very high degree temperature of molten metal. Permanent mould is design in such a way that casting can be withdrawn from mould without break the mould as same happening in expendable mould. Permanent mould can be used for repeating casting for illustration purpose a mould made by cast iron can produce 5,000 numbers of aluminium casting. Permanent moulds are various type such as

- **die casting mould,**
- **Pressure die casting mould,**
- **continuous casting mould,**
- **centrifugal casting mould, etc.**

7.5.1.3 Composite Mould

As the name depicts this type of mould is made by the composition of two different material. Such as sand graphite and metal. This type of mould is basically made to get the some **special properties** of the mould by combining of two different metals. Composite mould possesses very **good strength.**

Turbine impeller are casted by the composite shell mould.

Some other different types of mould used in industries are given below.

- **Green sand mould,**
- **Dry sand mould,**
- **Loam sand mould,**
- **Air dried mould,**
- **Plaster mould,**
- **CO_2-hardened mould,**
- **Core sand mould,**
- **Shell mould,**
- **Ceramic mould,**
- **Graphite mould, etc.**

7.5.1.4 Green Sand Mould

Green sand mould is referred as mould in moisture or damp condition, at the time when metal is pouring into the mould. Green sand mould consist the **silica, clay, sand, water and some additives.** As no baking of mould is required before pouring metal, therefore all ferrous and non-ferrous metal can be easily casted by green sand mould (small and medium size product). Green sand mould is a cheaper mould. Green sand mould **possesses poor strength and less hardness.**

For casting cast iron, green sand mould should have the following composition:

(i) River sand 50-60%, (ii) Water 4-10%,
(iii) Bentonite 2-20%, (iv) Clay 12-15%,
(v) Coal dust 5-10%

7.5.1.5 Dry Sand Mould

Dry sand mould is defind as completely dried sand mould before pouring of molten metal into mould (cavity) and this mould cavity dried by baking the mould in a furnace or oven before the molten metal is poured into the cavity for casting purpose. As we know that at the time of preparing the mould sand mixture have enough moisture content (by mixing the molasses, dextrine, etc.) to easy making and shaping the mould, by latterly moisture contents is completely evaporated into atmosphere by baking process in oven and finally convert into dry sand mould. Baking is done to provide the **more strength** and **hardened** to mould as the baked sand consist less moisture therefore less gases formation during the pouring process, hence result a defect free casting.

Baking of mould is done at 150-300°C. Dry sand mould is suitable for large casting of steel and cast iron.

For a typical casting of cast iron, composition of sand mixture:

(i) Floor sand–50%,
(ii) Water–10%,
(iii) Saw dust–10%,
(iv) River Sand–25%,
(v) Clay–5%.

7.5.1.6 *Loam Sand Mould*

This type of moulds are widely used to casting very large products which have symmetrical shape like, **conical container/cylindrical container.** Figure 7.6 shows the loam sand mould, which is daubed over on the bricks. Refractory brick is given to mould face and the mould dried to get a strong and hard mould.

A Loam and Mortar is prepared by adding, **Clay, Coarse Slica Sand, Chopped Straws, Fire Clay** with water. Loam Sand Mould is basically used to produce product such as **large size cylinder, gear blanks,** etc:

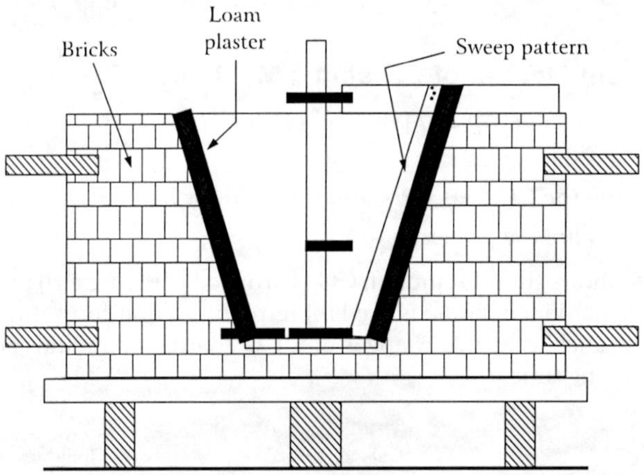

Fig. 7.6 Preparing a loam sand mould using a sweep pattern

7.5.1.7 *Air Dried Mould*

Air Dried Mould is quite similar to skin dried mould, mould face is dried and hardened by air.

7.5.1.8 *Plaster Mould*

In this type of system mould is made out of gypsum slurry. For plaster, mould two halves (mould) are prepared which are clamping together and molten metal is poured in the two halves. Gypsum basically consist **gypsum plaster, silica flour, silica sand** and **water.**

Plaster moulds are used for casting of **Aluminium, Zinc, Magnesium and Copper base** alloys with fine surface. This mould have low porosity therefore special cares are required for escaping the gases during casting process.

7.5.1.9 *CO_2 Hardened Mould*

Basically **carbon-dioxide** is used to **hardening** and **strengthening** to moulds and cores. Moulds are prepared by adding **dry silica sand, sodium silicate, liquid base binder and moisture** in appropriate proportion.

Some percentage of graphide is added in mould sand to increases the collapsibility. Sand mixture is packed in moulding flask with pattern and then CO_2 gas is passed about it for 30-40 sec before withdrawing the pattern from the sand. CO_2 react with sodium silicate, precipitating SiO_2 and giving strength and hardness to the mould.

$$Na_2SiO_3 + CO_2 + H_2O \rightarrow Na_2CO_3 + \underbrace{SiO_2 + H_2O}_{\text{Silica jel}}$$

7.5.2 Different Method of Gassing a Mould by CO_2

(i) By using rubber strip and plywood cover.

(ii) By using a shower curtain.

(iii) By passing the CO_2 gas throughout mould for harden.

(iv) By using a hollow pattern.

Figure 7.7 shows the CO_2 mould. CO_2 hardened cores are also produce by the same way. CO_2 hardened mould is used for ferrous and non-ferrous metals casting for thin section such as heat exchanger fins. CO_2 hardened moulds and cores can be stored for a long time.

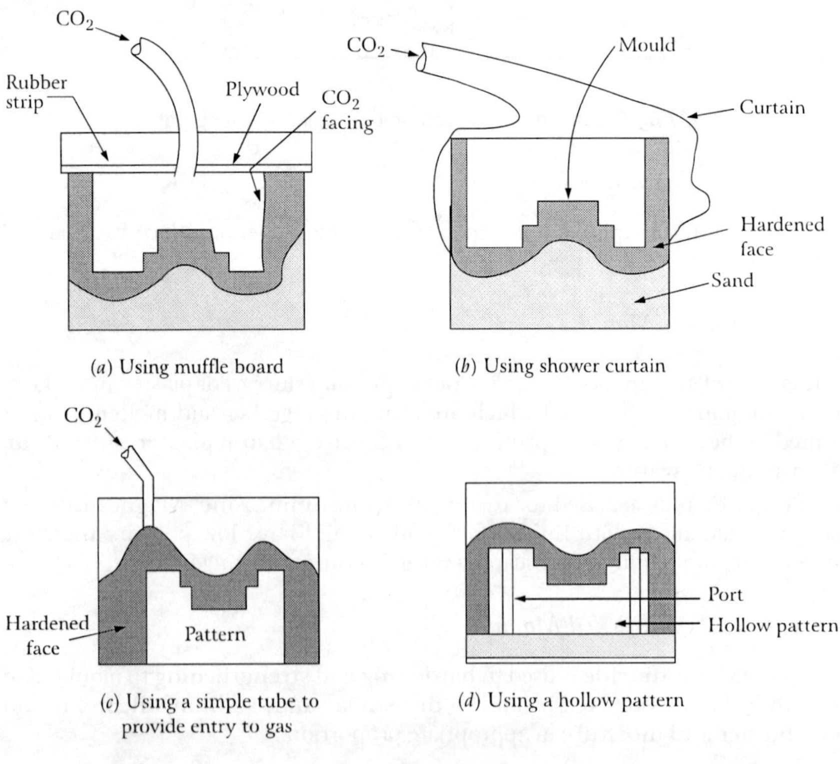

(*a*) Using muffle board

(*b*) Using shower curtain

(*c*) Using a simple tube to provide entry to gas

(*d*) Using a hollow pattern

Fig. 7.7 CO_2-Moulding

Advantages of CO_2 hardened mould

(i) Surface finish of casting is better
(ii) A mould or core can be harded easily by CO_2 and no need to baked the same.
(iii) Very thin chance of mould or core distortion during the hardening process.
(iv) No need to skill person to perform the hardening.
(v) Permeability and flow ability of sand is improved.
(vi) Same sand is used for both cores and moulds.

Limitations of CO_2 hardened mould

(i) Relatively short storage life of CO_2 hardened mould and core.
(ii) Process is more expensive as compared to other mould.
(iii) Poor colapsibility problem some times.

❑❑❑

8

CHAPTER

GATES

8.1 GATING SYSTEM

Gating is defined as all the channels or passages through which the molten metal enters into mould cavity.

A gating system should be design in such a way that it can take account of the weight and shape of casting, fluidity of molten metal and relative susceptibility of metal oxidation. Size and location of gates should be such that the mould cavity should be filled in quickly with a minimum amount of cutting of the mould surfaces by the flowing metal. In a good gating system, gates should be a location from where they can be easily removed without damaging casting.

8.1.1 Functions of a Gating System

A good gating system should be:

(i) fill the molten metal into mould cavity before solidification of metal.

(ii) fill the mould cavity with molten metal without any turbulence.

(iii) assist to promote the temperature gradient favourable for systematic directional solidification.

(iv) control the flow rate of molten metal at which it enter into mould cavity.

(v) should incorporate traps for separation of non-metallic inclusions.

(vi) system should consume least amount of metal.

(vii) keep the metal velocity sufficiently low to avoid mould and core erosion.

8.1.2 Requirement while Designing the Gating System

(i) Molten metal should flow smoothly into mould cavity without any trouble problem.

(ii) Unwanted metal like dross, slag, etc. should not come into the mould cavity.

(iii) Metal entering into mould cavity should flow in a control speed.

(iv) Metal flows should be maintained in such a way that no gating or mould erosion take place.

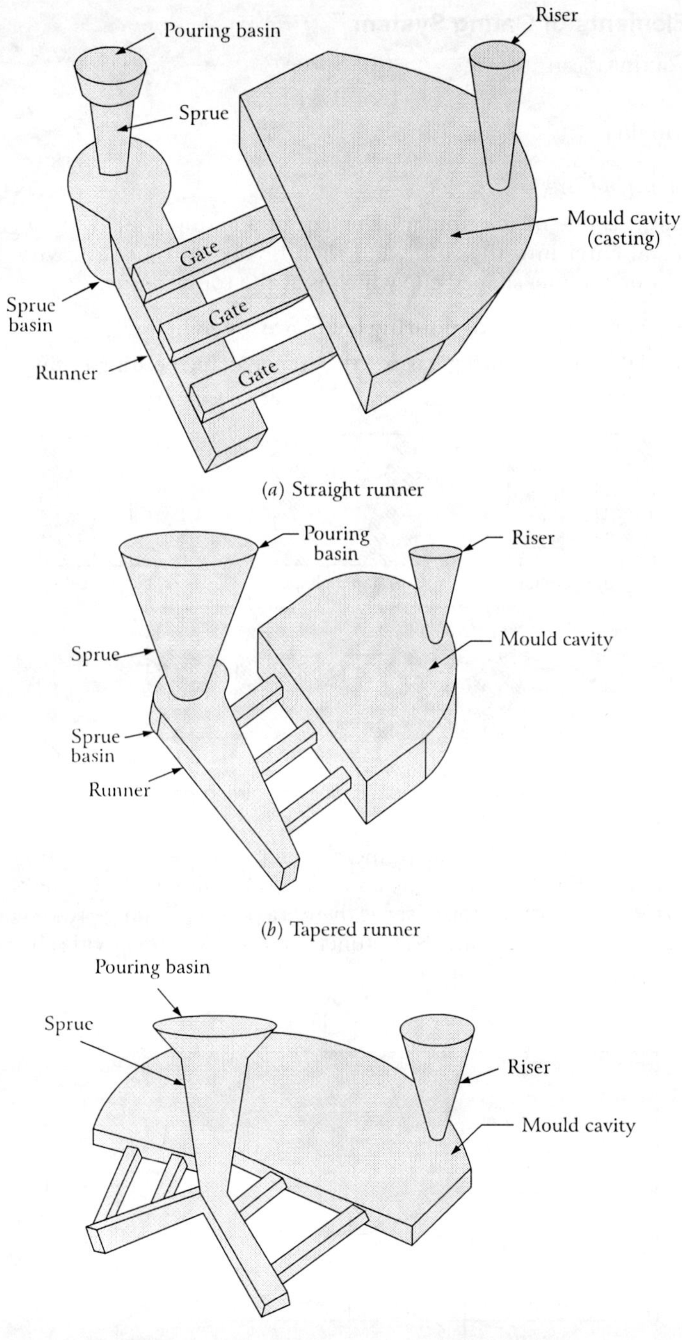

Fig. 8.1 Various component of a gating system (three type getting system)

8.1.3 Elements of Gating System

(i) Pouring basin (ii) Sprue
(iii) Runner (iv) Riser
(v) Mould cavity

8.1.3.1 Pouring Basin

Pouring basin is one of the primary element of a gating system, through which molten metal enter into the sprue and then finally into mould cavity. Pouring basin is just like funnel shaped cup which is at the top of sprue.

8.1.3.1.1 **Various shape of pouring basin are given below:**

(i) Straight shape pouring basin (ii) Taper shape pouring basin.
(iii) Straight P/B (iv) Positive taper sprue pouring basin.

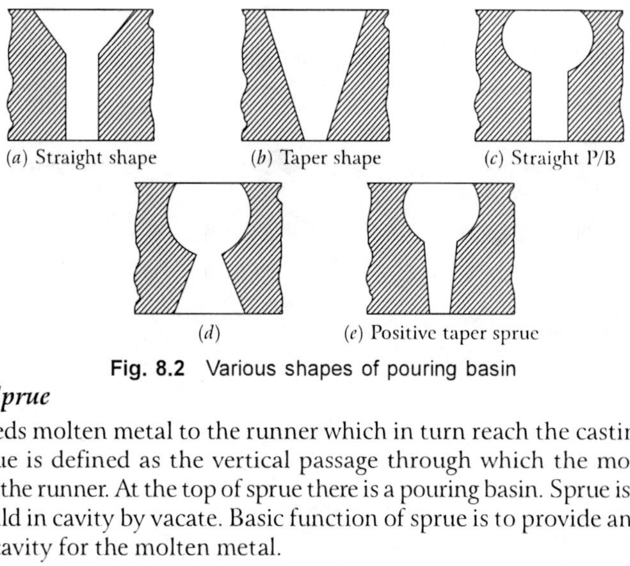

Fig. 8.2 Various shapes of pouring basin

8.1.3.2 Sprue

A sprue feeds molten metal to the runner which in turn reach the casting through gates. Sprue is defined as the vertical passage through which the molten metal enters into the runner. At the top of sprue there is a pouring basin. Sprue is connected to the mould in cavity by vacate. Basic function of sprue is to provide an enterance to mould cavity for the molten metal.

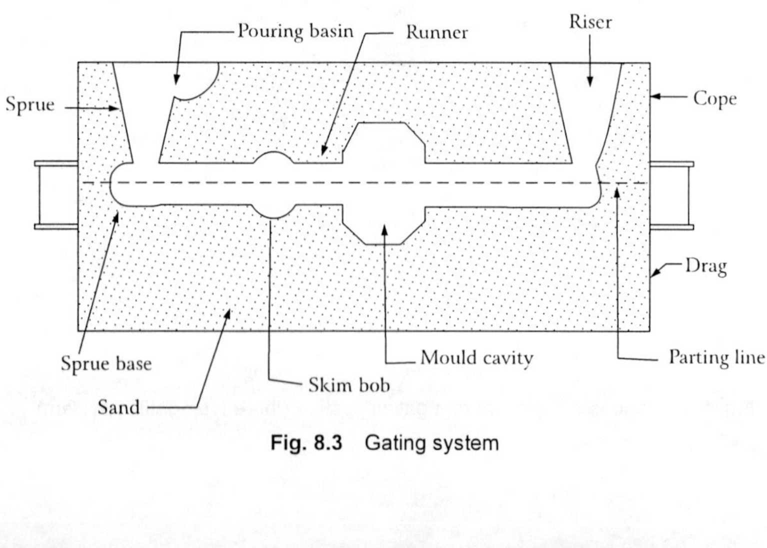

Fig. 8.3 Gating system

8.1.3.3 Riser

Risers are known as *reservoirs*. They are designed and located to feed the molten metal to the solidifying casting to compensate for solidification shrinkage. Riser is basically provide to check and ensure filling up of mould cavity.

Objects (Functions) of a Riser

(i) It allows to escape the gases and air as the mould cavity is being filled with molten metal.

(ii) It possesses in directional solidification.

(iii) Riser should cover entire casting section that require feeding.

(iv) A full of riser with molten metal indicates that entire mould cavity is also filled with molten metal.

(v) Its volume must be sufficient to compensate for metal shrinkage within the casting (mould cavity).

(vi) It should be effective in producing a temperature gradient within the casting so that the casting will solidify directionally towards the riser direction.

8.1.3.3.1 Types of Riser

(i) Open Riser
(ii) Blind Riser
(iii) Side Riser
(iv) Top Riser

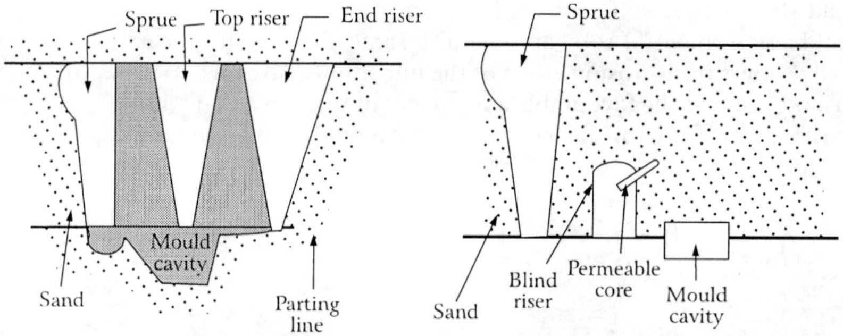

Fig. 8.4 Different type of riser

8.1.4 Differentiate between Open Riser and Blind Riser

Sl. No.	*Open Riser*	*Blind Riser*
1.	Open riser are generally cylindrical in shape.	Where as blind riser are in a rounded cavity, it shows the least practical ratio of surface area to volume.
2.	Portion of open riser is open in atmosphere.	Where as blind riser is closed at its top portion.
3.	Open risers are bigger in size as compared to blind riser.	A blind riser may be smaller in size as compared to open riser, hence more yield can be obtained.
4.	Cannot be removed so easily from casting.	Can be removed more easily from casting than open riser.

Gates

In casting process gates are the channels or passages which finally leads molten metal from runner into the mould cavity. In casting, gates should be so located that cracks do not develop after the solidification of molten metal. Gate should be so designed that neigther it is near at the mould cavity nor it is for away from mould cavity.

8.2 INTRODUCTION

As we know in the casting process, gate is define as opening or passage through which molten metal enter into the mould cavity.

Gates are various types such as:

- top gates,
- bottom gates,
- side gates,
- step gates,
- parting gates,
- ring gates,
- whirl gates, etc.

8.2.1 Top Gates

Top gate is also referred as drop gates as molten metal just enter into the mould cavity from the top. As molten metal poured directly into the mould cavity hence mould should have sufficent hard and strong to resist erosion by the dropping metal basic element of top gates are also same as the bottom gate. This system usually suffers from erosion effect of the unconfined stream and oxidation due to splashing. One of the favourable advantages of top gates that all the metal enters into the casting from top, hence metal in the mould cavity is in hottest position. In this type of gates them-self work as riser.

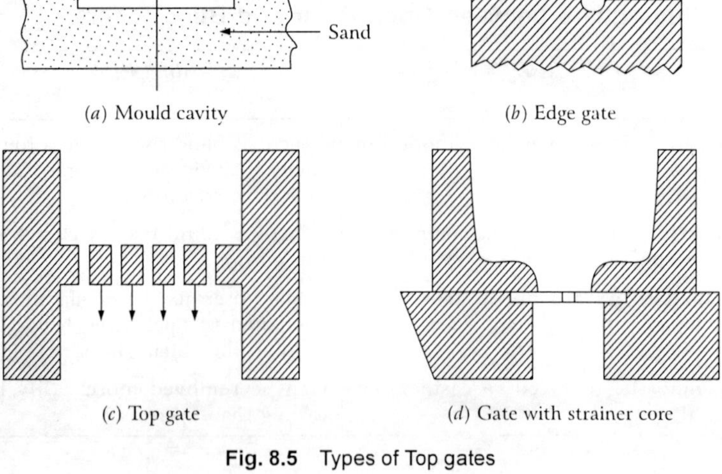

(*a*) Mould cavity (*b*) Edge gate

(*c*) Top gate (*d*) Gate with strainer core

Fig. 8.5 Types of Top gates

8.2.2 Bottom Gate

Bottom gate is associated with the drag portion of mould. In this type of the gating system, molten metal first comes at the bottom portion and then start to enter in the mould cavity. In bottom gate system, metal is hottest position at bottom and coolest at top portion due to this phenomenon there may be problem of directional solidification of molten metal. In bottom gate, less chance of mould erosion.

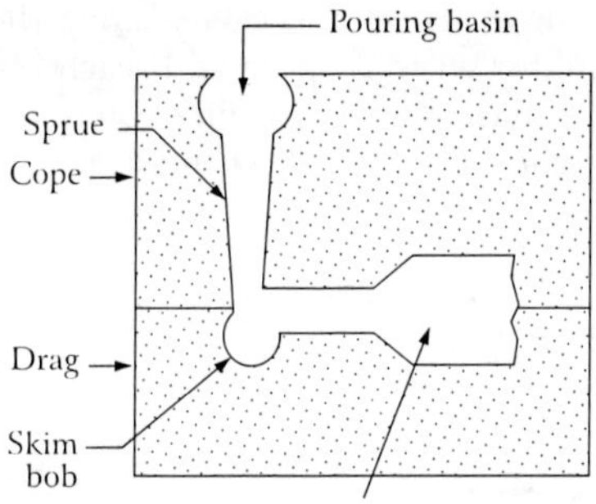

(*a*) Mould cavity

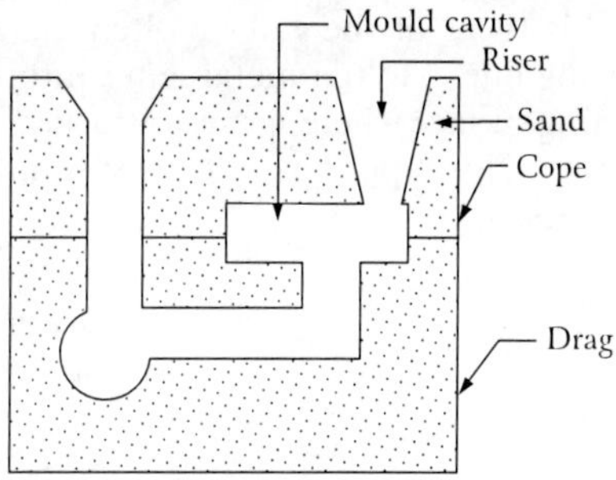

(*b*) Simple bottom gate

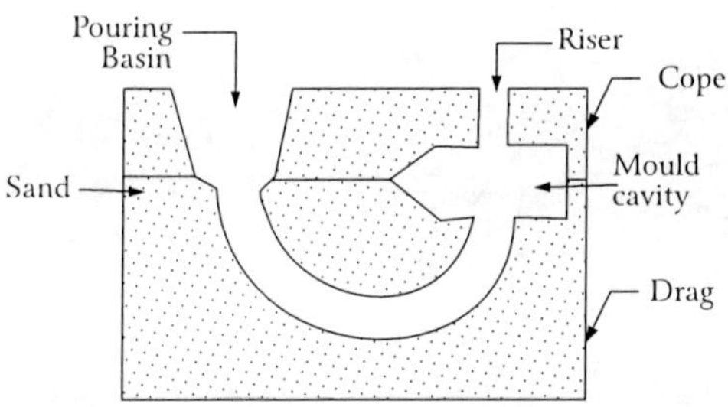

(*c*) Horn type bottom gate

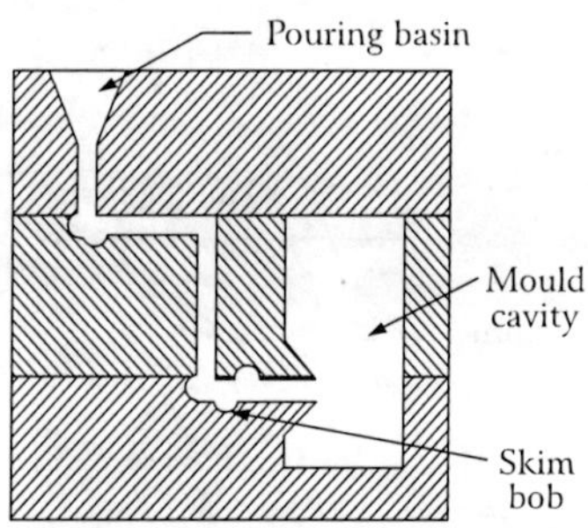

(*d*) Gate with skim bob

Fig. 8.6 Bottom gate

8.2.3 Step Gate

In the step gates, molten metal is enter upto mould cavity through a numbers of in gates. These gates are arrange in the vertical manner as shown in Fig. 8.7.

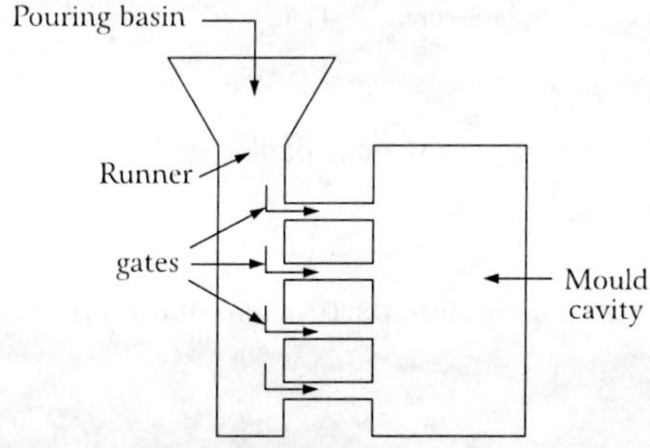

Fig. 8.7 Step gate

Size of these gates are increasing from top to bottom. It is only for easy entering the liquid metal from bottom to top. It ensure the gradual filling of molten metal for a sound casting.

This types of gates used for large and heavy casting.

8.2.4 Parting Gate

In this types of the gates molten metal enter in to the mould cavity along the parting line. Molten metal enter into the mould cavity via sprue and reach the parting surface where sprue is attached with gate in the horizontal direction to the casting. This type of gates consist skim bob or relief sprue to collect the slag.

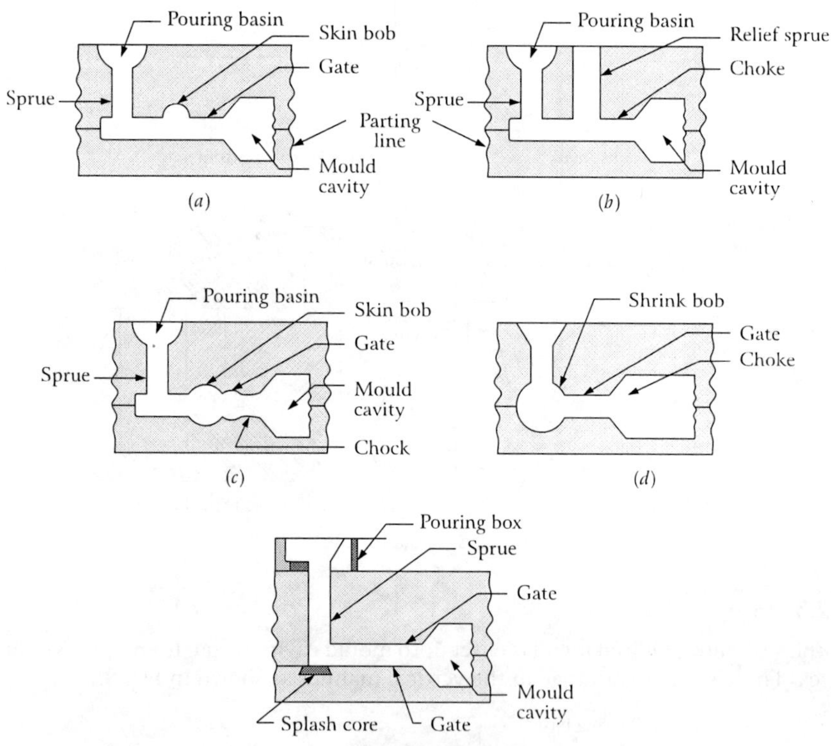

Fig. 8.8 Various parting gates

8.2.5 Whirl Gate

Purpose of whirl gate some time, like parting gate which is shown Fig. 8.9.

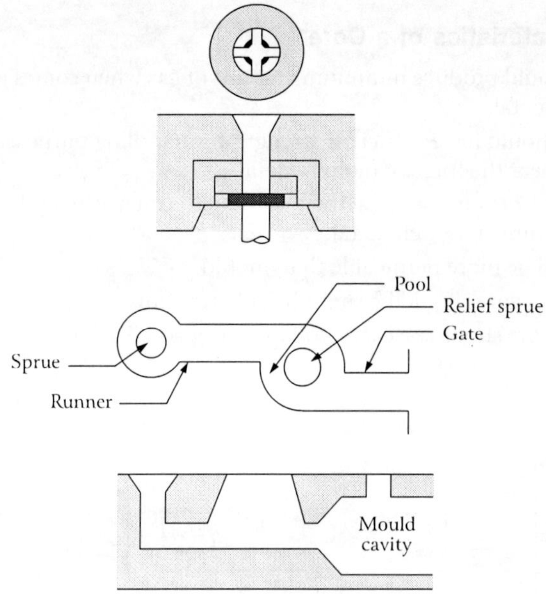

Fig. 8.9 Whirl gate

8.2.6 Branch Gate

Branch gate is also another type bottom gate. It is design either to feed a single casting at several points or a numbers of individual castings.

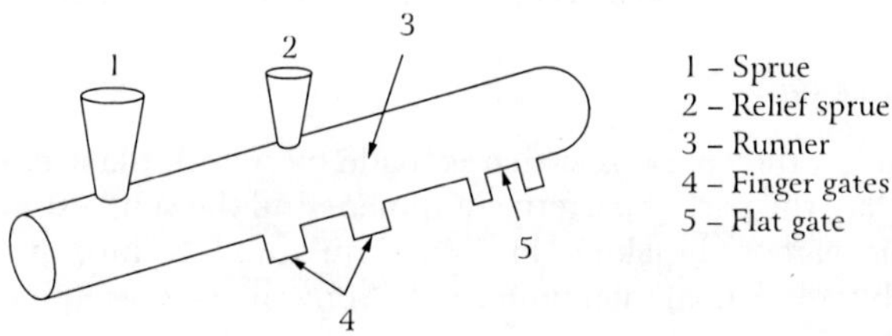

Fig. 8.10 Branch gate

8.3 CORE

Sometimes casting requires holes, cavities, etc. internally. This internally hole in casting body possible only by core. Therefore core is a body which is made out of some refractory material such as **sand, metal**, etc. which is kept in the prepared mould before casting and pouring mould cavity.

Therefore the **basic object** of core is to produce a hole in the casting object. As core is surrounded by the molten metal, hence having some special mechanical and thermal effect than the mould. Core eliminate by machining process such as drilling, boring, etc.

8.3.1 Characteristics of a Core

(*a*) Core should produce minimum amount of gas when comes in contact with molten metal.

(*b*) A core should have sufficient strength for handling purpose. Core must be able to bear the force of molten metal.

(*c*) It should be highly refractory so that it can withstand at high degree temperature of molten metal.

(*d*) It should be more permeable than mould.

(*e*) Surface of core should be smooth for better finish of casting.

(*f*) A good core should be easily collapse after solidification of molten metal.

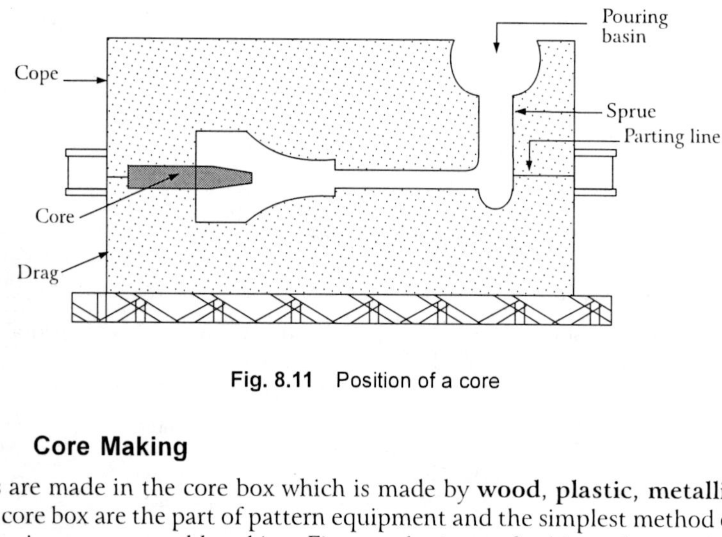

Fig. 8.11 Position of a core

8.3.2 Core Making

Cores are made in the core box which is made by **wood**, **plastic**, **metallic**, etc. these core box are the part of pattern equipment and the simplest method of core making is same as mould making. First sand mixture feed into the core box and then rammed by wooden rammer uniformely. Sometimes core requires wire or nails to provide internal support for easy handling purpose. Venting and other operations are performed during the preparation of core as usual.

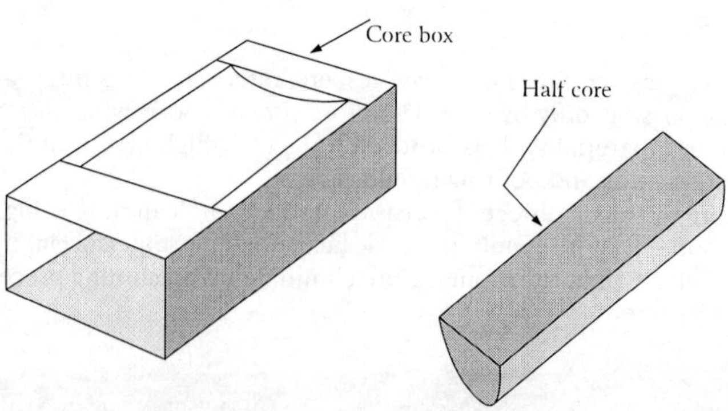

Fig. 8.12 Half core box and Half core

Core is withdrawn from the core box placed at a metal tray and baked in an oven at a temperature from 150–450°C for appropriate time.

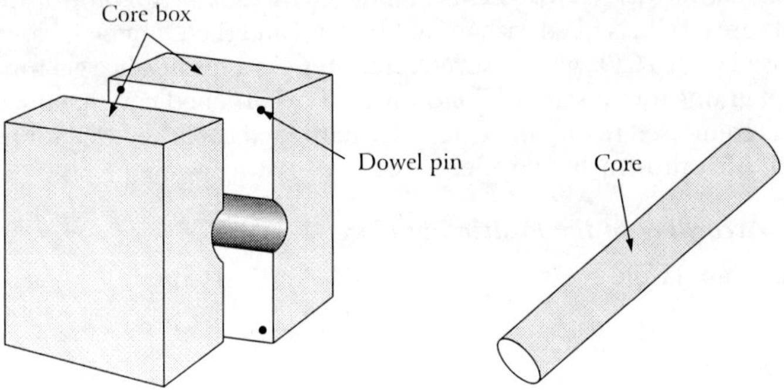

Fig. 8.13 Split core-box

8.3.3 Classification of Cores

Core can be classified on the basis of the following points:

8.3.3.1 According to the State of Core

(*a*) Green Sand Core
(*b*) Dry Sand Core
(*c*) Oil Sand Core
(*d*) CO_2 Hardened Core
(*e*) Shell Mould Core

8.3.3.1.1 Green Sand Core

This type of cores are basically used for light casting purpose as they are weaker in strength. Mixture:

- silica sand – 75%,
- dung – 10%,
- molases – 15% and water.

8.3.3.1.2 Dry Sand Core

They are baked core with refractory coating for proper strength. Dry sand core consist coarse grained **silica sand, moulding sand, silica flour**. They are baked until perfect dry.

8.3.3.1.3 Oil Sand Core

Oil sand cores are commonly used as they possesses very good collapsibility. These type core do not absorb the water.

Constituent of oil sand core:

- silica sand – 95%
- cornflour – 1%
- bentonite – 1%
- linseed oil – 2%

8.3.3.1.4 CO_2 Hardened Core

As the name shows this type of cores are hardened by CO_2 gas for proper strengthing. First of all core mould is filled with moulding sand and then rammed. Now rammed core is gassed with a CO_2 gas for several minutes. As a result silica gel forms which binds sand grains into a strong. Solid form CO_2 hardened core having the more strength as composed to oil sand core. CO_2 hardened cores are used for producing **C.I.**, **steel**, **Aluminium** and **copper** base alloys casting.

8.3.3.2 According to the Position of Core

(*a*) Horizontal Core
(*b*) Vertical Core
(*c*) Balanced Core
(*d*) Hanging Core
(*e*) Drop Core
(*f*) Kiss Core

8.3.3.2.1 Horizontal Core

This type of core is used in the horizontal position and kept in between the parting line of two mould box as shown in the Fig. 8.13 horizontal core may be in circular or any other shape. Horizontal core commonly used in casting.

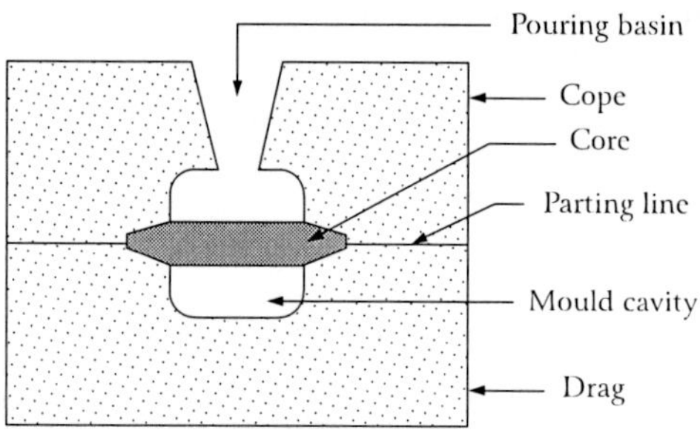

Fig. 8.14 Horizontal core

8.3.3.2.2 Vertical Core

Vertical core always used in the vertical position. Two ends of the vertical core are supported in core seat in moulding box maximum part of the core within the drag portion.

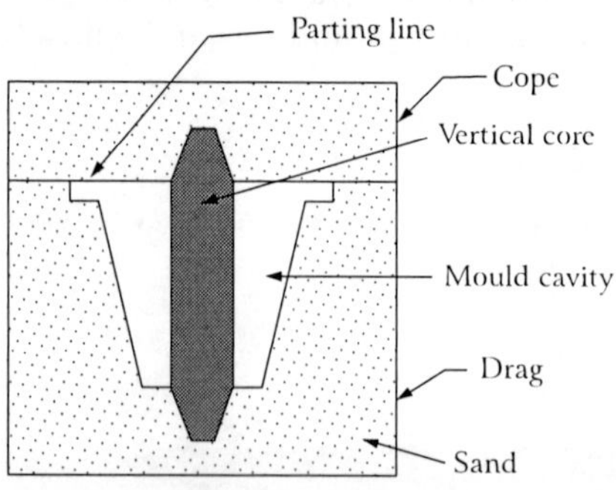

Fig. 8.15 Vertical core

8.3.3.2.3 Balanced Core

A balanced core is one which is supported and balanced from its one end only. It is often used to produce a blind hole in casting. Core print heavy and large enough to balance the weight of the core in the mould.

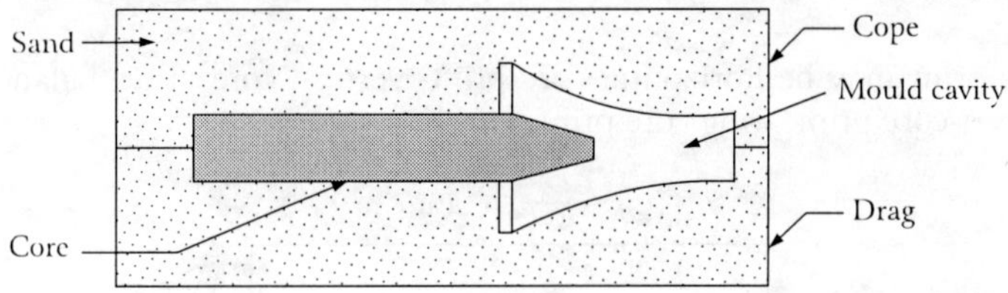

Fig. 8.16 Balanced core

8.3.3.2.4 Kiss Core

A kiss core generally do not require any seat for supporting purpose, A large number of holes can be produce by using a large number kiss cores. Core is lie held in position between cope and drag. Due to pressure exerted by cope on the drag.

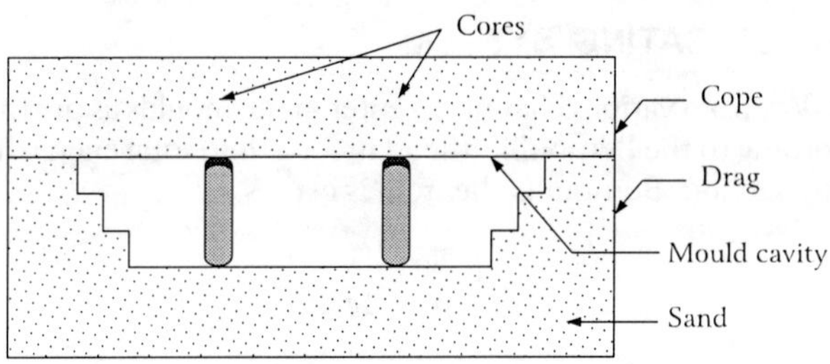

Fig. 8.17 Kiss core

8.3.3.2.5 Ram-up Core

This type of core is placed in the sand alongwith pattern before ramming operation of sand. This type of core cannot be kept in the mould after mould has been rammed.

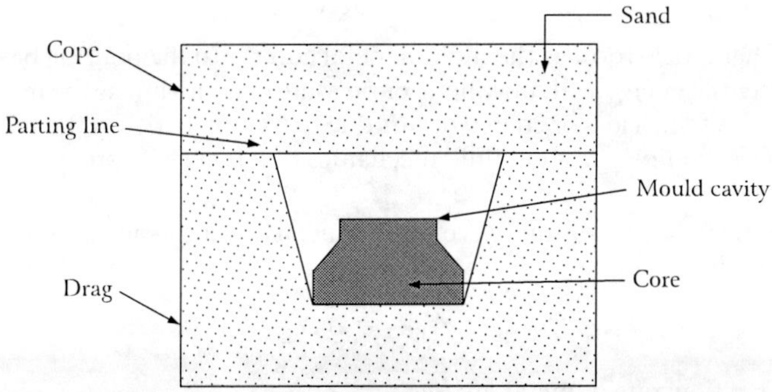

Fig. 8.18 Ram-up core

8.4 CORE PRINT

To support the core in the mould cavity, an imprison in the form of a recess is made in the mould with the help of a projection on pattern. This projection is known as *core print* or other world core print is define as extra projection provided on the pattern.

Core print may be horizontal core print, vertical core print, balanced core print, cover core print, wing core print, etc.

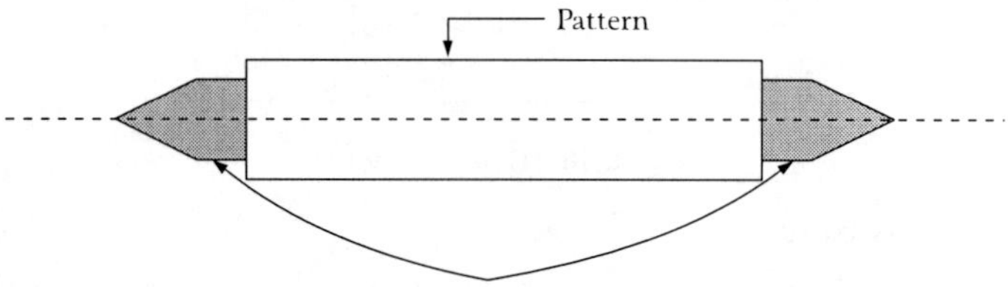

Fig. 8.19 Core print

8.5 DESIGN OF GATING SYSTEM

Molten metal flow from various channel to enter in the moulds as per the Bernoulli's theorem. According to the Bernoulli's theorem for a fluid total energy head remains constant at any section. Bernoullis theorem is expressed as

$$\frac{V_1^2}{2g} + h_1 + \frac{p_1}{\rho} = \frac{V_2^2}{2g} + h_2 + \frac{p_2}{\rho}$$

$$h + \frac{p}{\rho} + \frac{v^2}{2g} = \text{constant}$$

where, h = potential head, m

P = pressure, Pa

ρ = density of liquid, N/m^3

V = velocity of liquid, m/sec

g = gravitation constant = 9.8 m/sec^2

In mathematical form as the molten metal enter into the pouring basin, it has highest potential energy with no kinetic or pressure energies but as the metal moves from gating system, a loss or energy occurs due to friction between the mould walls and *Law of Continuity* is another fluid mechanics that is very helpful to understand the gating system.

According to law of continuity, volume of metal flowing at any passage (section) in the mould is constant.

$$Q = A_1 V_1 = A_2 V_2$$

where, Q = flow rate of metal, m^3/sec

A = area of cross-section, m^2

V = velocity of metal flow, m/sec

When metal is in pouring cup, it consist maximum potential energy and zero velocity.

Potential energy of molten metal convert into kinetic energy as molten metal moves down into sprue.

From Fig. 8.20 velocity at point B can be evaluated by Bernoulli's equation:

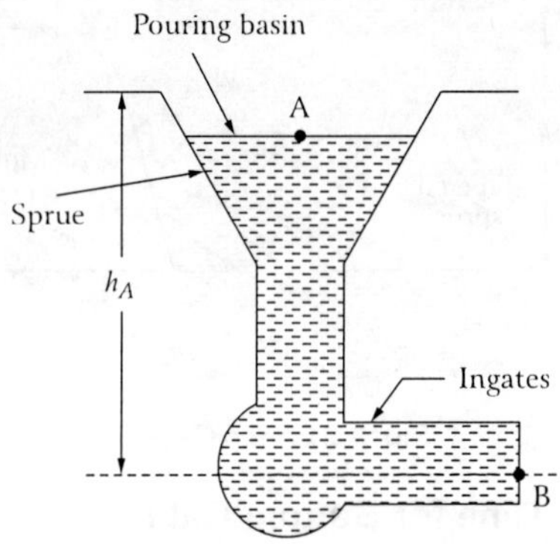

Fig. 8.20 Flow velocity at Ingate

$$\frac{V_A^2}{2g} + h_A + \frac{p_A}{\rho} = \frac{V_B^2}{2g} + h_B + \frac{p_B}{\rho}$$

but $V_A = 0$ and $h_B = 0$ then

$$h_A + \frac{p_A}{\rho} = \frac{V_B^2}{2g} + \frac{p_B}{\rho} \quad \text{since } p_A = p_B = 1$$

then $$h_A = \frac{V_B^2}{2g} \quad \text{or } V_B = \sqrt{2gh}$$

8.6 MOULD FILLING TIME

Basically rate of mould cavity filling by molten metal is depends upon the cross-section area of gates and this mould filling time is different for different type gating system such as **top gate**, **bottom gate**, etc.

8.6.1 Mould Filling Time for Top Gate

Time taken to fill the mould (t_m) is given by

$$t_m = \frac{V_m}{A_g V_g}$$

where, V_m = volume of mould

V_g = liquid metal velocity at the top of gate

A_g = cross-section area of gate

But velocity $V_g = \sqrt{2gh}$

where, H = height liquid in the pouring basin from gate

g = Gravitation constant

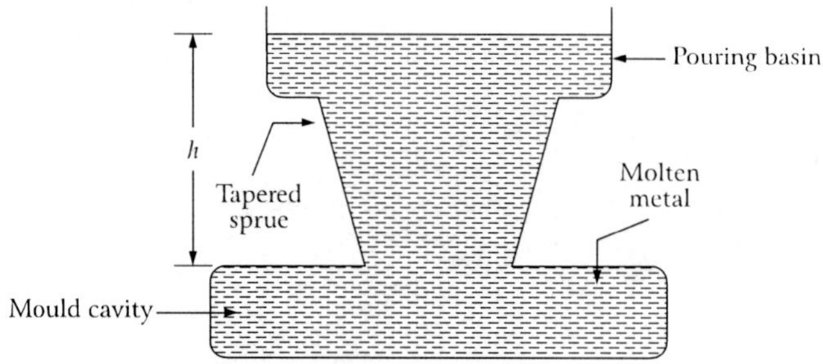

Fig. 8.21 Top Gate

8.6.2 Mould Filling Time for Bottom Gate

Time taken to fill up the mould cavity

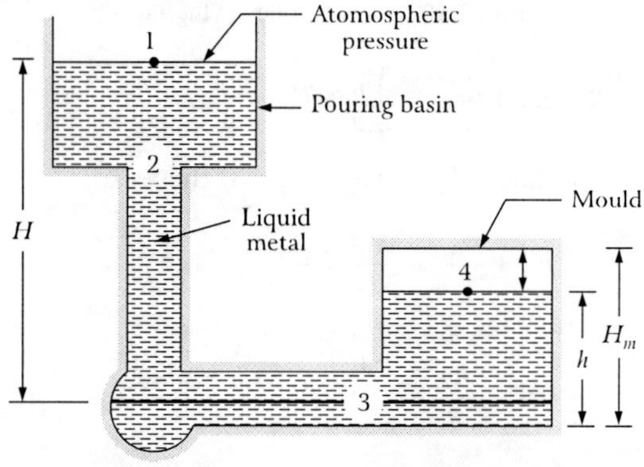

Fig. 8.22 Bottom gate

Applying Bernoulli's equation between surface 1 and 3, we get

$$M.g.H = \frac{MP_3}{w} + \frac{MV_3^2}{2}$$

$$gH = \frac{P_3}{w} + \frac{V_3^2}{2} \qquad \text{...(I)}$$

where P_3 = gauge pressure at surface 3

w = density of liquid metal

V_3 = velocity at gate

Similarly applying Bernoulli's equation between 3 and 4

$$\frac{MP_3}{w} = M.g.h \text{ or } \frac{P_3}{w} = g.h \qquad \text{...(II)}$$

From equations (I) and (II), we get

$$gH = gh + \frac{V_3^2}{2}$$

$$g(H-h) = \frac{V_3^2}{2}$$

or
$$V_3^2 = 2g(H-h)$$

$$V_3 = \sqrt{2g(H-h)}$$

Let A_m = Area of cross-section of mould.

Assume the metal level in the mould move up through a height dh in a time interval dt then by continuity equation

$$A_m \times \frac{dh}{dt} = A \times V_3$$

A = cross-section area of gate

Putting the value of V_3 in above equation

$$A_m \times \frac{dh}{dt} = A \times \sqrt{2g(H-h)}$$

$$dt = \frac{A_m}{A} \times \frac{dh}{\sqrt{2g(H-h)}}$$ Taking integration both sides

$$\int_0^T dt = \frac{A_m}{A} \int_0^{H_m} \frac{dh}{\sqrt{2g(H-h)}}$$

$$T = \frac{A_m}{A} \times \frac{\sqrt{2}\left(\sqrt{H} - \sqrt{H-H_m}\right)}{\sqrt{g}}$$

$$t_m = \frac{A_m}{A_g}\sqrt{\frac{2}{g}}(\sqrt{H} - \sqrt{H-h_m})$$

where, A_m = area of mould cross-section area
h_m = height of the mould

Velocity of liquid metal at the gate is given by

$$v_g = \sqrt{2g(H-h_l)}$$

where $H - h_l$ = effective head
h_l = height of liquid column in the mould

If riser is used, then pouring time (t_m) include the time requires to fill up riser too.

$$t_r = \frac{A_r}{A_g}\sqrt{\frac{2H}{g}}$$

where A_r = riser cross-section area
t_r = time take to fill the riser

Example 8.1

A mould 60 cm × 40 cm × 15 cm is to be filled by liquid metal during sand casting process. Determine the time taken to fill up mould by

(i) Top gate system

(ii) Bottom gate system

Cross-section area of gate is 6 cm^2.

Solution:

(i) For top gate system

$$V_m = 60 \times 40 \times 15 = 36000 \text{ cm}^3$$

Area of gate A_g = 6 cm^2, Head = 15 cm

Velocity of molten metal

$$V_g = \sqrt{2gH} = \sqrt{2 \times 9.8 \times 15}$$

$$= \sqrt{29400} = 171.46 \text{ cm/sec.}$$

$$t_m = \frac{V_m}{A_g V_g} = \frac{36000}{6 \times 171.46}$$

$$= 34.99 = 35 \text{ sec.} \quad \textbf{Ans.}$$

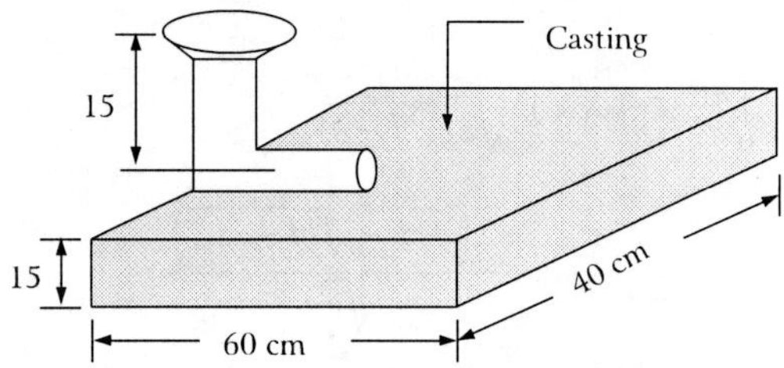

Fig. 8.23 Top gate system

(ii) **For bottom gate**

Area of cross-section of mould

$$A_m = 40 \times 60 = 2400 \text{ cm}^2$$

Height of mould h_m = 15 cm, Head = 15 cm

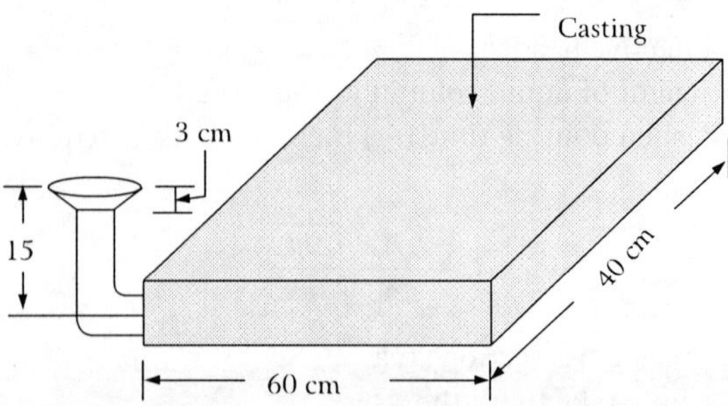

Fig. 8.24 Bottom gate system

$$t_m = \frac{A_m}{A_g}\sqrt{\frac{2}{g}}\left(\sqrt{H} - \sqrt{H - h_m}\right)$$

$$= \frac{2400}{6} \times \sqrt{\frac{2}{980}} \times (\sqrt{15} - \sqrt{15 - 15})$$

$$= 69.98 = 70 \text{ sec.} \quad \textbf{Ans.}$$

8.7 DESIGN OF SPRUE

Sprue should be always in taper form as if sprue is straight or having uniform cross-sectional area molten metal contracts inwards and is pulled away by wall of sprue this create the turbulence exact tapering can be obtained by the continuity equation.

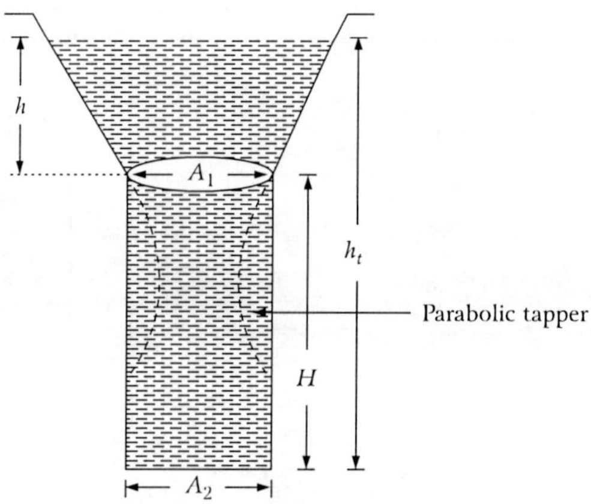

Fig. 8.25 Sprue

If $\quad t$ = top section of sprue

c = choke section of sprue

then

$$A_t V_t = A_c V_c$$

$$A_t = \frac{A_c \times V_c}{V_t}$$

as the velocities are proportionals then we can find heads by Bernoulli's equation

$$A_t = A_c\sqrt{\frac{h_c}{h_t}}$$

where, H = actual sprue height

$h_t = h + H$

Choke

Choke is smallest cross-sectional area of a gating system.

8.8 GATING RATIO

Gating ratio is defined as the proportion of the cross-sectional area between the sprue, runner and ingates. Usually it is denoted as sprue area: runner area; ingate area. As per the choke area gating system are of two types:

(i) Pressurised gating system,

(ii) Non-pressurised gating system.

Example 8.2

Show that the exit to inlet area ratio. R of a down delivery sprue for avoiding aspiration is given by

$$R = \frac{A_3}{A_2} = \sqrt{\frac{h_2}{h_3}}$$

where A_2 and A_3 are the cross-sectional area of the sprue at the sprue wall and pouring basin ends respectively and h_2, h_3 are head of liquid metal at the pouring basin and sprue well respectively. [*UPTU 2004*]

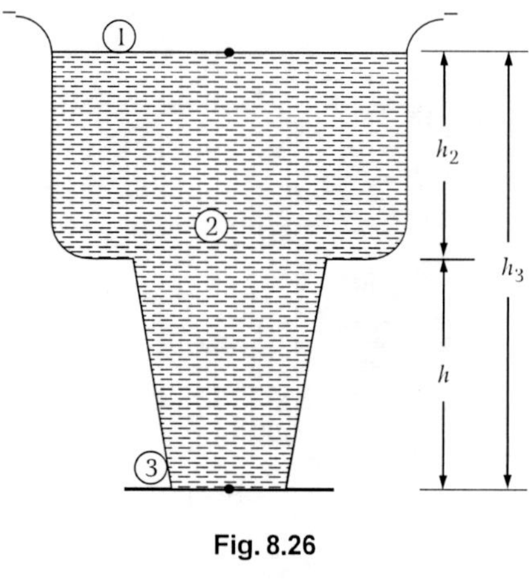

Fig. 8.26

Solution: According to Bernoulli's theorem for point 2 and 3

$$h + \frac{P_2}{\rho} + \frac{V_2^2}{2g} = o + \frac{P_3}{\rho} + \frac{V_3^2}{2g}$$

We know that sprue is a passage through which the molten metal feed from pouring basin to runner. Sprue should be design properly. If any air bubbles trapped in sprue will go in mould cavity and result blow holes and other defects. Sprue should be tapered to eliminate the problem of **low pressure** and to remove air bubbles trapping.

At point 2-3, continuity equation

$$A_2V_3 = A_3V_3$$

where, A_2 = Sprue area at point 2,

A_3 = Sprue area at point 3

V_2 = Velocity of molten metal at point 2,

V_3 = Velocity of molten metal at point 3.

But we know

$$V = \sqrt{2gh},\ V_2 = \sqrt{2gh_2} \text{ and } V_3 = \sqrt{2gh_3}$$

$$A_2 \times \sqrt{2gh_2} = A_3 \times \sqrt{2gh_3}$$

$$\frac{A_2}{A_3} = \sqrt{\frac{2gh_3}{2gh_2}} = \sqrt{\frac{h_3}{h_2}}$$

Ratio $$R = \frac{A_3}{A_2} = \sqrt{\frac{h_2}{h_3}}$$ **Proved.**

8.9 RISER DESIGN

Basic objective of the riser is to fill the mould cavity during solidification so that no shrinkage cavities are formed. Riser design basically depends upon the following factors such as **type** of molten **metal poured** and **complexity of casting**, etc.

Different material shrinks at different rates such as **gray cast iron**, requires very **little feeding** as it expands during shrinkage.

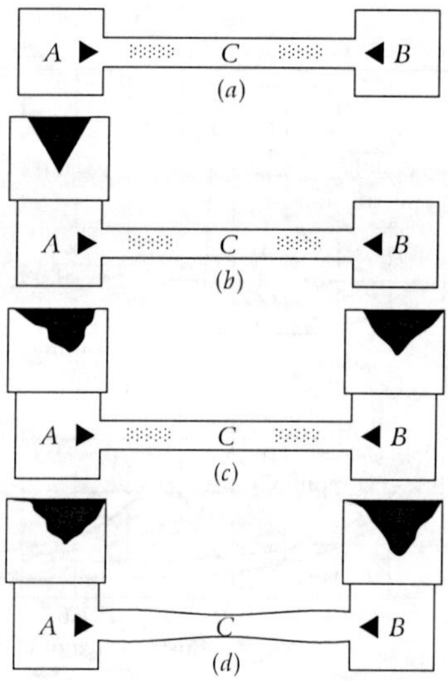

Fig. 8.27 Development (a to d) of a riser and padding system to ensure casting. Shaded area represent microporosity (from J.B. Caine)

On the other hand cast iron, steel and many other ferrous alloys requires more feeding as their higher shrinkage rate during solidification.

Directional solidification basically used in casting process it means that solidification of the molten metal start at the remotest point of the casting from the feeder.

Chvorinvov's rule: Chvorinvovs provide a rule regarding the solidification time of the casting which state that freezing time is proportional to square ratio of volume to area, i.e., $\left(\frac{V}{A}\right)^2$.

It means that if metal remains liquid in riser for a long time $\frac{V}{A}$ should be large or $\frac{A}{V}$ ratio should be small. [*IES 2006*]

$$t \propto \left(\frac{V}{A}\right)^2 \quad \text{or} \quad t = k\left(\frac{V}{A}\right)^2$$

where k = freezing constant
t = freezing time
v = volume
A = Area

8.9.1 Caine's Rule [*UPTU 2010-11*]

Caine's rule basically determine the size of riser based on experimentally determined hyperbolic relationship between relative freezing times and volume of the casting and riser.

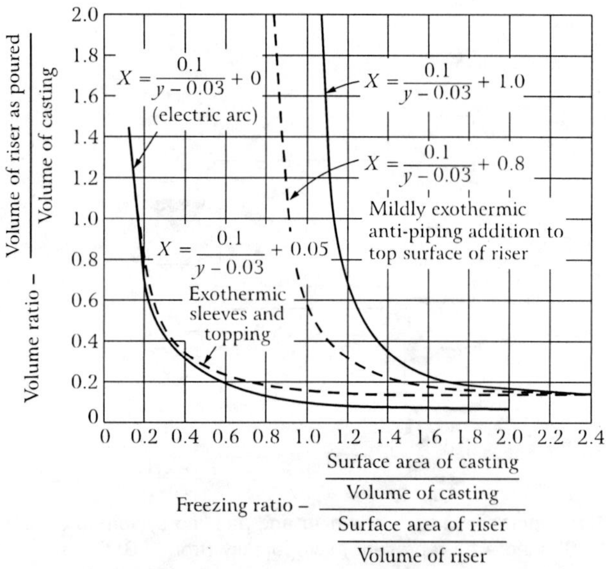

Fig. 8.28 Plot of riser-volume to casting-volume ratio vs. freezing ratio $\left(\frac{\text{casting area/casting volume}}{\text{riser area/riser volume}}\right)$ for various conditions existing in the riser. (From J.F. Wallace)

As per Caine's rule if the casting solidities infinitely rapidly, the riser (feeder) volume should be equal to the solidification shrinkage of the casting and if the feeder and casting solidity at the same rate, the feeder should be infinitely large.

Freezing ratio or freezing time

$$R_F = \frac{(A/V)_{\text{casting}}}{(A/V)_{\text{riser}}}$$

$$R_V = \frac{V_{\text{riser}}}{V_{\text{casting}}}$$

$$\text{Caine's formula}\, x = \frac{a}{y-b} + C$$

where, a = freezing constant of metal = 0.1 for steel

b = contraction ratio from liquid of solid = 0.03 for steel

C = relative freezing rate of riser and casting.

hence $C = 1$ if the medium around the mould and the riser is same.

Example 8.3

Calculate the size of cylindrical riser ($h = d$) necessary to feed a steel slab casting 25 × 25 × 5 cm with a side riser, casting poured horizontal into the mould.

Solution: $V = 35 \times 25 \times 5 = 4375 \text{ cm}^3$

Surface area of casting $= 2 \times 35 \times 25 + 4 \times 25 \times 5 = 2250 \text{ cm}^3$

$$\text{Volume of riser} = \frac{\pi D^3}{4}$$

$$\text{Surface area of riser} = \pi D^2 + \frac{\pi D^2}{4} = 1.25\,\pi D^2$$

$$X = \frac{\text{Surface area of casting/volume of casting}}{\text{Surface area of riser/volume of riser}}$$

Freezing ratio

$$X = \frac{2250 / 4375}{1.25\pi D^2 / 0.25\pi D^3} = \frac{0.5142D}{5} = 0.1028D$$

$$Y = \frac{\text{Volume of riser}}{\text{Volume of casting}} = \frac{0.25\pi D^3}{4375} = 0.000179\, D^3$$

$$\approx 0.00018D^3$$

Caine's equation for steel

$$X = \frac{a}{Y-b} + C \qquad \text{for steel } a = 0.1,\; b = 0.03 \text{ and } c = 1.0$$

$$0.103\, D = \frac{0.10}{0.000251\, D^3 - 0.03} + 1.0$$

On solving we get

$D^4 - 8.9286\, D^3 - 119.52D = 2490$

by errors and Trial method we get

$D = 11.44 \approx 12$ cm **Ans.**

Example 8.4

Prove that the volume/area ratio for most compact economical cylindrical side riser of height h and diameter d is $\frac{d}{6}$ and $\frac{d}{h}$ ratio is 1. *[UPTU 2001-02, 2003-04]*

Solution: h = height of cylindrical riser

d = dia. of cylindrical riser

⇒ **For side riser**

Surface area of riser $= \pi dh + \frac{\pi}{4}d^2 + \frac{\pi}{4}d^2$

$$A = \pi dh + \frac{\pi}{2}d^2 \qquad ...(i)$$

Volume of cylinder $\quad V = \frac{\pi}{4}d^2h,\ h = \frac{4V}{\pi d^2} \qquad ...(ii)$

Put value of h in above equation, we get

$$A = \pi d \times \frac{4V}{\pi d^2} + \frac{\pi}{2}d^2 = \frac{4V}{d} + \frac{\pi}{2}d^2$$

Diff. w.r.t. d

$$\frac{\partial A}{\partial d} = -\frac{4V}{d^2} + \pi d \quad \frac{\partial A}{\partial d} = 0 \text{ for least}$$

$$0 = \frac{-4V}{d^2} + \pi d$$

or

$$d^3 = \frac{4V}{\pi}$$

$$h = \frac{-4V}{\pi d^2} \text{ or } \frac{4Vd}{\pi d^3} = \frac{4Vd \cdot \pi}{\pi \cdot 4V} = d \text{ or } \frac{h}{d} = 1$$

Putting $h = d$ in equation (*i*) and (*ii*), we get

$$\frac{V}{A} = \frac{d}{6}$$

Proved.

Example 8.5

What will be the solidification time for a 1000 mm diameter and 30 mm thickness casting of aluminium if the mould constant is 2.2 sec/mm^2.

Solution: Given d = 1000 mm, h = 30 mm, c = 2.2 sec/min^2

Volume of casting $\quad V = \frac{\pi}{4}d^2h$

$$= \frac{\pi}{4} \times 1000^2 \times 30$$

$$= 23550000$$

Surface area of casting $\quad A = \frac{\pi}{4}d^2 \times 2 = \frac{\pi}{4} \times 1000^2 \times 2 = 1570000$

According to Chvorinov's rule

$$t = C\left(\frac{V}{A}\right)^2$$

$$t = 2.2 \times \left(\frac{23550000}{157000}\right)^2$$

$$= 495 \text{ sec. or } 8.25 \text{ min.}$$ **Ans.**

Example 8.6

Prove that the volume/area ratio for most compact economical cylindrical side riser of height h and diameter is $\frac{d}{6}$ and $\frac{d}{h}$ ratio is 1. [*CCSU 2010; UPTU 2003-04*]

Solution: h = height of cylindrical riser

d = diameter of cylindrical riser

For side riser surface area $= \pi dh + \frac{\pi}{4}d^2 + \frac{\pi}{4}d^2$

$$= \pi dh + \frac{\pi}{2}d^2 \quad ...(i)$$

$$A = \pi dh + \frac{\pi}{2}d^2$$

Volume of cylinder $\quad V = \frac{\pi}{4}d^2 h \quad ...(ii)$

$$h = \frac{4V}{\pi d^2}$$

Putting value of h in above equation

$$A = \frac{\pi}{4} \times d \times \frac{4V}{\pi d^2} + \frac{\pi}{4}d^2$$

or $\quad A = \frac{4V}{d} + \frac{\pi}{2}d^2$

For A minimum $\quad \frac{\partial A}{\partial d} = \frac{-4V}{d^2} + \frac{\pi}{2} \times 2d = \frac{-4V}{d^2} + \pi d$.

$$\frac{\partial A}{\partial d} = 0$$

$$\frac{-4V}{d^2} + \pi d = 0$$

$$\frac{4V}{d^2} = \pi d$$

$$d^3 = \frac{4V}{\pi}$$

Putting this value in h, $\quad h = \frac{4V \times d \times \pi}{\pi \times 4V}\; h = d$ or $\frac{h}{d} = 1$

Putting $h = d$ in Equs. (i) and (ii), we get

$$\frac{V}{A} = \frac{d}{6}$$

Proved.

Example 8.7

Two casting of same metal having the same surface area. One casting is sphere type and other is cube type. What is the ratio of the solidification time for the sphere to that of a cube. [*GATE-2007*]

Solution: A cube according to Chvorinov's rule $t \propto \left(\frac{V}{A}\right)^2$

$$A_{sphere} = A_{cube}$$

$$\frac{t_{sphere}}{t_{cube}} = \left(\frac{V_{sphere}}{V_{cube}}\right)^2 = \left(\frac{\left(\frac{4}{3}\pi r^3\right)}{a^3}\right)^2 = \frac{16\pi^2 r^6}{9a^2}.$$

Example 8.8

Two solid workpiece (*i*) sphere with radius *R* (*ii*) a cylinder with diameter equal to height have to be sand cast. Both object have same volume. Show that the cylindrical object will solidity faster than the spherical object. [*GATE 2004*]

Solution: Volume of sphere $V_S = \frac{4}{3}\pi R^3$,

Volume of cylinder $V_C = \frac{\pi}{4}d^2 \times h = \frac{\pi}{4}d^3$

Given $d = h$

$$V_{\text{sphere}} = V_{\text{cylinder}}$$

$$\frac{4}{3}\pi R^3 = \frac{\pi}{4}d^3 = \left(\frac{16}{3}\right)^{\frac{1}{3}} R$$

According to Chvorinov's rule

$$t \propto \left(\frac{V}{A}\right)^2$$

$$\left(\frac{V}{A}\right)_{\text{sphere}} = \frac{4/3 \times \pi R^3}{4\pi R^2} = \frac{R}{3} = 0.33\,R$$

$$\left(\frac{V}{R}\right)_{\text{cylinderical}} = \frac{\frac{\pi}{4}d^2h}{\pi dh + 2\frac{\pi}{4}d^2} = \frac{d}{6} = \frac{1}{6} \times \left(\frac{16}{3}\right)^{1/3} R = 0.29R$$

Thus cylindrical object will solidify faster than spherical object.

Example 8.9

Which one of the following casting shape would have least solidification time?

(i) **Diameter of sphere = 25 min.**

(ii) **Cylinder with diameter and height *h* = 25 mm.**

(iii) **Cube with a length of side *a* = 25 mm.**

Solution: According to Chvorinov's rule $t \propto \left(\frac{V}{A}\right)^2$

(i) $$t_{\text{sphere}} = c\left(\frac{\frac{\pi}{6}D^3}{\pi D^2}\right)^2 = \frac{C \times D^2}{36} = \frac{C \times 25^2}{36} = 17.36\,\text{C}$$

Here $h = d$

(ii) $$t_{\text{cylinder}} = \text{C}\left(\frac{\frac{\pi}{4}d^2h}{2 \times \frac{\pi}{4}d^2 + \pi dh}\right)^2 = \text{C}\left(\frac{\frac{d^3}{4}}{\frac{d^2}{2} + d^2}\right) = \text{C}\left(\frac{d}{6}\right)^2 = \text{C}\left(\frac{25}{2}\right)^2 = 17.36\,\text{C}$$

(iii) $$t_{\text{cube}} = \text{C}\left(\frac{a^3}{6a^2}\right)^2 = \text{C}\left(\frac{a}{6}\right)^2 = \text{C}\left(\frac{25}{6}\right)^2$$

as all casting have the same time, therefore, all casting will have the same solidification time.

Example 8.10

Compare the solidification time for casting of three different shapes of same volume: cube, cylindrical ($h = d$) and sphere.

Solution: Let

V = volume of each casting (say unity)

A = surface area of casting.

According to Chvorinov's rule $t \propto \left(\frac{V}{A}\right)^2$, C = casting constant

$$t = C\left(\frac{V}{A}\right)^2$$

(i) Cube casting

$$V = a^3 = 1 \qquad \text{as, } a = 1$$

$$A = 6a^2 = 6 \times 1^2 = 6 \text{ units}$$

$$t = C\left(\frac{V}{6}\right)^2 = C\left(\frac{1}{6}\right)^2 = 0.0278\,C$$

(ii) Cylinderical casting

$$h = 2r \qquad V = \pi r^2 h$$

$$= \pi r^2 \times (2r) = 1$$

$$2\pi r^3 = 1 \quad \text{or} \quad r = 0.5419 \text{ unit}$$

$$\text{Surface area} = 2\pi r^2 + 2\pi rh = 6\pi r^2$$

$$= 6\pi(0.5419)^2$$

$$= 5.53 \text{ unit}$$

$$t = C\left(\frac{1}{5.53}\right)^2 = 0.0327\,C$$

(iii) Spherical casting $\quad V = \frac{4}{3}\pi r^3 = 1$

or $\quad r^3 = \frac{3}{4}\pi$

$$r = 0.62 \text{ unit}$$

$$\text{Surface area} = 4\pi r^2 = 4\pi(0.62)^2 = 4.83 \text{ units}$$

$$t = C\left(\frac{V}{A}\right)^2 = \left(\frac{1}{4.83}\right)^2 = 0.0429\,C$$

Cubic casting have least solidification time where as spherical casting have maximum solidification time.

Example 8.11

Calculate the ratio of solidification of two steel cylinderical riser of size 36 cm in dia. by 72 cm height, 72 cm in dia. by 36 cm in height subjected to identical conditions of cooling.

Solution: $d_1 = 36$ cm, $h_1 = 72$ cm, $d_2 = 72$ cm, $h_1 = 36$ cm

by Chvorinov's rule $t \propto \left(\frac{V}{A}\right)^2, \quad t = C\left(\frac{V}{A}\right)^2$

For Riser Ist
$$= \frac{V}{A} = \frac{\frac{\pi}{4}d^2 h_1}{\pi d_1 h_1 + 2\frac{\pi}{4}d_1^2}$$

$$= \frac{\frac{d_1}{4}h_1}{d_1 h_1 + \frac{d_1^2}{2}} = \frac{\frac{36^2}{4}\times 72}{36\times 72 + \frac{36^2}{2}} = 7.2$$

For Riser IInd
$$= \frac{V}{A} = \frac{\frac{72^2}{4}\times 36}{72\times 36 + \frac{72^2}{2}} = 9$$

$$\frac{t_{Riser\ I}}{t_{Riser\ II}} = \frac{7.2}{9} = 0.64 \quad \text{Ans.}$$

Example 8.12

Calculate the size of a cylindrical rise with *d/h* ratio 1, necessary to feed a steel slab casting 25 × 25 × 5 cm^3. Assume the volume shrinkage on solidification is 3% for steel and that the volume of riser is three times that dictated by shrinkage considerations alone. [*UPTU 2003-04*]

Solution: Volume of cylindrical casting $V_c = 25 \times 25 \times 5 = 3125\ \text{cm}^3$

Surface area of cylindrical casting $SA_c = 2(lb + hb + hl)$

$$= 2(25 \times 25 + 25 \times 5 + 25 \times 5)$$
$$= 1750\ \text{cm}^3$$

Value of riser
$$V_r = \frac{\pi}{4}d^3$$

$$\frac{d}{h} = 1 \quad \text{(given)}$$

Minimum volume necessary for the riser is

$$V_r = 3 \times 0.03 V_C$$
$$= 3 \times 0.03 \times 3125 = 281.25\ \text{cm}^3$$

but
$$d = \left(\frac{4V_r}{\pi}\right)^{1/3}$$

$$= \left(\frac{4\times 281.25}{3.15}\right)^{1/3} = 7.10\ \text{cm}$$

for most compact economical cylindrical riser

$$\left(\frac{S_A}{V}\right)_r = \frac{6}{d} = 4\times\frac{6}{7.10}\text{cm}^{-1} = 0.85 \text{ cm}^{-1}$$

$$\left(\frac{S_A}{V}\right)_c = \frac{1750}{3125}\text{cm}^{-1} = 0.56 \text{ cm}^{-1}$$

dimensions of the riser are recalculated as follows for

$$\left(\frac{S_A}{V}\right)_r \le \left(\frac{S_A}{V}\right)_c \frac{6}{d} \le 0.56 \text{ cm}^{-1}$$

$$d \ge 10.71 \text{ cm}$$

with minium value of d

$$V_r = \frac{\pi}{4}d^3 = \frac{\pi}{4}\times(10.71)^3$$

$$= 964.85 \text{ cm}^3$$

This volume is much more than minium V_r necessary

$$h = d = 10.71 \simeq 11 \text{ cm} \quad \textbf{Ans.}$$

Example 8.13

A down sprue of 150 mm length has a diameter 20 mm at its top end. The liquid metal in the pouring basin maintained at 50 mm height. Calculate the theoretical diameter for down sprue at its lower end to ensure that there is no aspiration effect.

Solution: To avoid aspiration effect

$$h_1 = 50 \text{ mm}$$

$$h_2 = 150 \text{ mm}$$

$$\frac{A_1}{A_2} = \sqrt{\frac{h_2}{h_1}} \quad \text{or} \quad \frac{A_2}{A_1} = \sqrt{\frac{h_1}{h_2}}$$

$$\frac{A_2}{A_1} = \sqrt{\frac{50}{(150+50)}} = \sqrt{\frac{50}{200}} \qquad A_2 = A_1 \times 0.5$$

$$\frac{\pi}{4}d_2^2 = \frac{\pi(20)^2}{4}\times 0.5$$

$$d_2 = 14.15 \text{ mm} \quad \textbf{Ans.}$$

Example 8.14

A casting of 200 × 100 × 50 mm^3 size take 10 minutes for solidifies. Calculate the solidification time for 200 × 100 × 20 mm^3. Under similar conditions

Solution:

$$\text{Volume} = 200 \times 100 \times 50 = 10 \times 10^5 \text{ mm}^3$$

$$\text{Surface area} = 2(Lb + bh + Lh)$$

$$= 2(200 \times 100 + 100 \times 50 + 50 \times 200)$$

$$= 2 \times 35000 = 70 \times 10^3 \text{ mm}^3$$

According to Chvorinov's rule

$$t \propto \left(\frac{V}{A}\right)^2 \quad t = k\left(\frac{V}{A}\right)^2$$

$$t_1 = k\left(\frac{10 \times 10^5}{70 \times 10^3}\right)^2$$

Case-II

$$\text{Volume} = 200 \times 100 \times 20 = 4 \times 10^5 \text{ mm}^3$$

$$\text{Surface area} = 2(lb + bh + hl)$$

$$= 2(200 \times 100 + 100 \times 20 + 20 \times 200)$$

$$= 2 \times 26000 = 52000 \text{ mm}^2$$

Solidification time $t_2 = k\left(\frac{4 \times 10^5}{52 \times 10^3}\right)^2$ on solving we get

8.10 SOLIDIFICATION OF CASTING

It is very important to a foundry engineer to understand the mechanism of solidification of casting and its control to produce a sound casting. Once the solidification of casting is understand by an engineer then a defect free casting is possible to produce.

Once the mould cavity is filled by the molten metal and molten metal is allowed to solidity into desired shape. Cooling of a casting in a systematic manner is referred as solidification of casting.

Solidification of casting is basically affected by the:

(i) Grain size, shape and orientation.

(ii) Distribution of alloying element.

(iii) Underlying crystal structure and its imperfections.

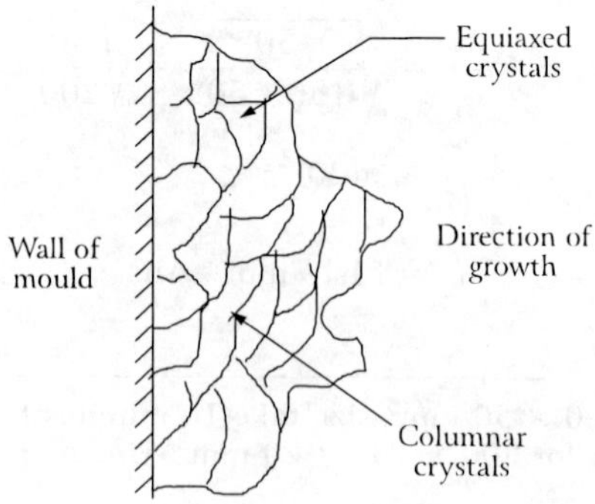

Fig. 8.29 Solidification mechanism

Mechanism of solidification is shown in Fig.8.29, it is necessary to predict how the casting will solidity which avoids casting defects. In case of casting after pouring of the molten metal in the mould, temperature falls until freezing commences at a particular points during solidification temperature remains constant due to release of latent heat. In fact those may be slight increase in temperature if super cooling has occured. As all alloys does not have a sharply define freezing temperature. Solidification of different alloys take place at different range of temperature (heat loss are very high near the mould cavity walls as compared to other place I[st] submicroscopic metal crystallinity refer as Nuclei form here). Crystal growth occurs in a dendritic manner. On slow cooling is responsible for long grow of dendrites where as on fast cooling short dendrites growth produce dendrite arms grow as atoms of metals attach each other to solid dendrite and the arrangement of atoms in a three dimensional manner is referred as unit cell.

Type of unit cell:

(i) B.C.C. Structure,

(ii) F.C.C. Structure,

(iii) H.C.P. Structure.

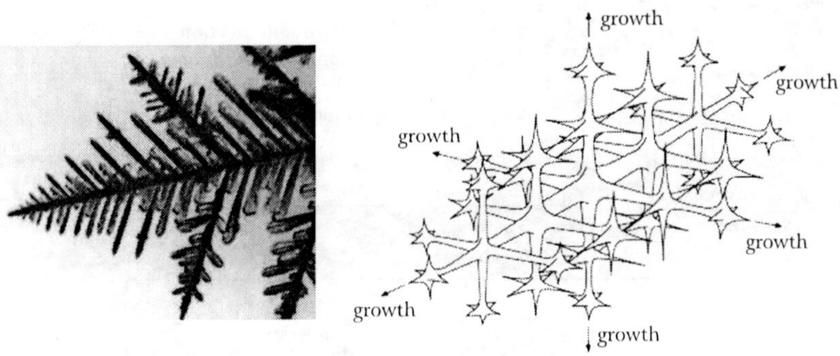

Fig. 8.30 Classical concept of Dendrite (metal) growth

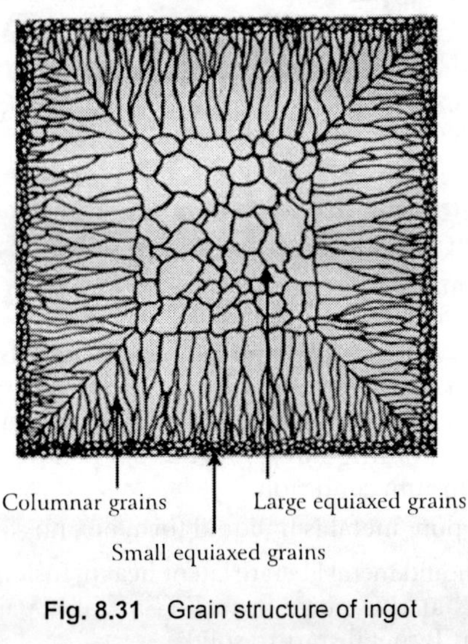

Fig. 8.31 Grain structure of ingot

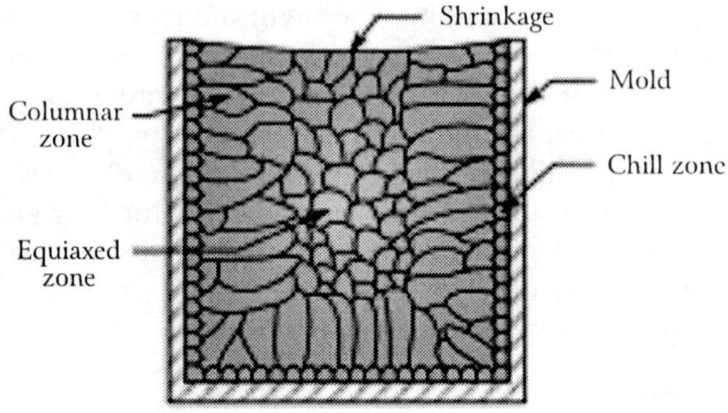

Fig. 8.32 Grain structure in cast metal

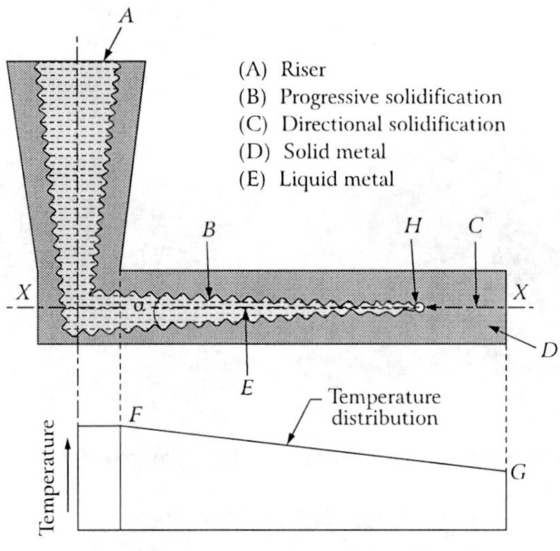

Fig. 8.33 Progressive and directional solidification

8.10.1 Solidification of Pure Metal

Generally pure metal like **copper** and **aluminium** are good conductor of electricity and having good thermal conductivity. They possesses also higher melting point, good ductility and low tensile strength but have better corrosion resistance. Pure metal melts and solidify at a single temperature which is also referred its freezing point or melting point. Solidification of pure metal take place at the constant temperature which can be easily understand by the time-temperature curve. This curve is also known as **cooling curve**. Figure 8.34 shows a cooling curve for a metal under the equilibrium condition.

- From point *A-B* pure metal is in liquid form i.e., no solid metal.
- From point *B-C* liquid metal liberate latent heat of fusion, temperature remains constant. Means at this range from *B-C* metal is partly liquid and partly solid. From point B metal start to solid.

- Now from point C metal start to cool and tends to convert completely solid at point D.

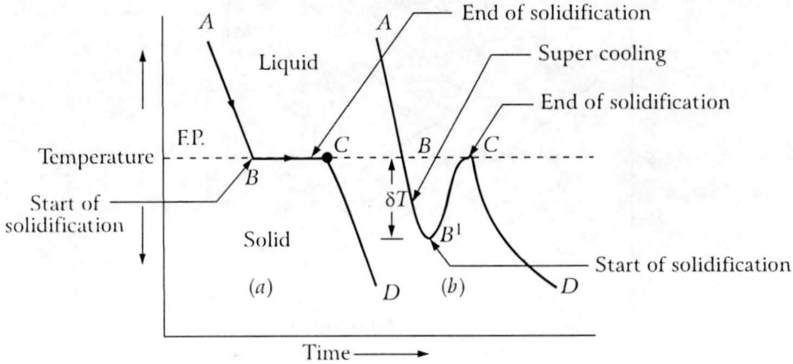

(a) Equilibrium cooling curve (b) Super cooling curve

Fig. 8.34 Cooling curve of a solidifying pure metal

Phase diagram:

Phase diagram is associated with solidification of alloys during casting process. They also known as equilibrium curve. Phase diagrams are differents for different metals. Phase diagram consist three phase i.e., **liquid stage, solid + liquid stage, solid stage**. Phase diagram shows on cooling (sudden/gradually) of liquid metal, what atomic structure produced and they basically depends upon the rate of cooling. Every phase diagram shows two phases, one is liquid metal solution and other is solid solution. In phase diagram upper line is liquidus line while lower line is solidus line. Liquidus line refers at this line liquid is not total liquid. It is some part of liquid and solid, while solidus line shows at this stage metal is not perfectly solid. It is in between solid and liquid stage.

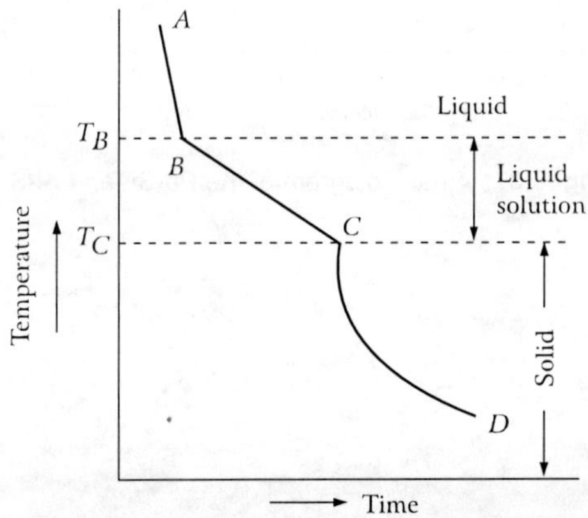

Fig. 8.35 Ideal cooling curve of alloy.

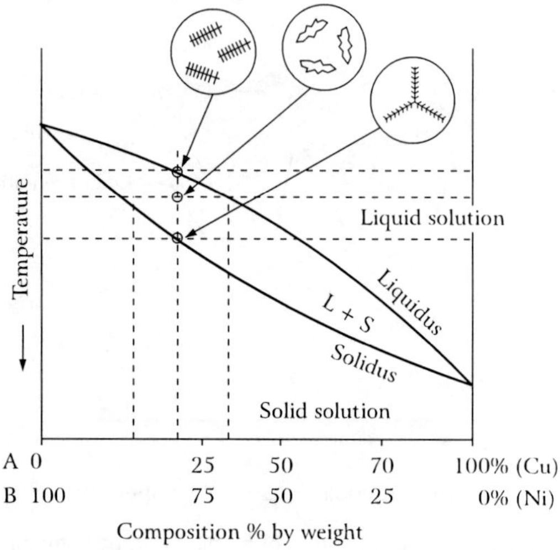

Fig. 8.36 Phase diagram

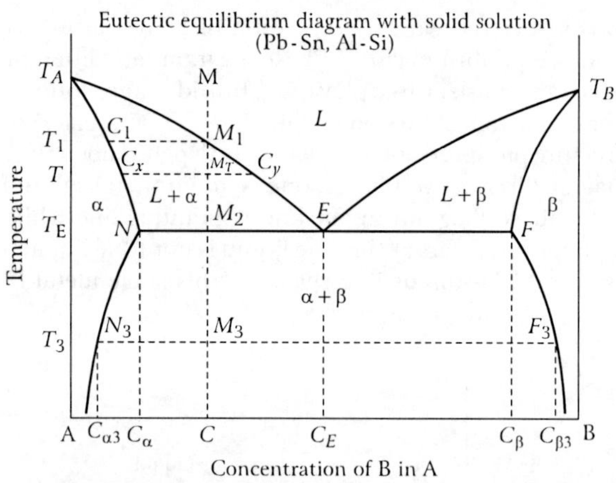

Fig. 8.37 Phase diagram of the Pb-Sn and Al-Si

❑❑❑

9

CHAPTER

CASTING DEFECTS

9.1 CASTING DEFECT

[*UPTU 2009-10*]

In sand casting process, large number of defects occurs. All casting defect can not be eliminated they can be only minimize.

Various factors are responsible for producing the casting defects such as:

(*a*) Casting design.
(*b*) Pattern design.
(*c*) Moulding and core making equipment.
(*d*) Core and mould material.
(*e*) Risering and gating system.
(*f*) Melting and pouring temperature.
(*g*) Method of mould and core making technique.
(*h*) Composition of metals.

9.1.1 Defect

If any part/component/product produced by casting not fulfilling the total requirement or product is not as per specification then it is referred as a casting defect.

Basically three types of casting defects are produced:

(i) Defects which are very minor and can be easily rectified.
(ii) Defects which are some difficult to rectified and invest some amount of money.
(iii) Defects which are very major and cannot be rectified at minimum cost in such condition product is completely useless.

9.1.2 Seven Basic Major, Categories Casting Defects

1. Cavities—include internal and external blow holes, pin holes, etc.
2. Metallic Projection—mismatch of pattern/moulding box, fins flash.
3. Incorrect Dimensions and Shape—shrinkage or war page.
4. Discontinuities–cracks, hot tear, cold tear, etc.
5. Defective Surface–fusion, excess or less of penetration, crushes, scabs, hard spots, scars, blister, etc.
6. Incomplete Casting—cold shut, pour short.
7. Inclusion.

9.2 DIFFERENT CASTING DEFECTS

9.2.1 Blow Holes

Blow holes are round holes and appears in the mould cavity. They are basically due to entrapped bubbles of gases in the metal.

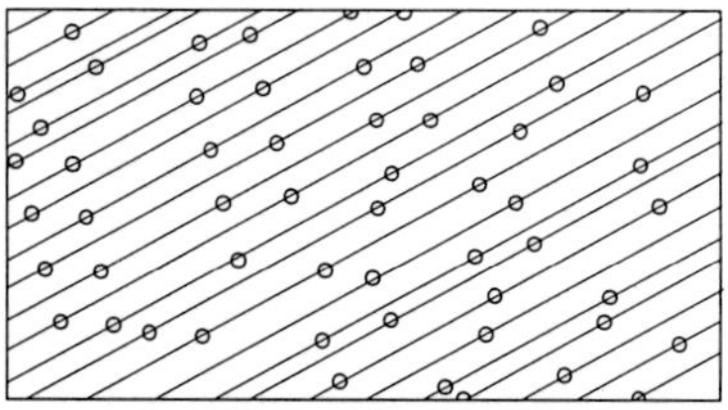

Fig. 9.1 Blow holes or gas holes

Causes	*Remedies*
• Excess moisture contents in moulding sand. • Super fine grain size and low permeability of casting sand. • Rusted chill and inserts. • Very hard ramming. • Mould not adequately vented.	• Moisture content should in proper proportion. • Permeability and grain size of sand should be proper. • Chill and inserts should be rust free. • Ramming should be proper. • Mould should be properly vented.

9.2.2 Misrun

Once the mould cavity is not fill completely then it is referred as misrun i.e., some part of mould cavity is empty (not filled by the molten metal).

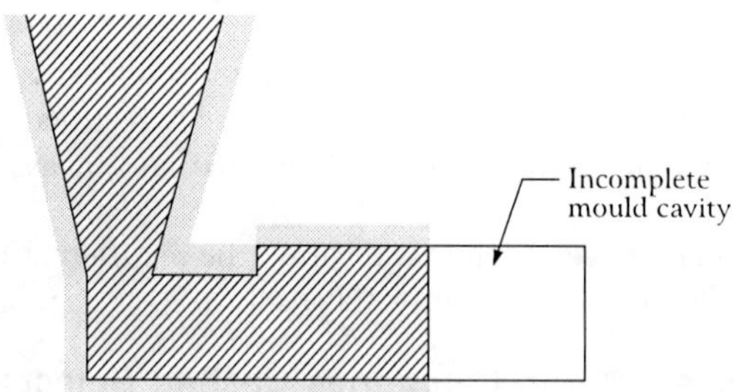

Fig. 9.2 Misrun

Causes	*Remedies*
• Inadequate metal supply. • Low pouring temperature. • Metal lacking in fluidity. • Improper gating system.	• **Metal supply** should be **proper.** • **Pouring temperature should be adequate.** • Gate size should be proper. • Use proper gating system.

9.2.3 Mismatch

Mismatch defect basically encountered when there is shifting of individual part of a casting with respect to each other.

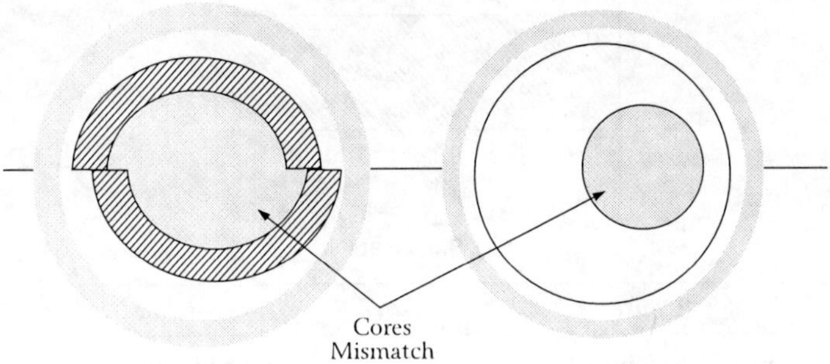

Fig. 9.3 Cores mismatch

Causes	*Remedies*
• Due to not proper assembly of lower and upper half of moulding box. • Worn out clamping pins. • Faulty core box. • Improper location of cores.	• Proper care should be taken while assembling of upper and lower half. • Provide adequate support to cores. • Use the proper core box. • Core should be properly located.

9.2.4 Hot Tears

Hot tear is also one of the type of cracks, having ragged edge due to high stress during solidification of casting.

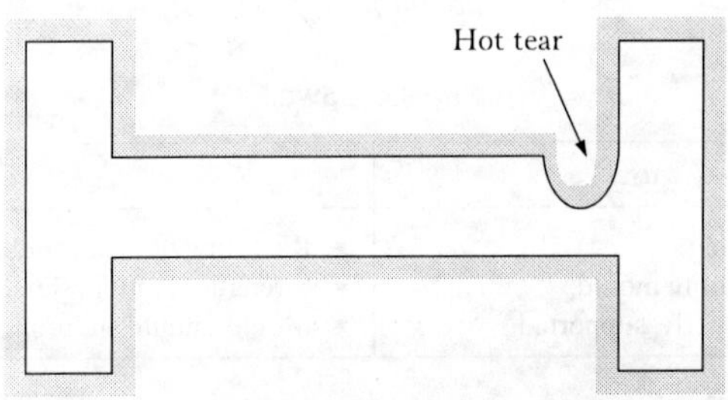

Fig. 9.4 Hot tear

Causes	*Remedies*
• Very hard ramming and excessive mould hardness. • High dry and hot strength of sand. • Very fast solidification of casting. • Very low pouring temperature. • Very high sulphur content.	• Ramming should be uniform throughout. • Strength of sand should be appropriate. • Solidification should be uniform. • Pouring temperature should be adequate. • Sulphur content should be low.

9.2.5 Rat Tails

Rat tail is a long shallow angular depression in the surface of casted part.

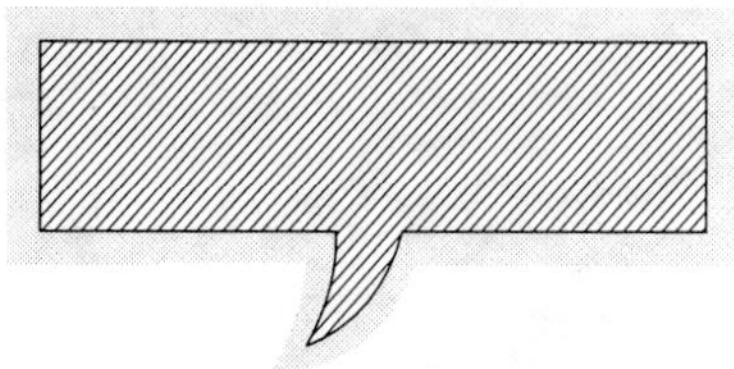

Fig. 9.5 Rat tail

Causes	*Remedies*
• Excessive mould hardness. • Very large flat surface of casting. • Continuous large flat surface on casting.	• Reduce **mould** hardness. • Apply grooving/depressions breaking large flat surface.

9.2.6 Swell

It is usually found on the vertical surface of casting. It is enlargement of mould cavity due to the pressure of molten metal to side and wall of mould cavity.

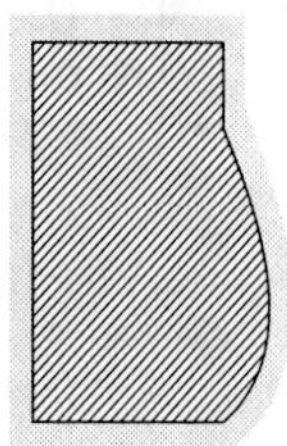

Fig. 9.6 Swell

Causes	*Remedies*
• Due to soft ramming. • Less strength of entire mould. • Mould is not properly supported.	• Ramming neither should be hard nor soft. • Strength of mould should be adequate. • Mould should be properly supported.

9.2.7 Cold Shut

If the molten metal is entering in the mould cavity through more than one gate and the molten metal is meeting physically in mould cavity but not fuse this is referred as cold shut.

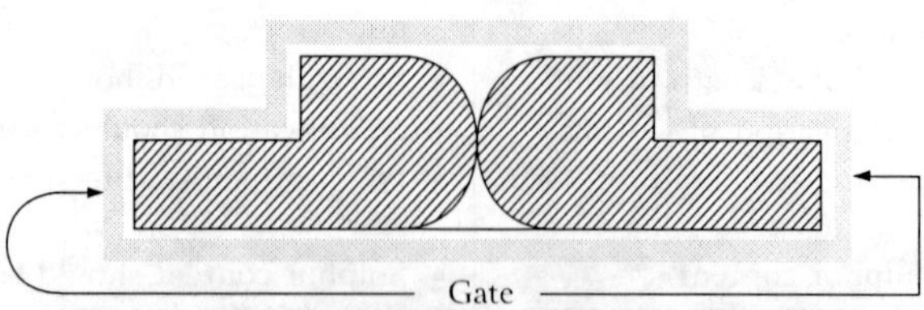

Fig. 9.7 Cold shut

Causes	*Remedies*
• Pouring temperature should not so high.	• Pouring temperature of molten metal should be adequate.
• Inadequate molten metal supply.	• Supply of molten metal should be uniform.
• Improper design of gates.	• Gate design should be full proof.
• Restrictions in the gate.	• Gate should be free from any restriction.

9.2.8 Gas Porosity/Gas Hole

Gas porosity is different with blow holes. Gas porosity encountered due to the low permeability of moulding sand basically hydrogen is responsible for gas porosity.

Causes	*Remedies*
• Low permeability of sand.	• Permeability of sand should be proper.
• Excessive moisture content in mould.	• Reduce the moisture contents from sand.
• High pouring temperature.	• Pouring temperature should be proper.
• Faulty metal	• Proper metal
• Slow rate of solidification.	• Solidification should be completed at appropriate rate.

9.2.9 Scrab (Metal Penetration)

These are rough, irregular projections on the casting surface with embedded sand. It occurs when molten metal penetration behind the surface layer of sand.

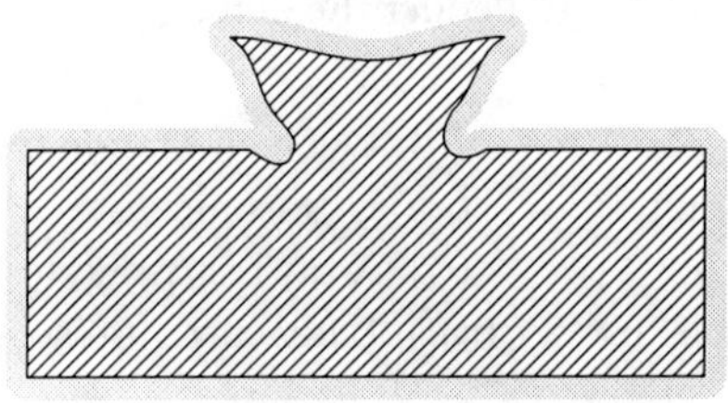

Fig. 9.8 Scrab

Causes	*Remedies*
• Too fine sand.	• Grain size of sand should not be too fine
• Low permeability of sand	• Sand should have proper permeability.
• High moisture constant in sand	• Moisture content should be in inappropri-ate ratio.
• Soft ramming	• Ramming should be uniform.
• Slow running of molten metal	• Molten metal should be run with uniform speed.

9.3 NDE

NDE is referred as non-destructive examination of any parts/components without its breaking.

9.3.1 Non-Destructive Examination of Casting

During the melting and pouring of a liquid metal into mould cavity and allow it solidified into mould cavity. This process have some defects. Detection of the same defects and rectification of same is very necessary.

In non-destructive testing after inspection part/component may be further reused as there is no change its shape and size. With the help of NDT (Non-Destructive Testing) several invisiable subsurface defect can be easily deduct easily. NDT make casting component more reliable and safe as NDT do not provide direct measurement.

9.3.2 Advantages of NDT

(a) This type of test requires low cost
(b) These tests gives the correct microscopic structure of object without destroyed the parts.
(c) These tests are faster and easy to handle.
(d) Reliability of these type tests are more
(e) Most test methods are rapid.
(f) In service testing is possible
(g) 100% checking of parts is possible
(h) Very little preparation is sufficient.

9.3.3 Several Non-destructive Inspection Methods

Various methods are available to deduct the defects from casting parts. They are following:

(*i*) Florescent test
(*ii*) Dye-penetration test
(*iii*) Visual inspection test
(*iv*) Ultrasonic test
(*v*) Cycle test
(*vi*) X-ray test/γ-rays test (radiography test)
(*vii*) Magnetic particle test
(*viii*) Eddy current test

9.3.3.1 Liquid Penetration Test

It is a method to detect the surface openings such as **surface porosity**, **cracks**, **inclusions** in **stainless steel**, **aluminium**, **brass** and other **non-ferrous metal casting**. It is based on the ability of a penetrating liquid to wet surface opening of a weld/cast defect and to drawn into it.

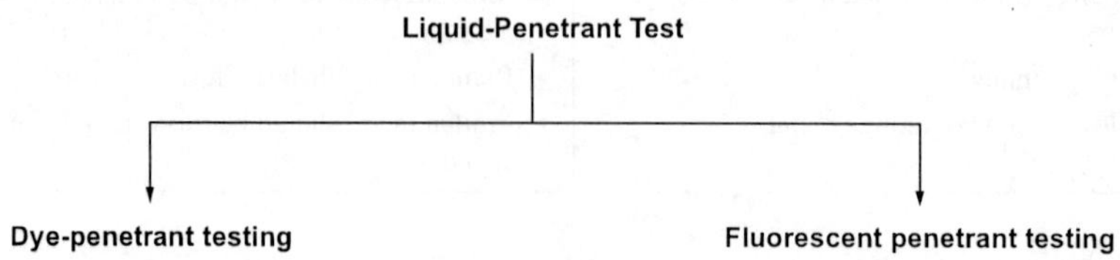

Once the penetrant enters into the cracks or opening the rest of the liquid can be wiped or washed away but the liquid in the cracks or opening remains thereby the application of a chalk like developer blatting paper action take place with the developer drawing the dyne from the crack and showing it on the surface of the developer in such a way flaw is detected.

This type of test basically involves the following steps:

(a) Preparation of surface
(b) Penetrant application
(c) Dwell time
(d) Excess penetration removal
(e) Developer application
(f) Surface inspection
(g) Post cleaning

9.3.3.1.1 Dye-Penetrant Test

Basic difference between these two types of test that dye-penetrant test utilises visible penetrant usually red colour that produces a contrasting indication against the white background of developer for the case of fluorescent testing medium used fluorescent liquid, *i.e.* a liquid which flows by ultra violet (UV) or black light of a wavelength (λ) of approximate 3651 Å. As compared to Dye-Penetrant Test flourescent penetrant test method is more expensive and it needs UV light source. Some very well internationally renowned liquid dye-penetrants are *Met-L-Check*, *Dy-Check*, etc.

(a) Surface Preparation

Basic need of such type test that surface should be free from contaminants that inhibit flow of penetrant into cracks and other surface openings. Such inhibiting contaminants may be **dirt, grease, scab, acids, chromate, rust, casting flux, spatters, paints**, etc. There are several methods are available for surface cleaning such as solvent cleaning; it is very popular method. Solvent should be dilute so that it would not harmful for weld/cast surface, acetone, isopropyl alchol, CH_3Cl and perchlorethylene are some usefull solvent. Shot-blasting or sand blasting is not recommended as surface cleaning process as they peens-over the cracks resulting in thier closure if needed drying may be done by working the parts in drying ovens, with infrared lamps or forced hot air.

(b) Dwell Time

Dwell time is the time for which the weld/cast structure will dip in liquid for examination purpose by varying penetration time, optimum time for a particular size and kind of defects can be determined and time range may be from 90 to 180 sec.

(c) Procedure for Penetrant testing:

Following steps are performed in sequence to perform the penetrant test:

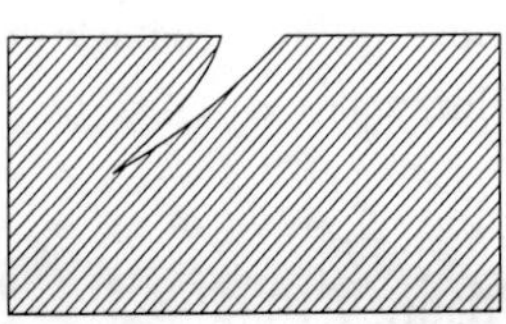

(a) Open defect

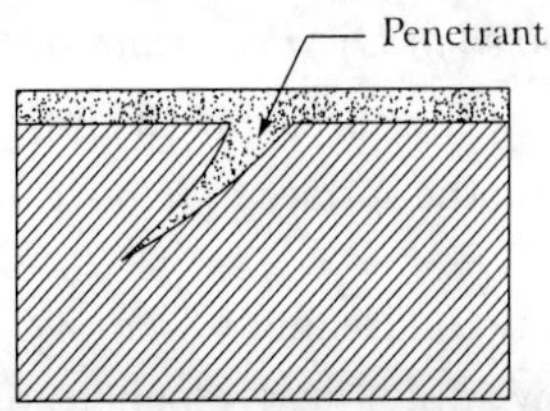

(b) Applied penetrant in the crack.

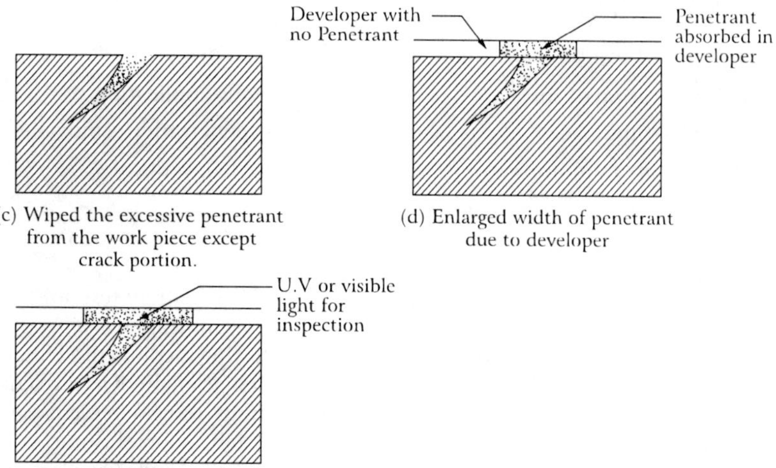

(c) Wiped the excessive penetrant from the work piece except crack portion.

(d) Enlarged width of penetrant due to developer

(e) Detection of defect (crack) by visible inspection

Fig. 9.9 Sequence of steps in liquid penetrant Test

1. Cleaning
2. Penetrant application
3. Wipe of excessive penetrant from workpiece
4. Applying developer
5. Inspection and Evaluation.

1. Cleaning

It is one of the most important step before starting the liquid penetrant test. In cleaning process workpiece is clean for any dust, scale, grease, paint and other chemical that are not cleaned from the surface. A satisfactory combination of solvents, brushes, etchants etc. must be chosen for a particular inspection.

2. Applying penetrant

Second most important step in this test is proper apply penetrant fluid to the cleaned surface. dwel time, which is the amount required for penetrant to enter into the crack will vary depending on the shape and size of crack and environmental condition too. dwel time may be 20 to 30 min.

Steps Involved in Dye Penetrant Test

(a) Clean the surface with cloth to avoid any dust and dirt particles.

(b) Brush the surface to remove any grease, oil film, scales, rust, etc. by a smooth hand.

(c) Spray the cleaner to remove oil, etc.

(d) Now apply the dye-penetrant as shown in figure 9.10 (a).

(e) Allow it 3 to 5 minute to penetrate into the cracks.

(f) Wipe off the excess penetrant from the surface.

(g) Spray the developer as shown in figure 9.10 (c) with very thin layer. Layer absorbs the penetrate from the cracks and red spots on the casted structure to give a very clear indication of flow.

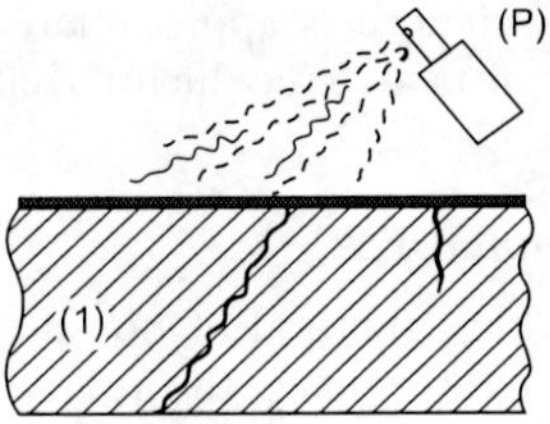

(a) Dye Penetrant (P)

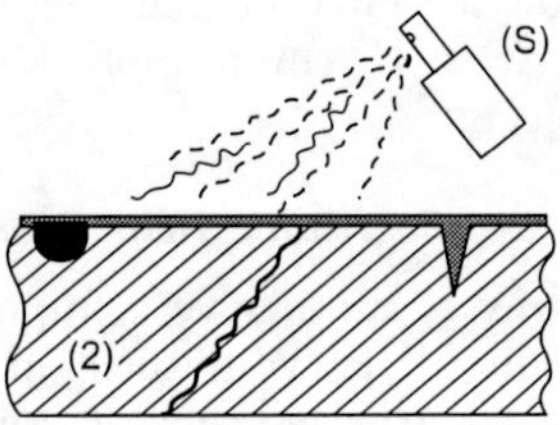

(b) Solvent (S)

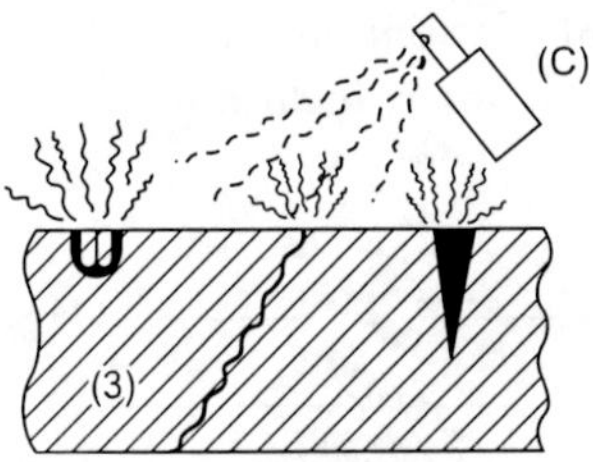

(c) Chalk developer (C)

Fig. 9.10 Steps in Dye-penetrant Test

9.3.3.1.2 Fluorescent-Penetrant Test

It is also another method to detect small surface cracks on casted/welded parts. It is same as magnetic particle inspection test. This test can be applied for ferrous and non-ferrous metal (casted/welded parts). **Zyglo** is the trade mark of magnaflux corporation which is applied to its equipment and material for fluorescent penetrant test.

Steps Carried out in Florescent Inspection

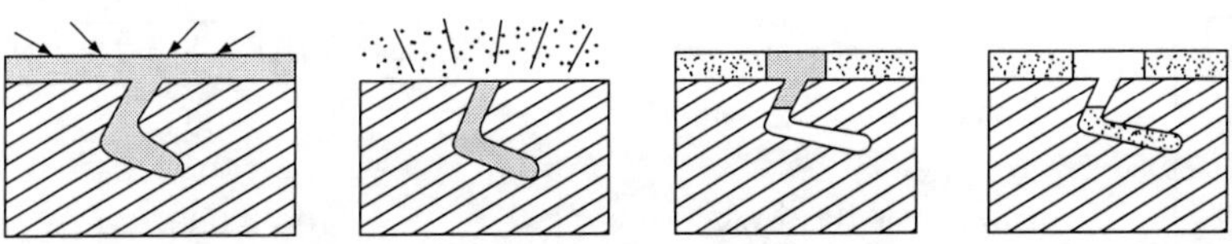

1 - Penetrate 2 - Cleaning 3 - Applying developer 4 - Inspect

Fig. 9.11 Steps in Fluorescent-Penetrant Test

Procedure

(i) First of all clean the surface that to be inspected.

(ii) Now apply a layer of fluorescent penetrant by brush or by dipping and allow it to penetrant approx. for 60 minutes.

(iii) Now clean the penetrant surface by water spray to remove the penetrant layer from surface that to be inspected.

(iv) Now it is time to apply developer. A developer act as blatter to draw penetrant out of cracks, etc. and enlarge the size of cross-section area of penetrant indication.

(v) Now a source of black light whose wavelength is approximate 3650 Å (1 Å = 10^{-8} cm) is applied on developer surface to see the internal surface defects.

9.3.3.2 Radiographic Inspection (Radiographic Test)

Radiography is another very important and oldest method of non-destructive examination (NDE). With the help of NDE it is easy to detect the internal defects of any casted structure such as **cranks**, **porosity**, **inclusions**, etc. When a test piece (casted/welded object) is exposed to a penetrating radiation like X-ray or γ-ray variation in the amount of variation transmitted through it depend upon:

(*a*) relative dansity of metal or any inclusions

(*b*) thickness variation

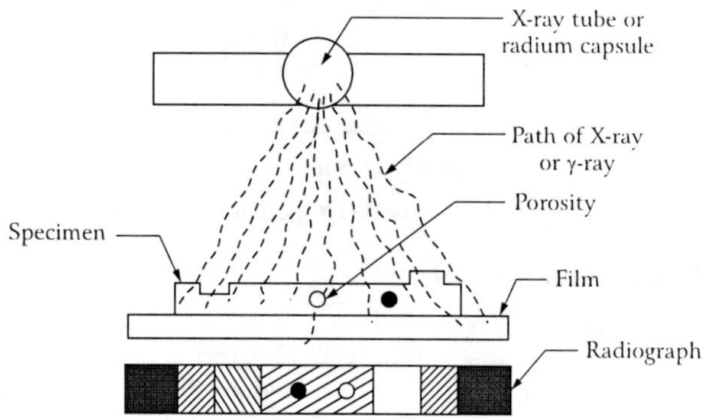

Fig. 9.12 Effect of metal thickness and inclusion on X-ray absorption

X-ray and γ-rays comprises part of electromagnetic spectrum that can be divided on the basis of wavelength.

S. No.	*Nature of rays*	*Wavelength (Å) 1 Å = 10^{-8} cm*
1.	γ-rays	0.1 – 0.005
2.	X-rays	5 – 0.004
3.	Ultraviolet (UV) rays	10 – 4,000
4.	Infrared rays	7500 – 4 × 10^6
5.	Radiowaves	10^6 – 10^{13}

Applications of Radiographic Test:

Radiographic test have the following applications

1. Radiographic test extensively used for casted, forged and weldments part when these is a critical need to ensure that parts/object is free from any internal flows.
2. This method can be used to inspect both Ferrous and non-ferrous solid materials.

3. This test is also suitable to non-metallic and composites material.
4. This test can be used to inspect the condition and proper placement of components for liquid, level measurement in sealed parts etc.
5. This method (Radiographic) is also suitable to inspect of semiconductor device for checking the cracks, broken wire, forgoing material and misplaced components.

Characteristic of X-ray and γ-rays

1. They are invisible electromagnetic variations.
2. They are differentially absorbed.
3. They can penetrate any material easily.
4. They can travel very easy in a straight line.
5. They can produce photochemical effect in photographic emulsions.
6. They can easily ionise the gases through which they pass.
7. They are not easily affected by electromagnetic field.
8. They can travel with the speed of 3×10^{10} cm/sec.
9. They are capable to liberating photoelectrons.

Basic Needs/Requirements of a Radiography Test

Requirements of a radiography test are followings:

(*i*) A penetrating source such as X-ray machine or a radioactive isotope.
(*ii*) Casted/welded object that to be inspected.
(*iii*) A recording film or a viewing screen.
(*iv*) A well trained radiographer to operate X-ray or γ-ray machine.
(*v*) A means to process the exposed film or operate viewing system.
(*vi*) A skilled person for interpretation of radiographs.

9.3.3.2.1 X-rays

When high speed of electron in the form of a beam known as cathode ray strike on a metal target (weld object or casted) placed in an X-ray tube, the X-ray are produced at which velocity the electron strike the target, is determined by the tube voltage where potential difference (is the tube voltage) between the source electrons and target energy of electron which is used to penetrating radiation is often expessed in terms of electron volts (eV), KeV, or MeV, *i.e.* million electron volts.

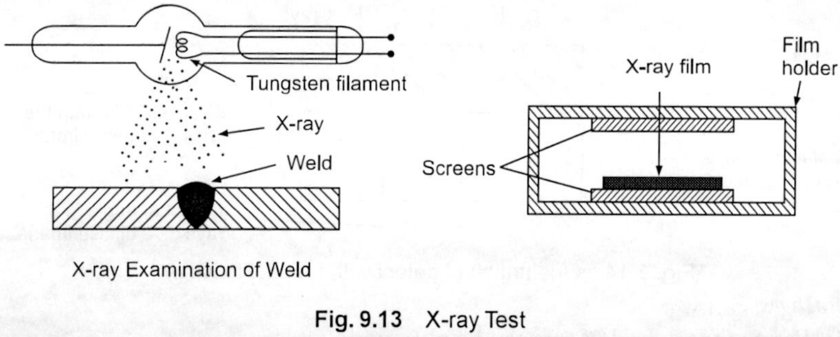

Fig. 9.13 X-ray Test

Energy of X-ray produced at the target is proportional to the square of the mean velocity of the electrons and shorter the wavelength of X-ray, greater their penetrating power.

Kinetic energy is equal to work done in accelerating the electron and it is given by

$$K.E. = \frac{1}{2} MV^2 = 1.6 \times 10^{-12} V.$$

where M = Mass of electron (9.1×10^{-28} g)

V = Tube voltage

v = Velocity of electron cm/sec

Efficiency of X-ray producing can be determined by the following:

$E = 1.4 \times 10^{-7} \times ZV$

Z = Atomic number of target

E = Efficiency

V = Tube voltage

There are basically two types of X-ray tubes are available:

(i) gas tube

(ii) coolidge tube.

In radiography it is expected to have a source of X-ray which approaches an ideal point source finer the focal spot better the radiographic image. Power required for X-ray machine range from 50 to 24000 kV.

A 2400 kV X-ray machine is capable to photographing approx. 500 mm thickness.

Test Procedure

X-rays are allowed to fall upon the test specimen. Some X-ray absorbed by test piece. Extent of this absorption, as already stated, that depends on the presence of voids, foreign inclusions cracks in the casted part/weld object as the result X-ray passing (orientation of defect with respect to X-ray) through the cast part and falling upon a photographic plate or film behind it will produce area differing in optical density dark spot corresponding to narrow area and light spot to some defect in interior surface of cast structure. If the intensity of radiation source is good then no doubt quality of (X-ray) picture is also very good, it also depends upon angle of X-ray inclination, type and thickness of metal. This test is very effective to find out the **cracks, lack of fusion, undercutting, slag inclusion, porosity, pinholes, blowholes**, etc.

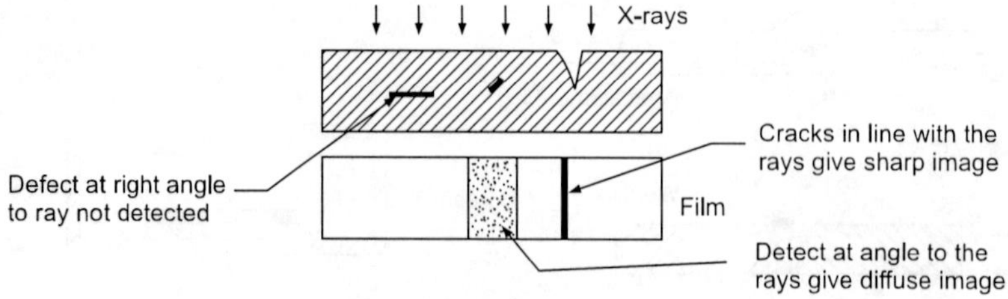

Fig. 9.14 Orientation of detect with respect to X-ray

Advantages

(i) Permanent record of test can be obtained.

(ii) Reference standards for defects are available.

Limitation

(i) Slow and expensive NDE process.
(ii) Trained operator is needed.
(iii) Harmful radiation generate during process.
(iv) Costly equipment.

9.3.3.2.2 *γ-rays*

γ-rays are also another NDE process to detect the internal defects. γ-rays are quite different with the X-rays. These rays are generated with the help of **cobalt-60** where X-rays are produce by **X-ray** tube. γ-rays are used to find out the internal defect which cannot be detected by X-ray very easily. γ-rays do not have a continuous spectrum as X-rays but consist of one or more descrete energies. γ-rays produced by natural source such as Radium Ra^{226} and Radon, Rn^{222}. One of the big characteristic of γ-rays that at a time it can inspect a numbers of casted/welded object. γ-rays are not recommended to detect the small defect. γ-rays from its source are emitted in all direction therefore a number of separate casted parts/welded object having cassette containing film a fastened to the back of each object are dissipated in a circle around the source placed in a central position.

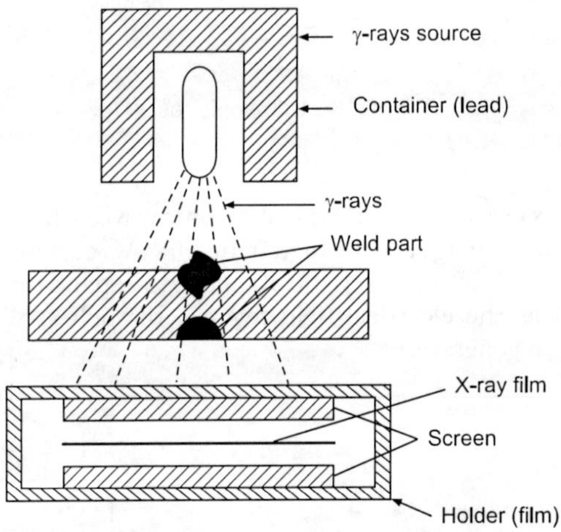

Fig 9.15 γ-rays Procedure

9.3.3.3 *Ultrasonic Testing*

Ultrasonic testing is also another most popular non-destructive testing to find out the internal defect of a casted parts/weld object. It is quite different with radiograph testing technique. There are no harmful ultravoilet rays generate in ultrasonic testing as generated in radiographic technique. A frequency that is hearing above the normal human range ear is called ultrasonic waves. This term generally include all waves having a frequency greater than 20,000 Hz to 25,000,000 Hz or cycle per sec. In this process ultrasonic waves are introduce into a test piece to detect and locate surface and internal defects.

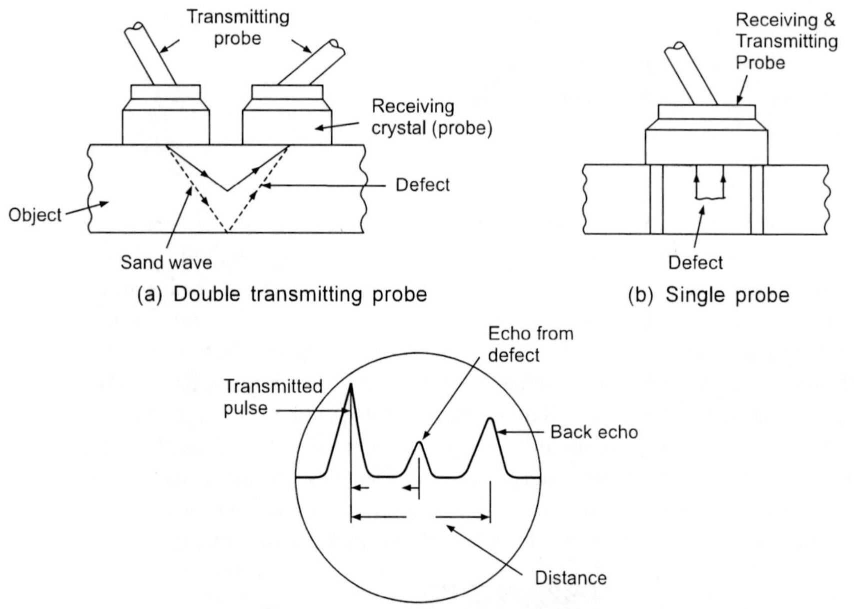

(a) Double transmitting probe (b) Single probe

(c)

Fig. 9.16 Ultrasonic test

Test Procedure

Piezoelectric effect is used to generate the ultrasonic waves which convert electrical energy into mechanical energy. As the high frequency AC current passed across the faces of quartz crystal. It is characteristic of quartz to expand during I[st] half of the cycle and contract as the electrical field reverse. In such as way sound waves/ mechanical vibration generate.

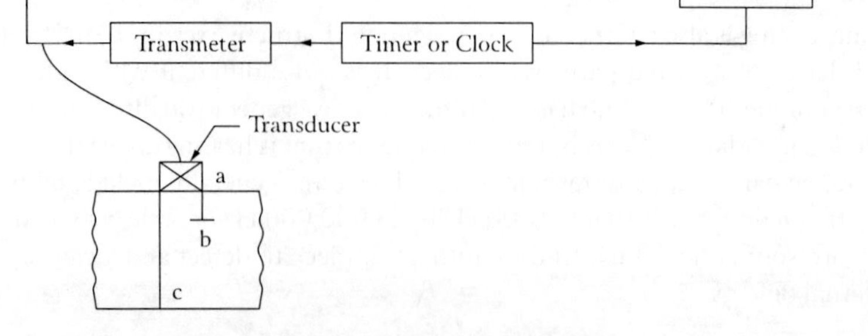

Fig. 9.17 Ultrasonic flaw dector

The casted welded part that should be inspected for surface defects and internal defect by ultrasonic testing machine should be smooth surface and free from the dust particles by any suitable method. In such a way that ultrasonic waves can transmitted from probe (crystal) to casted/weld part.

Usually three type of different systems are employed for testing.

(i) pulse ecosystem.

(ii) through transmission system.

(iii) resonance system.

A oil film should be there in between the probe and object surface. For proper contact to better transmission of waves from probe (crystal) to casted part/weld object.

For operation, ultrasonic wave is introduced into the metal and time internal between transmission of the outgoing and reception of incoming signals are measured with a cathode ray oscilloscope (CRO). In case of defect exist in between the top and bottom surfaces, most of beam striking their defect will get reflected from the defect, reach the receiving probe and in dictate a pip (echo) on the CRO screen before the pip given by the waves striking the bar end of the job and returning.

Advantages

(i) Very fast and reliable process.

(ii) Process is more sensitive than radiography.

(iii) Cost of inspection is very low and high speed of operation.

(iv) Sensitivity is very high of flow detection.

(v) Big casted parts/weldments can be systematically scanned for initial detection of major defect.

Limitations

(i) Surface that to be tested must be smooth and clean.

(ii) Trained and equalified operator is need to perform the operation.

(iii) A very complex shape and complicated structure casted part/weld metal object cannot be detected by this process.

Applications of Ultrasonic test:

- For non-destructive testing pulse-echo technique is most widely used. Weldment and cast structure is the largest application of ultrasonic testing.
- Bar, rod, rails, pipes etc. of ferrous and non-ferrous semi-finished products.
- Simple shapes parts which are large like castings are tested by ultrasonic testing.
- Complicated shapes parts are inspected for flaw dictating by this method.
- Most important application of ultrasonic testing to finding dangerous incipient defects i.e., fatigue crack and corrosion effect.
- Boiler components are inspected regularly for fatigue cracks in weld/cast for corrosion in certain area.

9.3.3.4 Magnetic Particle Inspection

Magnetic particle test is another non-destructive inspection method to detect surface or near surface defect or discontinuities in magnetic material, *i.e.*, ferrous materials. But deeper subsurfaces defects are not satisfactory detected by this process as the influence of the distorted lines of magnetic flux on magnetic particle. Defects that can be easily detected by this process are **cracks**, **thermal cracks**, **grinding cracks**,

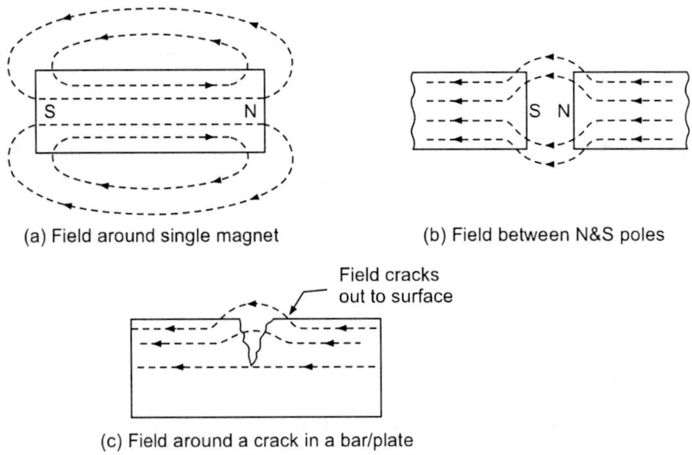

Fig. 9.18 Magnetic field around bar magnet and a crack

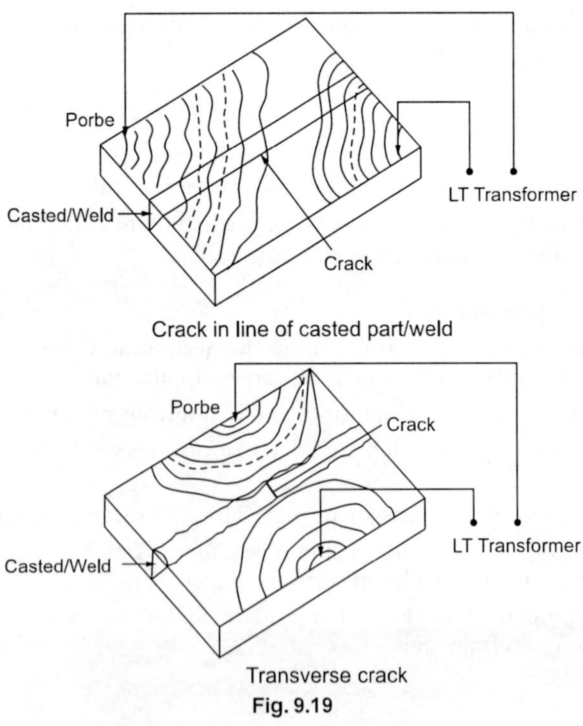

Fig. 9.19

laps, seams, over laps, fatigue crack, hot tear etc. It is a very simple and very inexpensive testing method. Any type of shape and size of weld/cast structure object can be detected for defects. This **test do not work** on aluminium, copper, zinc, austenitic steel as all materials are non-magnetic.

Test Procedure

Following steps are performed to ensure satisfactory detection of defects:

(a) Surface preparation of component before testing
(b) Initial demagnetization
(c) Cleaning and Degreasing
(d) Magnetization of the component.

Two basic method are commonly used one is **dry method**. In which dry magnetic power is spinkled on the workpiece and other is the ***wet method***. In which liquid suspended magnetic particles is employed. Permanent or electromagnetic can be used to create a magnetic field in the casted part. More commonly a current is passed through the area being examined as shown in Fig. 9.20 detecting fluid or ink is usually particles of magnetic iron oxide suspended in kerosene. Particles may be coated with a fluorescent compound for viewing under ultra-violet light. In case where there are objections to use of kerosene the magnetic particles can be dispersed from a '**puffer**' to from a cloud above the surface of a weld. This alternative is particularly useful when inspecting root runs in 'V' joints since there is no need to clean the groove before depositing the next run of weld metal.

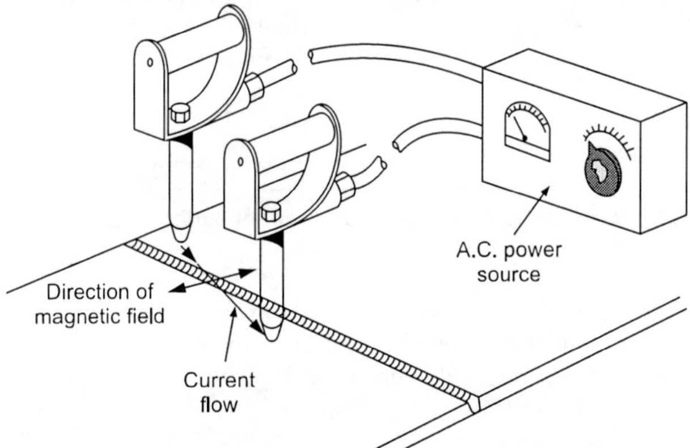

Fig. 9.20 Magnetic-particle inspection using alternating current

Limitations of Magnetic Particle Test:

- MPT having certain limitation.
- This process is only applicable only surface opening and subsurface defects in ferromagnetic materials.
- proper care is necessary to avoid local heating and burning of finished parts or surface at the points of electrode contact.
- another limitation is necessity to complete demagnetization of the parts after completing the test.

Applications

- This method is widely used in the field of casting.
- Nowadays several types of defects can be easily detected with the help of magnetic particular inspection such as **surface** and **subsurface cracks, incomplete fusion, slag inclusion, porosity undercuts, bonding, laminations** etc.
- It is widely used in **aircraft industry**. Such as landing gears, etc.
- This technique is also very helpful to detecting the beginning of the *tiny* fatigue crack.

9.3.3.5 Eddy Current Test

In this type of non-destructive examination of casted parts. An alternating current is passed through a coil which is placed in the proximity of casted parts. As the resulting of changing the current in coil an alternating magnetic field in the casted object produced. Varying magnetic field in the casted parts produces electrical eddy current. Now this eddy current which vary with the magnetic field create their own magnetic field which interacts with the initial field.

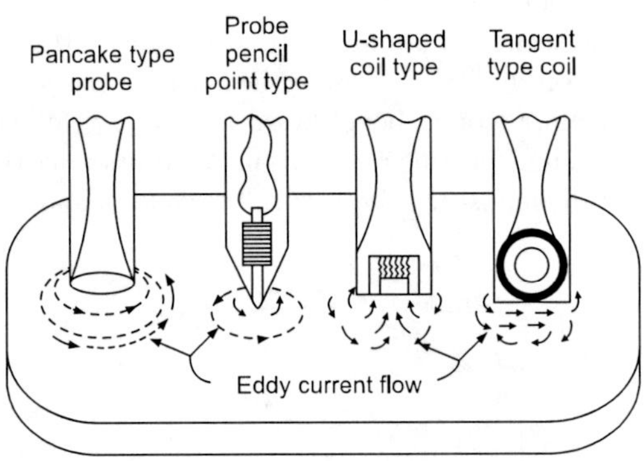

Fig. 9.21 Basic four type of used eddy current surface inspection probes.

Test Procedure

Test block (**casting part**) is placed in alternating current field for producing eddy current. Coil may be in the shape of probe or in the case of tabular shape may be wound to fit inside the tube. Eddy current also produce own magnetic field in the

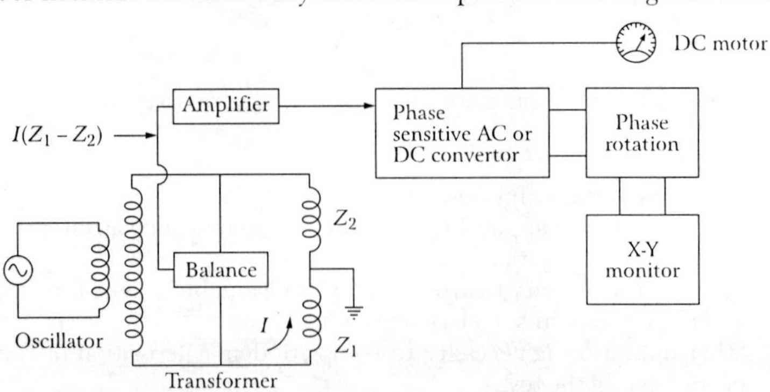

Fig. 9.22 Block diagram of eddy current test instrument.

metal test piece (casted object) which opposes the original impedance of the exiting coil/second coil coupled to the Ist and inclare primary proximity to the test piece is affected by the presence of induced eddy current if there is any crack in casted part discontinuity must be oriented nearly normal to the eddy current flow to disturb it eddy current decrease with depth.

Advantages

(*i*) No need to contact of prob with surface to be tested.
(*ii*) Eddy current test is applicable for ferrous and non-ferrous metal.

Limitation

(*i*) Test is suitable only for good electrical conductor.

Applications of eddy current test method:

In industries eddy current test is widely used to inspect a wide range of products. **ECT** can be applied to **flat**, **round**, **cylindrical** and **irregular** shaped conductive objects.

(a) Conductivity (varies with material characteristics)
(b) Hardness (hardness change as conductivity change)
(c) Heat treatment (variation in heat treatment causes variations in conductivity)
(d) Dis-continuities (cracks, inclusion etc. cause change in output impedance).
(e) Coating thickness (non-conductive coating varies the lift-off between the coil and the conductive surface of the object).

For inspection of **seamless** and **welded piping** and **tubing during** inspection. For ferromagnetic **steel**, **austenitic steel**, **copper alloys**, etc.

❑❑❑

10

CHAPTER

MELTING FURNACES

10.1 MELTING FURNACE

Furnace is a device which is used to melting the material at specific temperature for casting purpose. Furnaces are various types. Selection of a furnace for melting the material depends upon the following factors:

(a) Type of fuel or energy used to melt the material.
(b) Quantity required for melting.
(c) Size and shape of metal charge that to be melted.
(d) Degree of refining required and control over the process.
(e) Compositional range of charge and superheat required.
(f) Initial cost of furnace.
(g) Pouring method desired.
(h) Operation cost and maintenance cost
(i) Flexible to melt different melting temperature material.
(j) Pouring temperature of material required.

Different furnaces for melting are:

- Crucible furnace – from 10 to 200 kg melting of material.
- Oper hearth furnace – suitable for melting – 100 to 500 tonnes
- Cupola furnace – suitable for melting –1 to 100 tonnes
- Electric arc furnace – suitable for melting –0.5 to 2 tonnes.
- Pit furnace – suitable for melting – upto 100 tonnes.

10.2 CUPOLA FURNACE

One of the oldest method of produced the cast Iron only by cupola furnace. It is a vertical type furnace consisting a steel barrel (hollow from both ends) lined with refractory material from inside (to avoid the heat transfer) having tuyeres and wind box. Cast iron is produced by the melting of pig iron, coke, flux and steel scrap in cupola furnace.

10.2.1 Description of Cupola

Cupola furnace having the following parts:

1. **Steel Barrel:** As cupola is a vertical type furnace having a steel with the length 7-9 m long and diameter 1-2 m Long and this steel shell is made out of 6-12 mm thickness sheet and this steel barrel consisting the refractory lining from inside (SiO_2 and Al_2O_3).
2. **Legs:** Entire cupola furnace is mounted on the bricks foundation or on steel columns. Bottom of steel barrel having drop bottom door through which slag can be discharge at the end of melting process.
3. **Tuyeres:** For combustion of cupola air is needed hence tuyeres is a device through which the air supplied for combustion of fuel and the position of tuyeres is 0.6-1.2 meter height above working bottom.
4. **Slag Hole:** There is a provision of slag hole in the cupola furnace through which the slag can be removed at regular interval and the position of slag hole 25 mm below the centre of tuyeres.
5. **Wind Box:** It is a device through which the air is delivered to the tuyeres for combustions purpose of fuel. It is just like a duct made by steel and mounted on the outer shell of cupola.
6. **Blower:** Blower is used to delivered a high pressure air to the wind box via blast pipe.
7. **Chimney:** Every cupola furnace consist a chimney or stack. Chimney is fitted at the top of the steel shell.
8. **Charging Door:** As the name depict it is one of the type door. Through this door the metal charge, coke, flux are feed into the furnace and the position of charging door is 3-4 m above the tuyeres.

10.2.2 Cupola Zones

Cupola zones are shown in Fig. 10.1 they are different type:

1. **Combustion Zone Oxidation Zone:** This zone is basically 1.5 m 3.0 m above the tuyeres. A large amount of heat is generated due to exothermic reaction.

$$C + O_2 = CO_2 + \text{Heat}$$

$$Si + O_2 = SiO_2 + \text{Heat}$$

$$2Mn + O_2 = 2MnO + \text{Heat}$$

$$Fe + O = FeO + \text{Heat}$$

2. **Reducing Zone:** Position of reducing zone is just above of combustion zone and reducing zone is upto the top of coke bad. CO_2 and CO reduction encountered in this zone and it has the temperature of 1200°C.

$$CO_2 + C(\text{coke}) - 2CO - \text{heat}$$

3. **Melting Zone:** Region of melting zone starts at the first layer of metal charge and melting zone extend upto the 9 m heights. Temperature of this zone is approximate 1600°C.

   Following reaction take place in melting zone.

$$3Fe + 2CO - Fe_3C + CO_2$$

4. **Preheating Zone:** Preheating zone starts from the end of the melting zone and its goes up to the bottom of the charging door. In this zone, there is an

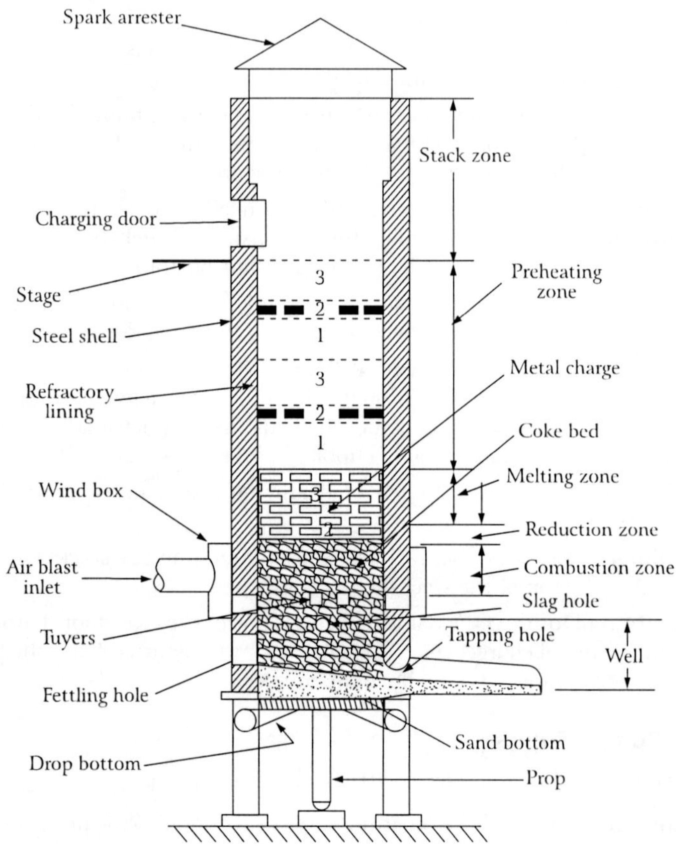

Fig. 10.1 Cupola furnace

alternative layer of coke, limestone and metal. Charging metals are introduced in this zone and get preheat. Temperature of this zone is round of 1100°C. Preheated charge door gradually down in the melting zone.

5. **Stack Zone:** Region of this zone starts from the charging zone and extends up to the top of the cupola. Gases produced during Cupola operations are escaped in the atmosphere by this zone.

10.2.3 Efficiency of Cupola Furnace

$$\text{Melting efficiency} = \frac{\text{Heat utilized in preheating, melting and super heating}}{\text{Heat input due to heat in the coke + Heat evolved due to oxidation of Xc, Fe, Si, M + Heat in air blast}}$$

10.2.4 Cupola Operations

Following steps are involved in cupola operations:

- Preparation of cupola
- Firing of cupola

- Charging of cupola
- Melting of charge
- Opening of air blast
- Shut downing the cupola

(i) **Preparation of Cupola:** First of all open the drop bottom of cupola for cleaning the slag (unburned coke, etc.) and repair from the inside of refractory lining if any. Now bottom door is closed and a sand bed is prepared. Sand bed having the some degree of taper towards the tap hole. Slag hole is also make to remove the slag.

(ii) **Firing of Cupola:** Cupola is started three hours before the melting of metal as it requires the heat for heat up. For firing purpose soft and dry wood piece is placed on the sand bed and now coke is placed over the wooden piece up to tuyeres now wooden piece is ignited via tap hole. Air for combustion purpose enter through tuyeres. Once the coke is proper burning additional amount of coke is feed to the desired height (approximately 75 mm).

(iii) **Charging of Cupola:** Charging of cupola means an alternative layer of limestone, coke, pig iron up to the charging door in cupola operation flux is added to remove the impurities as well as to prevent from oxidation. Generally limestone ($CaCO_3$) is used as a flux. As all flux are basic by nature hence cannot be added in greater quantity they can be add up to 2-4% by weight of metal and fuel used for charging of cupola should be low sulphur grade.

(iv) **Opening of Air Blast:** As the soaking period is over, air blast is open. Tap hole is off to accumulate the sufficient amount of metal.

(v) **Shutdown of Cupola:** Shutdown of cupola means closing or off the cupola furnace. Once the melting process is over now no need to feed the more feeding charges. Supply of air is cut off and prop is removed, dropped cupola remains quenched with water immediately and the metal and coke are recovered from the same for use the next time heat.

10.2.5 Advantages of Cupola Furnace

(i) Melting cost is low.
(ii) Better control of chemical composition.
(iii) Very easy to temperature control.
(iv) High degree of efficiency.
(v) Low initial installation cost as compared to other furnace of same capacity.
(vi) Suitable for mass production of cast iron.
(vii) Suitable for burning of inferior quality of fuel.
(viii) Less floor area required (as it is a vertical furnace) as compared to same capacity of other furnace.
(ix) Simple design and simple construction.
(x) Maintenance is easy.

10.2.6 Limitations of Cupola

(i) Cupola is not suitable to produce iron below 2.8%. Carbon hence duplex process is used to produce the white cast iron (% C < 2.7).
(ii) Difficult to control the closer dimensional temperature.

Electric furnace

Electric furnaces are the another most important and major method for melting in steel and iron based foundries. Now-a-days in modern foundries electric furnaces are widely used. One of the unique characteristic of electric furnace that good temperature control and it is suitable for mostly all types of metals and alloys.

10.2.7 Characteristics of Electric arc Furnace

1. High degree of temperature control
2. Less melting losses
3. High degree flexibility of operation
4. Greatest control over melting condition and pollution due to the absence of fuel consumption.

Electric furnaces are being used for melting and refining all types of steel, tool steel, etc. electric furnace is also used for melting to cast steel and plain steel. **Electric furnaces are categories in three parts.**

1. Direct arc furnace
2. Indirect arc furnace
3. Induction furnace.

10.2.7.1 Direct arc Furnace

Direct arc furnace is used to melting and refining the metal at contact temperature. Most well known direct arc furnace is **Heroult furnace.** Direct arc furnace is used to melting the material upto 3 to 20 tonnes, and current used for transformer vary from 850 kVA to 30,000 kVA depending upon the size of furnace.

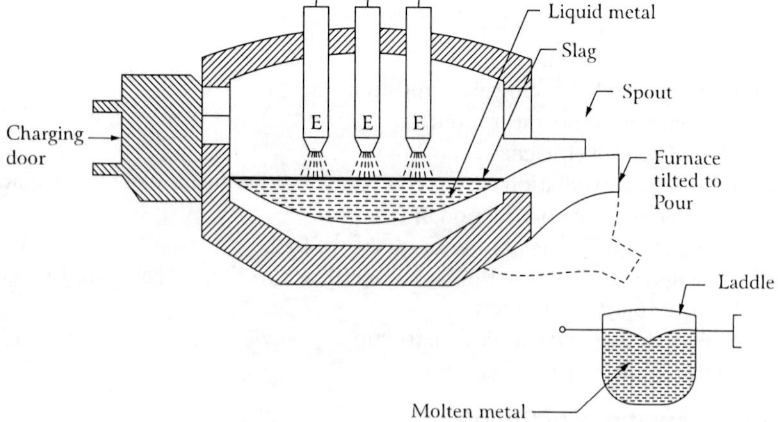

Fig. 10.2 Direct arc furnace

A direct arc furnace is shown in Fig. 10.2. Which consist a cylindrical steel shell having refractory bricks lining and top cover/roof containing three holes through which the three electrodes can be lowered or rised automatically by suitable electric or electrically controlled device or by hydraulic control with the support of servomotors. Distance between the molten change and electrodes is controlled automatically. Once metal charge is converted into liquid metal, furnace can be tilted at angle by some arrangement. Electrodes are surrounded by cooling box.

These are two spouts opposite to each other, one spout is used to remove the slag where as other spout is used to pouring the molten metal. There is also a provision of charging door which is known as pouring spout.

Furnace works on the principle that heat is produced when resistance is offered to flow of electricity. In direct arc furnace metal in the charge that provides the resistance to flow the current slag offers resistance to the current flow, when metal is in the molten state.

Advantages of direct arc furnace

(a) Rate of heat application can be controlled closely
(b) Furnace atmosphere above molten metal can be easily controlled
(c) Most alloying element such as Cr, W, Ni, etc. can be recovered from the scrap with very little loss.
(d) Steel may be made direct from steel scrap and pig iron, using the same method as for open hearth furnace.

10.2.7.2 Indirect arc Furnace

Indirect arc furnace is used to melting the all types of material but it is specially design for melting the non-ferrous metals such as ***Al***, ***Cu***, ***Brass***, ***Bronze***, etc. In this type of furnace carbon electrode are mounted in horizontal position and heat is generated by arcing between the electrodes. This type of furnace basically consist a barrel shaped drum, mounted horizontally and can be rotated at angle upto 180°. There are refractory lining inside the drum.

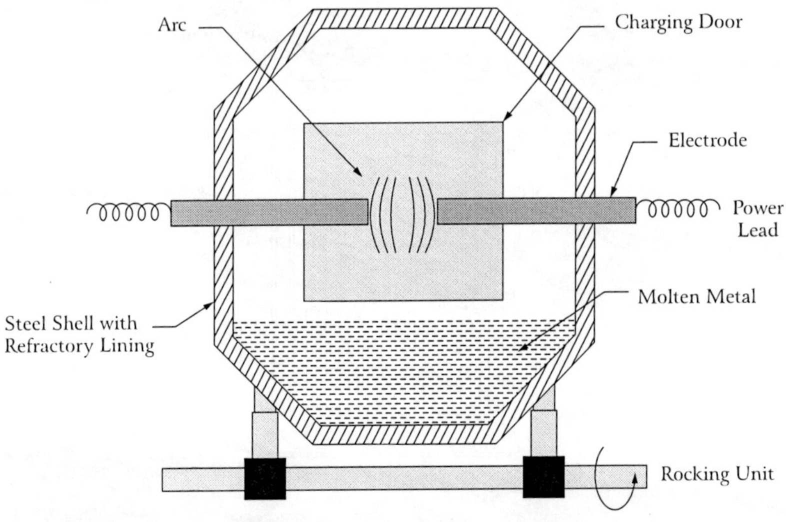

Fig. 10.3 Indirect arc furnace.

Indirect arc furnace ensure good mixing of melt and heat recovery from the refractory. Single phase power is required to run this type of furnace and best suitable to melting from 100 Kg to 5 tonnes.

10.2.7.3 Induction Furnace

Induction furnace is quite different as compared to arc furnace. In induction furnace all heat is generated in the charge it self. This type of furnace consist a crucible or a monolithic lining which is surrounded by water cool copper coil and it is all inside the shell. Fire bricks are placed at the bottom portion of shell, a high frequency current of 1000 cycle/second is supplied by a motor generator set, a spark-gap converter or an electronic solid state converter, by induction , secondary currents, called eddy current, are produced in the crucible charge. In this process melting is quite fast, so much so that there is only a slight loss of easily oxidized elements.

Advantages of induction furnace

(i) Melting process is simple.
(ii) Temperature can be easily control and same can be control quicker.
(iii) Melting cost is very low.
(iv) Process is also very suitable for melting of small quantity of any element composition.
(v) Metal losses are minimum.
(vi) metal and alloys melted in induction furnace are of high quality and free from gases.

Limitations of induction furnace

(i) Initial cost of furnace is high
(ii) As melting and quicker and fast hence bath sampling is difficult.

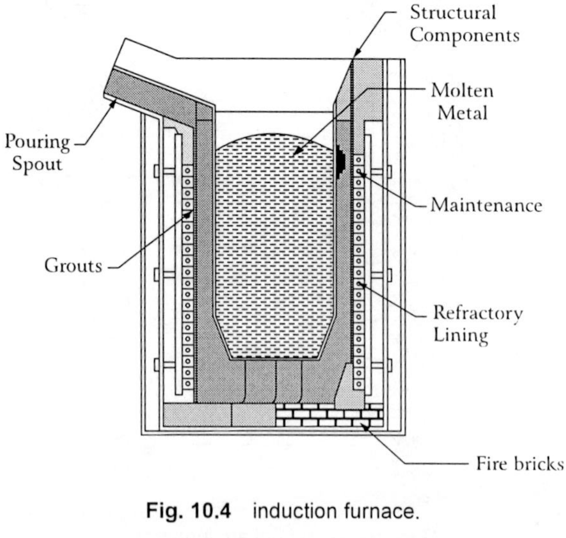

Fig. 10.4 induction furnace.

❑❑❑

11 CHAPTER

OTHER CASTING PROCESSES

11.1 PERMANENT MOULD CASTING

Permanent mould casting is also refered as **gravity die casting.** This permanent mould casting process consist a metal mould, process is quiet different from sand casting.

Metal mould casting have the following **advantages over sand casting:**

(i) Metal mould can produce a component with closer dimensional tolerance and smooth surface finish.

(ii) Metal mould can be housed repeatedly to produce different numbers of casting.

(iii) Process can be easily mechanized and automated to reduce the casting time and casting defect.

(iv) Metal mould casting process is suitable for mass production.

(v) Thermal conductivity of metal mould material facilitates the cooling rate and solidification resulting improve the mechanical properties.

11.1.1 Construction of Metal Mould

This type of mould is made by two halves (dies) which are hinged by a dowel pin for faster opening and closing. To creat a hole in the casting steel cores or grey cast iron cores are used. These cores are the design as the part of casting. Once the molten metal is solidified and cooled in the mould cavity then mould cavity opened by some hand lever, mechanical device or a hydraulic device.

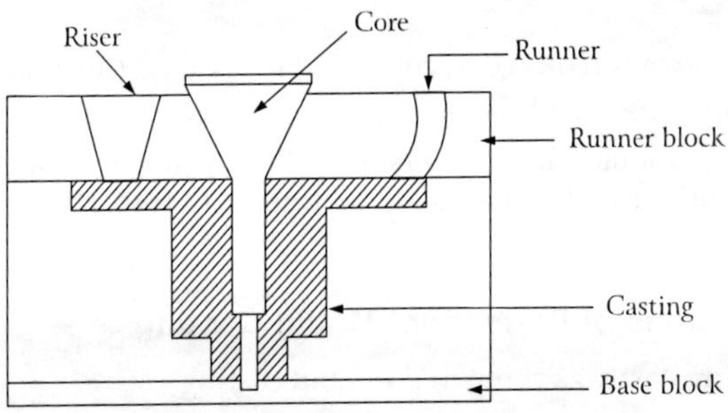

Fig. 11.1 Metal mould

11.2 DIE FOR GRAVITY CASTING OR PERMANENT CASTING

Generally dies (half) used for this process are in a pair forms. Dies are equipped to feeding and ejecting the molten metal. They possesses a hole for escaping the gases. Die may be air cooled or water cooled.

Proper taper allowances provided in the die surface for easy removal of casting. In order to increases the production rate, die may be water cooled or air cooled. **Four types of common dies are used**:

(i) **Two Piece Die** A two piece (part) die is split along a vertical line passing through the die casting.

(ii) **Recessed Die** Die is made hollow to reduce the weight and enhance the thermal balance. In die, there is a provision of ribs to provide the strength and stiffness.

(iii) **Hinged Die** Two halves open about a hinge. Small die for simple and shallow. Casting are easily opened and closed manually if they are hinged.

(iv) **Multi Stage Die** This type of die is made out of the several pieces. Which are either joined together firmly or kept loose to facilitate sliding or lifting. This types of die are basically suitable for complex shapes which are very difficult to obtained by two piece die.

11.2.1 Die Material

Selection of die material basically depends upon the required dimensional accuracy. Die is commonly made by high quality **cast iron** or **die steel, graphite, copper, aluminium**, etc. Steel die should stabilised first by heating at 550°C for 10 hours and then cooling slowly.

11.2.2 Advantages of Permanent Mould Casting

(i) Process is quit faster as compared to other casting process. Product requires 2-4 minutes for completing.

(ii) Less skill man power is required as compared to sand casting process.

(iii) Surface finish by this process is better than sand casting but inferior than pressure die casting.

(iv) Less chance of casting defect as compared to sand casting.

(v) Dimensionally tolerance of the ± 0.51 per 100 mm can be achieved by this process.

(vi) Less floor area is required as compared to sand casting for same capacity.

(vii) Fine grain structure of casting is possible by this process.

(viii) Permanent moulds have a very long life as much as one mould can be conveniently used for producing between 3,000 to 10,000 casting in cast iron.

11.2.3 Limitations of Permanent Mould Casting

(i) Moulds are costly as compared to sand casting.

(ii) Method is not suitable for a small quantity production.

(iii) Maximum weight of casting 120 kg can be made by this process.

(iv) Process is not suitable for very high melting alloys.

(v) Complicated shapes of casting may offer problem in design of mould.

(vi) Metal mould always likely to provide a chilling effect on the casting surface and **call** for greater precautions in this respect.

11.2.4 Metals and Alloys Cast in Permanent Mould

All cast metals, with the expception of steel, can be cast by the permanent mould method, which ensure better mechanical properties, better surface finish, resistance to corrosion. **Aluminium** alloys which are used for **piston, gear box housing, cylinder head, compressor part** are casted by this process.

Copper alloys which are sufficiently fluid in the molten state and used for **gear pinions, impellers** for small pumps can also be cast by this process. **Lead, zinc, gray cast iron** are casted by this process.

11.3 PRESSURE DIE-CASTING

Pressure die-casting is one of the fastest casting process with a very high degree of accuracy than that of conventional sand casting process. Pressure die-casting process is a advance version of permanent mould process. Pressure die-casting process consist a die in which the molten metal is allowed to eject at high pressure and allowed to solidified under extremely applied pressure. Process is suitable only for certain non-ferrous metal of small size.

Main advantage of pressure die-casting is the speed and surface finish of casting with better dimensional tolerance of casting. Process can be easily mechanized.

11.3.1 Type of Die-casting Process

They are basically four types:

(i) Hot chamber die-casting process,

(ii) Cold chamber die-casting process,

(iii) Vaccum die-casting process,

(iv) Air blow or goose-neck process.

Machine used in all the die-casting process have some common functions:

(i) Closing of two halves (part) of die.

(ii) Moulding the two halves of the die firmly together during pouring and solidification process.

(iii) Forcing the molten metal at high pressure into die.

(iv) Opening out the two halves of the die.

(v) Ejecting the casting.

1. Hot Chamber Die-casting Process

Figure 11.2 shows the main parts of hot chamber die-casting machine. This machine is basically operated by a hydraulic plunger, this plunger acts inside a cylinder formed at one end of the goose-neck type casting. Furnace is fitted within the machine for melting and holding the metal. Submerged below the surface of molten metal, a plunger operates within a cylinder. When plunger is rised it uncovers an opening in the wall of cylinder through which the molten metal spill into the cylinder. Once cylinder is filled, plunger is forced downwards, pneumatically or hydraulically closing, opening and then forcing the confined metal up through a channel and nozzle into die. Die is opened and solidified die-casting ejected. Pressure and speed of injection of molten metal are different for different metals and casting.

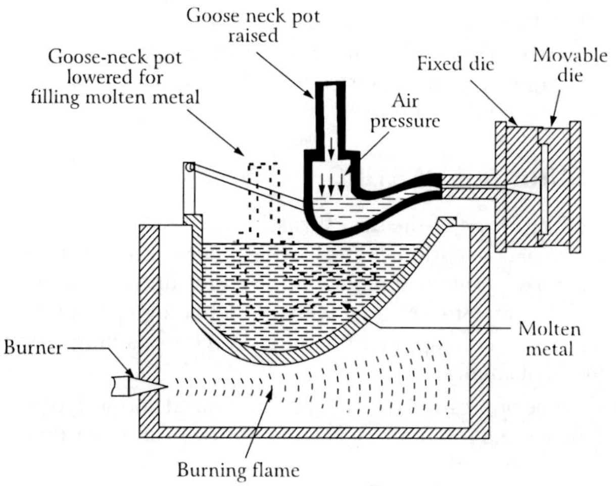

Fig. 11.2 Section through a goose neck type of air blown

These machine basically works at a pressure of **150 kg/km^2**. Casting is manually removed from the die by operator.

Hot chamber die-casting further divided into two type arrangement:

(i) Goose-neck or air injection type,

(ii) Submerged plunger type.

11.3.2 Advantages of Die-casting

(i) Less floor area is required.

(ii) Production rate is very high.

(iii) Most of non-ferrous metal can be casted by this process.

(iv) Very high degree dimensional tolerance can be achieved.

(v) Surface finish is also very high degree.

(vi) Die-casting dies retain their accuracy for a long time.

(vii) No need of skill worker.

(viii) Intricate shapes can be easily casted.

(ix) Very thin section can be casted. Minimum thickness that can be casted is 0.5 mm for Zn and 0.8 mm for Al and 1.27 mm for Mg.

(x) No further machining process is required.

(xi) Die-casting is less defective than sand casting.

11.3.3 Limitations of Die-casting

(i) Equipment initial cost and die cost is very high.
(ii) All types of metals and alloys cannot be casted.
(iii) If molten metal temperature is very high, life of die gradually decreases.
(iv) Size of casting is limited.
(v) Process is not economical for small quantity production.
(vi) Unless special precautions are adopted for evacuation of air from die cavity some air always entrapped in the casting causing porosity.
(vii) Special supervision and special maintenance is required for die.

11.3.4 Applications of Die-casting

(i) Process is suitable for many non-ferrous alloys such a **zinc, aluminium, copper, lead, tin**, etc.
(ii) Automobiles parts.
(iii) Home appliances.
(iv) Instruments.
(v) Parts of **regulator, washing machine, refrigerators**, etc.

11.4 CENTRIFUGAL CASTING

In centrifugal casting process, molten metal is poured into the mould while the casting are in revolving condition. In centrifugal casting process, centrifugal force play an important role in feeding and shaping of the casting.

Molten metal feed into the centre of mould at the axis of its rotation towards its periphery by the centrifugal force as impurities are lighter in weight are left behind at the centre. Casting is totally free from the porosity defect by the application of centrifugal force. In the centrifugal casting process large gates, cores, feeders are eliminated hence process is less expensive.

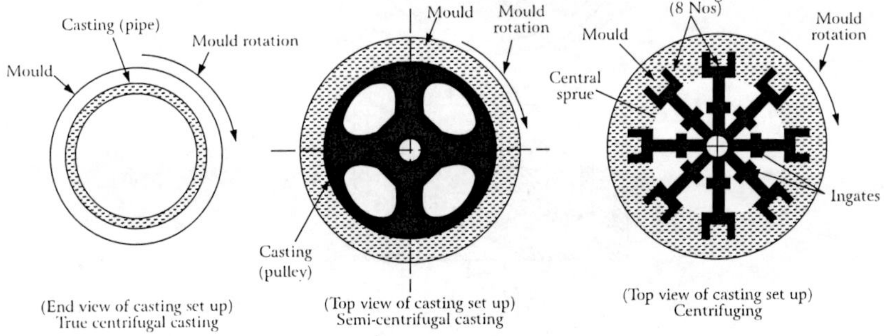

Fig. 11.3 Showing the Basic Difference between True Centrifugal Casting, Semi-centrifugal Casting and Centrifuging

11.4.1 Advantages of Centrifugal Casting

(i) Process is suitable for producing the high quality cylindrical (hollow) objects.
(ii) Usually in centrifugal casting process, no wastage of material in the form of gates.

(iii) Cast tube (hollow) produced by this process is very high degree quality.

(iv) Production is rapid with low production cost.

(v) As the centrifugal assist to feed the molten metal and also assist the directional solidification of molten metal resulting precipitating the gases. Thus casting is free from the gas and shrinkage cavity and porosity defect.

(vi) No need to use the cores for producing the hollow cylinder as rotating mould causes the liquid metal to rotate to periphery by rotating force.

11.4.2 Limitations of Centrifugal Casting

(i) All alloys cannot be casted by this process.

(ii) Inner surface diameter may be inaccurate.

(iii) Internal surface of casting contaminated with segregates and non-metallic inclusion and thus increased finishing allowance will be required.

11.4.3 Methods of Centrifugal Casting

They are basically of three types:

(i) Centrifuging,

(ii) Centrifugal casting,

(iii) True centrifugal casting.

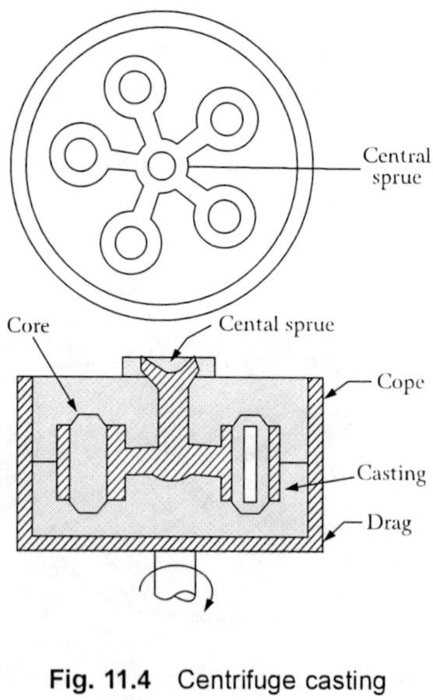

Fig. 11.4 Centrifuge casting

1. Centrifuging

In this casting process, several small sized patterns are placed in the mould and molten metal is feed into these cavity by radial gates from the centre. Pattern may be same or different shape and size, metal from a central pouring reservoir passes to different mould cavity that are offset from the axis of rotation. Centrifuging is possible only in vertical direction.

Advantages of Centrifugal Casting

(i) Process is quick and economical than other casting process.
(ii) Use of risers, feed heads, core are totally eliminated.
(iii) Good surface finish.
(iv) Different type of smaller shapes can be casted by this process.

Application of Centrifugal Casting

Following parts are casted by this process:

Valve body, pillow blocks, value bonnets, etc.

2. Semi-centrifugal Casting

It is also known as *profited centrifugal casting*. Semi-centrifugal casting is quite similar to true centrifugal casting with the only difference that a central core is used to form the inner surface. Casting is done in mould rotating about its vertical axis. No need high speed of spinning as required in centrifugal casting. By proper gating of casting directional solidification is possible. Parts produced by this process includes **gear, track wheels, fly wheels, etc.**

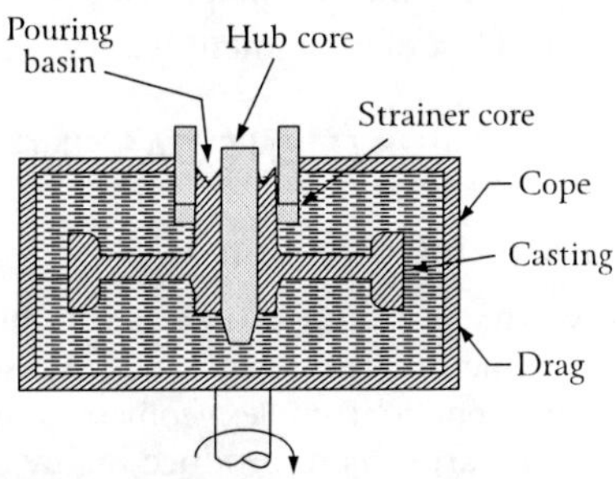

Fig. 11.5 Semi-centrifugal casting

3. True Centrifugal Casting

It is one of the most important, casting method. By true centrifugal casting process **hollow**, centrifugal parts such as **gun barrels**, **pipes**, lamp post, tubes, hollow bushes, etc. are produced as molten metal is always pushed outward due to centrifugal force hence no core is required for producing the concentric hole. In this process, molten metal is poured in rotating mould. Axis of rotation is usually horizontal but may be vertical for small object. Speed of mould rotating varies from 400-3000 r.p.m. mould may be out of steel, cast iron or graphite with coated refractory lining from inside to increase the mould life.

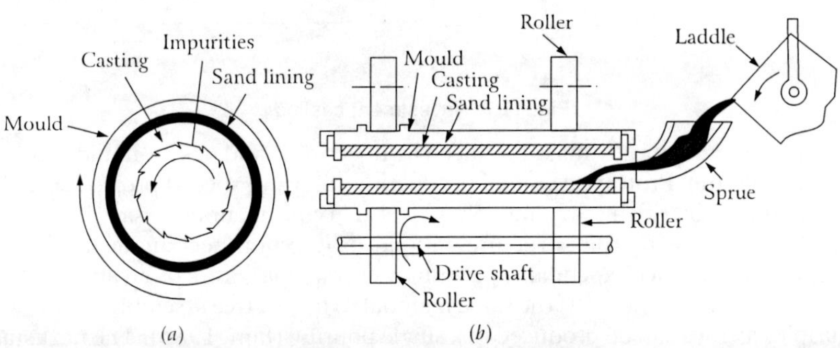

Fig. 11.6 Principle of centrifugal casting

Advantages of True Centrifugal Casting

(i) Mechanical properties of centrifugal cast object part are better as compared to other process as slag or oxides get segregated towards the centre and can easily removed by machining.

(ii) Cores are totally eliminated for producing concentric holes.

(iii) Gates, runners are eliminated which increases the casting yield.

(iv) Proper directional solidification can be obtained.

Limitations of True Centrifugal Casting

(i) Only limited shapes can be casted by this process.

(ii) Cost of equipment is very high hence suitable only for mass production.

11.5 INVESTMENT CASTING

[*UPTU 2009-10*]

11.5.1 Introduction

Investment casting produces very high surface quality and dimensional accuracy. Investment casting is commonly used for precision equipment such as **surgical equipment**, for complex geometries and for precious metals. This process is commonly used by artisans to produce highly detailed artwork. The first step is to produce and pattern or replica of the finished mould. Wax is most commonly used to form the pattern, although plastic is also used.

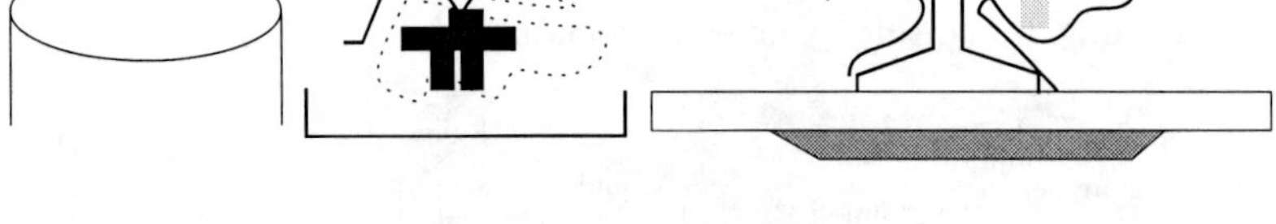

Fig. 11.7 Investment casting

Patterns are typically mass-produced by injecting liquid or semiliquid wax into a permanent die. Prototypes, small production runs and special projects can also be undertaken by carring wax models. Cores are typically unnecessary but can be used for complex internal structures. Rapid prototyping techniques have been developed to produce expendable patterns. Several replicas are often attached to a gating system constructed of the same material to form a tree assembly. In this way, multiple castings can be produced in a single pouring (Jain, 1986). The next stage is to create a one-piece destroyable mould around the pattern. This mould is built up around the wax pattern in stages by alternately coating the assembly with a specially formulated cut heat resistant refractory slurry mixture and then applying a granulated refractory 'stucco' shell.

The initial coats use a fine powder, which creates a vary smooth and dimensionally accurate negative of the pattern. Subsequent coals use a coarser refractory material to build up sufficient thickness. This material hardens around the assembly at room temperature. This investment shell casting method is the more common process.

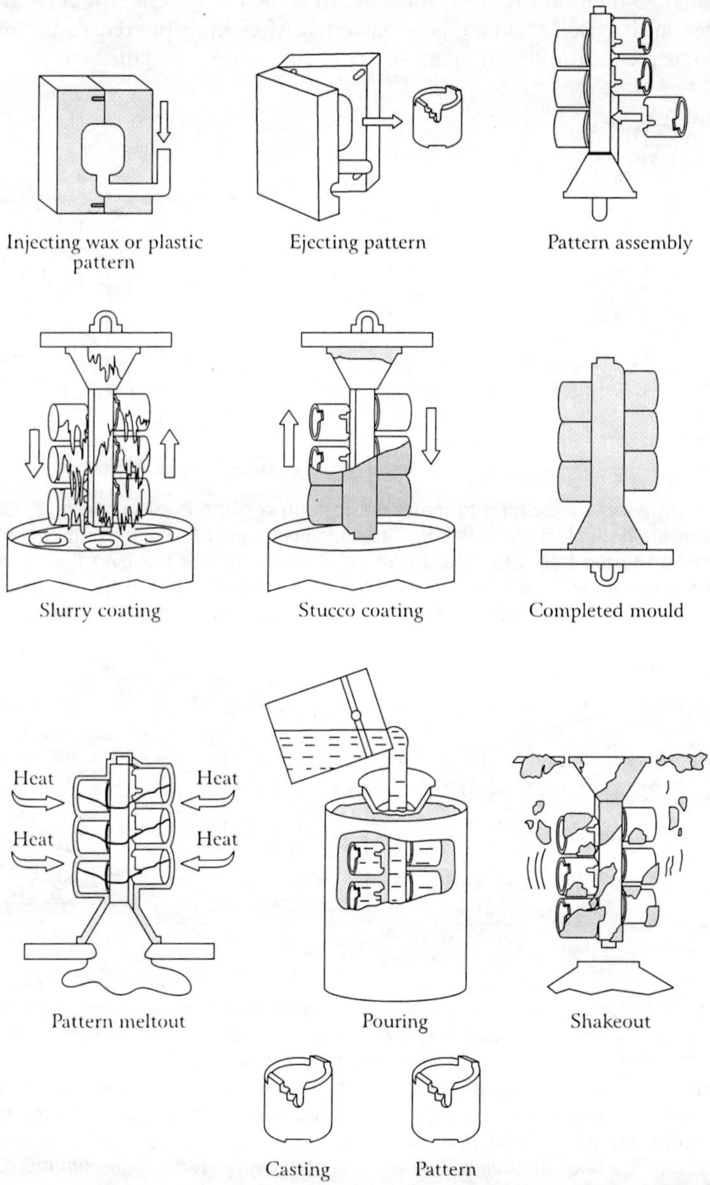

Fig. 11.8 Lost Wex Process or Investment casting (from ASM, Metals Handbook 1988)

An alternative process is to use an investment flask where sand is packed around the mould. This can be desirable where additional mould strength is required and also allows the casting size to be increased. In both shell and flask casting, the pattern is removed from the mould prior to the pouring stage. The mould is inverted and heated to melt and remove the wax. In some operations, the melted wax is recovered and reused to make new patterns. After multiple reuses the material needs to be reconditioned to maintain its purity, or disposed of.

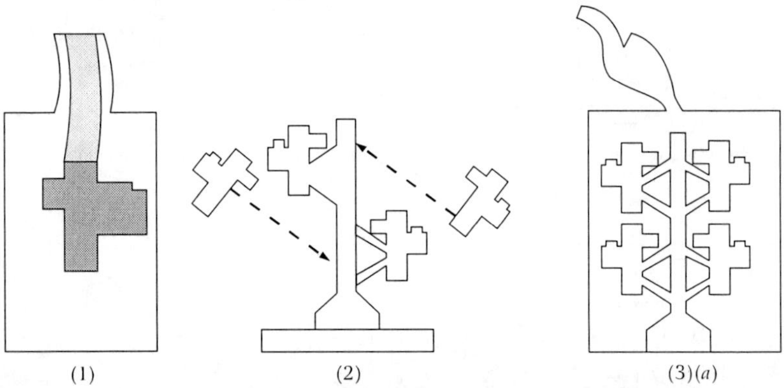

1. Pattern wax is injected into metal die to form a dispensable pattern. 2. Gating Patterns are gated to a sprue to form a tree or cluster and base of the pattern material is attached to the tree. 3(a). Solid Pattern: A metal flask is put over the cluster of pattern and sealed to a base plate to form a container. A hard setting moulding material is then poured into the flask completing the pattern cluster completely.

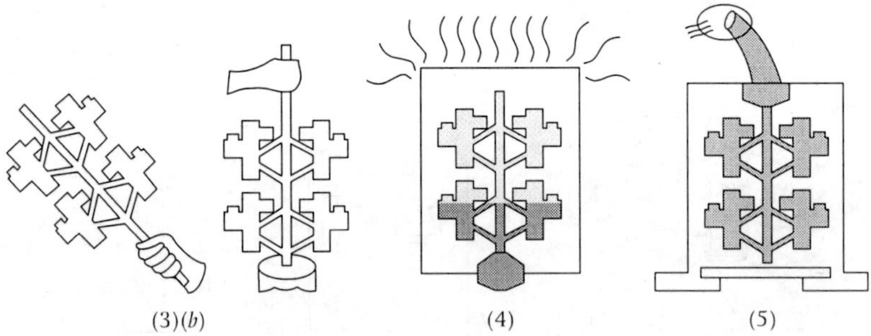

3(b) Shell Pattern: The ceramic shell is formed by dipped the clustered pattern in a ceramic slurry and then sprinkling them with a refractory grain. This procedure is repeated until the required thickness of the shell is achieved.
4. Pattern removal: the flask and shells are then placed in ovens to bake at a moderate temperature in order to slowly melt the embedded patterns. Cavity left in the mould will receive the poured molten metal.
5. Casting: the flask or shell is inverted and metal in poured into the hot mould.

Fig. 11.9 Investment casting process

The mould is then heated in an oven to remove any residual wax and to further cure and harden the mould. The temperature is raised to 980°C prior to pouring. This is a time and energy-consuming process: total heating time, from wax removal to pouring, can take up to 15 hours (Jain, 1986). Molten metal is then poured into the central cavity and flows into the individual moulds. After the metal has cooled, the mould material is removed.

Because of its very high dimensional accuracy the **Investment Casting Process** can achieve a net-shape cast requiring little or no machining. Great care is taken in the pattern making stage to remove any mould lines because it is more cost effective to remove unwanted material from the wax model than from the final cast.

11.6 ADVANTAGES OF INVESTMENT CASTING

- Very high dimensionals accuracy and surface finish can be achived by this process.
- Process is suitable for both ferrous and nun-ferrous precision pieces.
- Allows flexibility of design.
- The process can be adapted for mass production.
- Cores are typically eliminated.
- Can virtually eliminate the need for machining.
- Every high metal yields.
- Can produce castings that are impossible or difficult to produce by other casting methods and machining processes.
- Can be cost effective for repetitive casting and specialist jobbing applications.

11.7 LIMITATIONS OF INVESTMENT CASTING

- Investment casting is suitable for small size objects.
- Investment casting process is slow as compared to other casting process.
- Capital investment and operating cost of investment casting are very high in comparison to other casting process.
- Process became more difficult by using the core making.
- Pattern, die making cost very high and process requires some special tools and equipment.

11.8 APPLICATIONS OF INVESTMENT CASTING

Investment casting process is used to produce the following parts:

- Surgical parts.
- Sewing machine parts.
- Locks.
- Turbine: blades.
- Gears.
- Jewellery.
- Wear resistant.

As well as corrosion resistance parts which are used in chemical industries, diesel engine parts, etc.

❑❑❑

12

CHAPTER

FLASKLESS MOULDING

12.1 INTRODUCTION

Flaskless moulding is based on the principal that when high strength moulding sand (not less than 190 KPa in green state) properly bounded used and moulding sand ramming done at 15 MPa pressure, sand posseses a good strength to maintain the integrity of the mould.

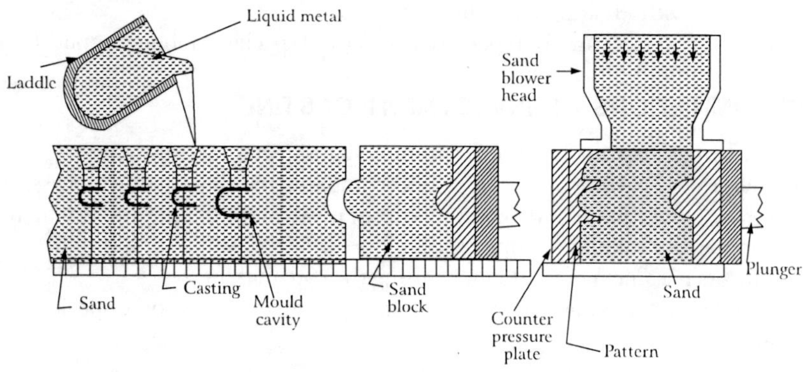

Fig. 12.1 Automatic flaskless moulding

12.2 PRINCIPLE

Basically this process consist vertical parted flaskless moulding machine. Two pattern plates, out of them one pattern plate is fixed with counter pressure plate and other plate is fixed with pressure plate.

First sand is feed through blower into moulding chamber under pressured air. Enterance of chamber is chopped off and sand feed under a pressure of 10-15 MPa by the action of plunger. Plunger now pushed sand block, forming a mould cavity. In such a way, a continuous moulding row if produced.

Advantages

- Light quality of casting can be produced.
- Process is automatic hence labour is eliminated.

- No casting flask is required.
- Process is suitable for mass production.
- High strength of mould can be achieved.

Applications

- Pipe fitting of malleable cast iron.
- Valve body of brass.
- Gas stove bodies, etc.

Fig. 12.2 Flask less moulding

Fig. 12.3 Molding [flask-less molding] machines

OBJECTIVE QUESTIONS

1. Max. quantity of moisture in moulding sand is
   (*a*) 2% (*b*) 4%
   (*c*) 8% (*d*) 12%
2. Draft is a ________ in casting.
   (*a*) machining allowance
   (*b*) shrinkage allowance
   (*c*) taper to facilitate its removal from mould
   (*d*) increases in size of cavity due to shrinkage
3. Gate commonly used in foundry is
   (*a*) top gate (*b*) bottom gate
   (*c*) side gate (*d*) step gate
   (*e*) parting gate
4. Which is not a casting defect?
   (*a*) decarbonization (b) blow hole
   (*c*) hot tear (*d*) scrab
5. To avoid the blow hole in a casting, mould must have adequate
   (*a*) hot strength (*b*) dry strength
   (*c*) permeability (*d*) green strength
6. Pistons are made by
   (*a*) centrifugal casting (*b*) shell moulding
   (*c*) permanent mould casting (*d*) continuous casting
7. Misrun defect occurs in casting due to
   (*a*) improper alignment of the mould flask
   (*b*) insufficient fluidity of the molten metal
   (*c*) absorption of gases by the liquid metal
   (*d*) very high temperature (pouring) of metal
8. Gas turbine blades are made by [*UPSC*]
   (*a*) die casting (*b*) shell moulding
   (*c*) centrifugal casting (*d*) lost wax casting
9. Carburetors body is made by
   (*a*) shell moulding (*b*) die casting
   (*c*) investment casting (*d*) centrifugal casting
10. Ornament (Jewellery) are cast by the following process
   (*a*) gravity casting (*b*) die casting
   (*c*) continuous casting (*d*) pressed casting
11. Investment casting is used for
   (*a*) complex shape (*b*) mild steel parts
   (*c*) mass production (*d*) difficult shapes
12. Large size of water pipe (hollow) are made by
   (*a*) sand casting (*b*) centrifugal casting
   (*c*) continuous casting (*d*) by joining (welding) [IES 2009]
13. Freezing time according to caine's, equation is
   (*a*) $(A_c/A_r)/(V_d/V_r)$ (*b*) $(A_r/V_r)\ (A_c/V_c)$
   (*c*) $(A_c/V_c)/(A_r/V_r)$ (*d*) $(V_c/A_c)\ (V_r/A_r)$

14. Function of riser is to
(*a*) feed the casting until complete solidification take place
(*b*) feed molten metal from pouring basin of gate
(*c*) feed the casting at a rate consistent with rate of solidification
(*d*) act as a reservoir for molten metal

15. When a pattern is made in three parts, the top part is known as a
(*a*) drag
(*b*) cheek
(*c*) cope
(*d*) any one of the above
(*e*) none of the above

16. When a pattern is made in three parts, the bottom part is known as a cope.
(*a*) True
(*b*) False

17. For smoothing and cleaning out depressions in the mould, a .... is used.
(*a*) slick
(*b*) lifter
(*c*) swab
(*d*) Bellow

18. Metal pattern require .... draft allowance than wooden patterns.
(*a*) more
(*b*) less

19. A taper provided on the pattern for its easy and clean withdrawl from the mould is known as
(*a*) machining allowance
(*b*) draft allowance
(*c*) shrinkage allowance
(*d*) distortion allowance

20. When a pattern is made in three parts, the .... is known as a drag.
(*a*) bottom part
(*b*) middle part
(*c*) top part

21. Shrinkage allowance is made by adding to external dimensions and subtracting from internal dimensions.
(*a*) Agree
(*b*) Disagree

22. Riddle is used for
(*a*) smoothing and cleaning out depressions in the mould
(*b*) cleaning the moulding sand
(*c*) moistening the sand around the edge before removing pattern
(*d*) reinforcement of sand in the top part of moulding box
(*e*) repairing and finishing the mould

23. Slick is used for repairing and finishing the mould.
(*a*) Correct
(*b*) Incorrect

24. Swab is used for
(*a*) smoothing and cleaning out depressions in the mould
(*b*) cleaning the moulding sand
(*c*) moistening the sand around the edge before removing pattern
(*d*) reinforcement of sand in the top part of the moulding box
(*e*) repairing and finishing the mould

25. Green sand is a mixture
(*a*) 30% sand and 70% clay
(*b*) 50% sand and 50% clay
(*c*) 70% sand and 30% clay
(*d*) 90% sand and 10% clay

26. Loam sand is a mixture of
(*a*) 30% sand and 70% clay
(*b*) 50% sand and 50% clay
(*c*) 70% sand and 30% clay
(*d*) 90% sand and 10% clay

27. The property of sand due to which it evolves a great amount of steam and other gases is called

(*a*) collapsibility
(*b*) permeability
(*c*) cohesiveness
(*d*) adhesiveness

28. A Jolt machine is used to
(*a*) ram the sand harder at the pattern face with decreasing hardness, towards the back of the mould
(*b*) ram the sand harder at the back of the mould and softer on the pattern face
(*c*) produce uniform sand hardness throughout the mould
(*d*) produce uniform packing of sand in the mould
(*e*) give uniform sand hardness throughout the mould
(*f*) draw the pattern from the mould

29. The property of sand due to which the sand grains stick together is called
(*a*) collapsibility
(*b*) permeability
(*c*) cohesiveness
(*d*) adhesiveness

30. The purpose of a gate is to
(*a*) deliver molten metal into the mould cavity
(*b*) act as a reservoir for the molten metal
(*c*) feed the molten metal to the casting in order to compensate for the shrinkage
(*d*) deliver molten metal from pouring basin to gate
(*e*) none of the above [*IES 2007*]

31. The sand used for making cores is
(*a*) green sand
(*b*) dry sand
(*c*) loam sand
(*d*) oil sand
(*e*) parting sand
(*f*) none of the above

32. A sand employed on the faces of the pattern before moulding is called
(*a*) green sand
(*b*) dry sand
(*c*) loam sand
(*d*) oil sand
(*e*) parting sand
(*f*) none of the above

33. The adhesiveness is the property of a sand due to which
(*a*) it evolves a great amount of steam and other gases
(*b*) the sand grains slick together
(*c*) it cling to the sides of a moulding box
(*d*) none of the above

34. In order to deliver molten metal from pouring basin to gate, a .... is used.
(*a*) riser
(*b*) sprue
(*c*) core
(*d*) runner

35. A pouring basin acts as a reservoir for the molten metal.
(*a*) Agree
(*b*) Disagree

36. The purpose of a riser is to
(*a*) deliver molten metal into the mould cavity
(*b*) act as a reservoir for the molten metal
(*c*) feed the molten metal to the casting in order to compensate for the shrinkage
(*d*) deliver the molten metal from pouring basin to gate

37. Cores are used to [*PH.D J.M.I*]
(a) form internal cavities in the casting
(*b*) improve mould surface
(*c*) form a part of a green sand mould
(*d*) all of the above
(*e*) none of the above

38. In a die casting method, the molten metal is forced into mould under high pressure.
(*a*) Correct
(*b*) Incorrect

39. When a hole or cavity to be cored is not in line with the parting surface, then a .... is used.
(*a*) horizontal core
(*b*) vertical core
(*c*) drop core
(*d*) balanced core

40. In a hot chamber die casting machine
(*a*) melting pot is separate from the machine
(*b*) melting pot is an integral part
(*c*) melting pot may have any location
(*d*) high temperature and pressure is used

41. In a cold chamber die-casting machine
(*a*) melting pot is separate from the machine
(*b*) melting pot is an integral part of the machine
(*c*) melting pot may have any location
(*d*) low temperature and pressure is used

42. In permanent mould casting method
(*a*) molten metal is poured in a metallic mould, retained in the mould long enough for the outer skin to solidify and finally mould is turned over to remove molten metal still in molten condition
(*b*) molten metal is poured and allowed to solidify while the mould is revolving
(*c*) molten metal is forced into mould under high pressure
(*d*) none of the above

43. In a ......, the molten metal is poured and allowed to solidify while the mould is revolving.
(*a*) die-casting method
(*b*) slush casting method
(*c*) permanent mould casting method
(*d*) centrifugal casting method

44. When the molten metal is fed in the cavity of a metallic mould by gravity, the method of casting is known as
(*a*) die-casting method
(*b*) slush casting method
(*c*) permanent mould casting method
(*d*) centrifugal casting method

45. The casting method adopted for ornaments and toys of non-ferrous alloys, is
(*a*) permanent mould casting
(*b*) slush casting
(*c*) die-casting
(*d*) centrifugal casting
(*e*) investment casting
(*e*) any one of the above

46. Gas, iron and steel pipes are produced by
(*a*) slush casting
(*b*) investment casting
(*c*) true centrifugal casting
(*d*) die-casting

47. In a cold chamber die-casting machine, a measured quantity of molten metal is brought in a ladle from the melting pot to a chamber and forced into the closed die section by applying hydraulic pressure upon the plunger.
(*a*) Right
(*b*) Wrong

48. Which of the following statement is wrong?
(*a*) The hot chamber die-casting machine is used for casting zinc, tin, lead and other low melting alloys,
(*b*) The cold chamber die-casting machine is used for casting aluminium, magnesium, copper-base alloys and other high melting alloys.
(*c*) The castings produced by centrifugal casting method have open and coarse grained structure.
(*d*) all of the above
(*e*) none of the above

49. In a centrifugal casting method
    (*a*) core is made of sand
    (*b*) core is made of ferrous metal
    (*c*) core is made of non-ferrous metal
    (*d*) core may be made of any metal
    (*e*) no core is used

[*IES 2007*]

50. In a hot chamber die-casting machine
    (*a*) ferrous alloys with low melting temperature are casted
    (*b*) ferrous alloys with high melting temperature are casted
    (*c*) non-ferrous alloys with low melting temperature are casted
    (*d*) non-ferrous alloys with high melting temperature are casted

51. Shift is a casting defect which
    (*a*) results in a mismatching of the top and bottom parts of a casting
    (*b*) results in general enlargement of a casting
    (*c*) occurs near the ingates rough lumps on the surface of a casting
    (*d*) occurs as sand patches on the upper surface of a casting
    (*e*) results excessive smooth depressions on the outer surface of a casting

52. A casting defect which results in general enlargement of a casting is known as
    (*a*) shift
    (*b*) sand wash
    (*c*) swell
    (*d*) scab
    (*e*) blow hole
    (*f*) none of the above

53. A casting defect which occurs near the ingates as rough lumps on the surface of a casting is known as
    (*a*) shift
    (*b*) sand wash
    (*c*) swell
    (*d*) scab
    (*e*) blow hole
    (*f*) none of the above

54. The operation of cutting a number of holes evenly spaced in a regular pattern on a sheet of metal is called *perforating*.
    (*a*) Agree
    (*b*) Disagree

55. In a cold chamber die-casting machine only non-ferrous alloys with .... are casted.
    (*a*) low melting temperature
    (*b*) high melting temperature

56. Blow holes are casting defects which
    (*a*) results in a mismatching of the top and bottom parts of a casting
    (*b*) results in general enlargement of the casting
    (*c*) occurs near the ingates as rough lumps on the surface of a casting
    (*d*) occurs as sand patches on the upper surface of a casting
    (*e*) none of the above

57. In a gating system, the ratio 1:2:4 represents [*GATE 2010*]
    (*a*) sprue base area, runner area, ingate area
    (*b*) runner area, ingates area, casting area
    (*c*) pouring basine area, ingate area, runner area
    (*d*) spure base area, ingate area, casting area

58. Match the item in Column (I) and Column (II) [*GATE 2009*]

Column (I)	Column (II)
P. Metallic Chills	1. Support for core
Q. Metallic Chaplates	2. Reservoirs the molten metal
R. Riser	3. Control cooling of critical sections
S. Exothermic Padding	4. Progressive solidification

    (*a*) P-4, Q-1, R-2, S-3
    (*b*) P-1, Q-3 R-2, S-4
    (*c*) P-1, Q-4 R-2, S-3
    (*d*) P-3, Q-4, R-2, S-1

59. Two streams of liquid metal which are not hot enough to fuse properly result into a casting defect known as [*GATE 2009*]
(*a*) cold shut (*b*) swell
(*c*) scab (*d*) sand wash

60. While cooling a cubical casting of side 40 mm undergoes 3%, 4% and 5% volume shrinkage during the liquid state, phase transition solid state, respectively. Volume of metal compensated from the riser is [*GATE 2008*]
(*a*) 2% (*b*) 7%
(*c*) 8% (*d*) 9%

61. A 200 mm long down sprue has an area of cross-section of 650 mm^2 where the pouring basin meets the down sprue (i.e., at the beginning of the down sprue). A constant heat of molten metal is maintained by the pouring basin-molten metal flow rate is 6.5×10^5 mm^3/sec. Considering the end of down sprue to be open to atmosphere and an acceleration due to gravity of 10^4 mm/sec^2. The area of down sprue in mm^2 at its end (avoiding aspiration effects) should be [GATE 2008]
(*a*) 650 (*b*) 350
(*c*) 290 (*d*) 190

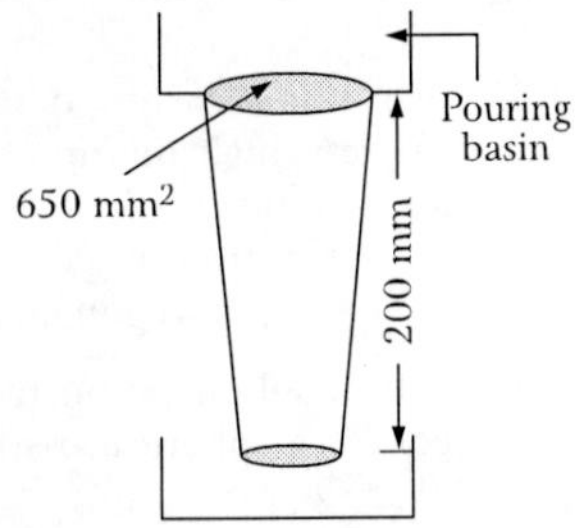

62. Which of the following engineering materials is the most suitable for hot chamber die-casting? [*GATE 2007*]
(*a*) low carbon steel (*b*) titanium
(*c*) tin (*d*) copper

63. Volume of a cube of side '*l*' and volume of sphere of radius '*r*' are equal. Both the cube and sphere are solid and of same material they are being cast, ratio of the solidification time of the cube to the same of sphere is [*GATE 2007*]

(*a*) $\left(\frac{4\pi}{6}\right)^2\left(\frac{r}{l}\right)^4$ (*b*) $\left(\frac{4\pi}{6}\right)^2\left(\frac{r}{l}\right)^3$

(*c*) $\left(\frac{4\pi}{6}\right)^3\left(\frac{r}{l}\right)^6$ (*d*) $\left(\frac{4\pi}{6}\right)\left(\frac{r}{l}\right)^2$

64. In a sand casting process, the total liquid head is maintained constant such that it is equal to the mould height time take to fill the mould with a top gate is t_A. If the same mould is filled with a bottom gate, then the time taken is t_B. Ignore the time required to fill the runner and frictional effects assume atmospheric pressure at the top molten metal surfaces the relation between t_A and t_B is

(*a*) $t_B = \sqrt{2}t_A$ (*b*) $t_B = 2\sqrt{2}t_A$

(*c*) $t_B = t_A\sqrt{2}$ (*d*) $t_B = 2t_A$ [*GATE 2006*]

65. An expandable pattern is used in
(*a*) slush casting (*b*) investment casting
(*c*) squeeze casting (*d*) centrifugal casting

[*GATE 2006*]

66. A mould has downsprue whose length is 20 cm and the cross-sectional area at the base of the downsprue is 1 cm^2. The down sprue feeds a horizontal runner leading into the mould cavity of 1000 cm^3 time required to fill the mould cavity will be
(*a*) 4.055 sec (*b*) 5.05 sec
(*c*) 6.05 sec (*d*) 7.25 sec [*GATE 2004*]

67. Match the items of List-I (equipment) with the item of List-II (process) the correct answers using the given codes.

	Equipment		Process
P.	Hot chamber machine	1.	Cleaning
Q.	Muller	2.	Core making
R.	Dielectric baker	3.	Die casting
S.	Sand blaster	4.	Annealing
		5,	Sand mixing

(*a*) P-2, Q-l, R-4, S-5 (*b*) P-3, Q-5, R-2, S-l
(*c*) P-4, Q-5, R-1, S-2 (*d*) P-4, Q-2, R-3, S-5

[*GATE 2004*]

68. Misrun is a casting defect which occurs due to
(*a*) very high pouring temperature of the metal
(*b*) insufficient fluidity of the molten metal
(*c*) absorption of gases by the liquid metal
(*d*) improper alignment of the mould flask [*GATE 2004*]

69. With a solidification factor of 0.97×10^6 s/m^2, find the solidification time (in sec) for a spherical casting of 200 mm diameter.
(*a*) 539 (*b*) 1078
(*c*) 4311 (*d*) 3233 [*GATE 2003*]

70. Match the following moulding/casting process with the product.

	Process		Product
A.	Slash casting	1.	Turbine blades
B.	Shell moulding	2.	Cylinder block
C.	Dry sand casting	3.	Hollow casting
D.	Centrifugal casting	4.	Rain water pipe
		5,	Machine tool bed

(*a*) A-2, B-3, C-1, D-5 (*b*) A-3, B-2, C-1, D-5
(*c*) A-5, B-2, C-4, D-1 (*d*) A-1, B-5, C-2, D-3 [*GATE 1992*]

71. Match List I (Product) with List-II (Casting Process) and select the correct answer using the code given below the list.

	List-I (Product)		List-II (Casting Process)
A.	Hollow status	1.	Centrifugal casting
B.	Dentures	2.	Investment casting
C.	Aluminium alloys pistons	3.	Slash casting
D.	Rocker arm	4.	Shell moulding
		5,	Gravity die-casting

Code	A	B	C	D
(*a*)	3	2	5	4
(*b*)	3	2	4	5
(*c*)	1	3	4	5
(*d*)	1	2	3	4

[*IES 2000*]

72. Which one of the following casting processes is best suited to make bigger size hollow symmetrical pipes.
(*a*) die casting (*b*) centrifugal casting
(*c*) shell moulding (*d*) investment casting

[*IES 2009*]

73. Metal moulds are used in
(*a*) dry sand mould
(*b*) die casting process
(*c*) green sand mould
(*d*) loam moulding [*IES 2006*]

74. Disposable pattern are made of
(*a*) wood
(*b*) metal
(*c*) rubber
(*d*) polystyrene [*GATE 2000*]

75. Main purpose of chaplets is
(*a*) to align the mould bones
(*b*) to provide efficient venting
(*c*) to support the cores
(*d*) to ensure the directional solidification [*IES 2001*]

76. Chills are used in casting mould to
(*a*) achieve directional solidification
(*b*) reduce the freezing time
(*c*) reduce possibility of blow holes
(*d*) increases the smoothness of cast surface [*IES 1994*]

77. Which one of the following material will require the largest size of riser for same size of casting?
(*a*) Copper
(*b*) Cast iron
(*c*) Aluminium
(*d*) Steel [*IES 1995*]

78. Directional solidification in casting can be improved by using
(*a*) Chills and Padding
(*b*) Chills and Chaplets
(*c*) Chills chaplets and Padding
(*d*) Chaplets and Padding [*IES 1995*]

79. In solidification of metal during casting, compensations for solid contraction is
(*a*) made by providing chills
(*b*) provided by the oversize pattern
(*c*) obtained by promoting direction solidification
(*d*) achieve by properly placed riser [*IES 1999*]

80. Shell moulding can be used for
(*a*) producing milling cutters
(*b*) producing thin casting
(*c*) make gold ornaments
(*d*) producing heavy and thick walled casting [*IES 2006*]

81. Consider the following advantage attributed to shell moulding.
1. Easier
2. Close dimensional tolerance
3. Low cost
4. Good surface finish, which of these attributes are correct
(*a*) 2, 3 and 4
(*b*) 1, 2 and 4
(*c*) 1, 2, 3 and 4
(*d*) 1, 3 and 4 [*IES 2010*]

82. Cast iron is produced in
(a) blast furnace
(b) open hearth furnace
(c) crucible furnace
(d) cupola furnace.

83. Properties of sand through which gases are escape in atmosphere is known as
(a) cohessiveness
(b) porosity
(c) green strength
(d) none of above.

84. Vertical surface of a pattern are provided a Taper known as
(a) shrinkage
(b) distortion
(c) machining
(d) draft.

85. Chaplets in foundry practice are used to
(a) support the cores
(a) provide efficient venting
(c) reinforce the sand in the top part of the moulding
(d) ensure the directional solidification.

86. According to Chvorinov's equation, the solidification time of a casting is proportional to
(a) V
(b) 1/V
(c) V^2
(d) $1/V^2$
Where V is the volume of casting *[IES-2006]*

87. Shrinkage rule is used for providing
(a) distortion allowance
(b) contraction allowance
(c) machining allowance
(d) draft allowance.

88. Match four correct pairs between List I and List II.

List I	List II
A. Aluminium brake shoe	1. Deep drawing
B. Plastic water bottle	2. Blow moulding
C. Stainless steel cups	3. Sand casting
D. Soft drink can	4. Centrifugal casting
	5. Impact extrusion

(*a*) A-3, B-1, C-2, D-4
(*b*) A-3, B-2, C-1, D-5
(*c*) A-5, B-2, C-4, D-1
(*d*) A-1, B-5, C-2, D-3 *[GATE 1998]*

89. Match the following moulding processes with the product.

Process	Product
A. Slush casting	1. Turbine blades
B. Shell moulding	2. Cylinder block
C. Dry sand moulding	3. Hollow casting
D. Centrifugal casting	4. Rain water pipe
	5. Machine tool bed

(*a*) A-3, B-2, C-1, D-5
(*b*) A-3, B-1, C-2, D-4
(*c*) A-5, B-2, C-4, D-1
(*d*) A-1, B-5, C-2, D-3 *[GATE 1992]*

90. Match the correct combination between the casting defects in column A with their meanings given in column B

Column A	Column B
A. Swell	1. Mismatch of top and bottom part of the casting
B. Shift	2. Rough lumps on the surface of casting
C. Hot tear	3. General enlargement of castings
D. Sand wash	4. Cracks in the casting

(*a*) A-3, B-2, C-1, D-5
(*b*) A-3, B-1, C-2, D-4
(*c*) A-3, B-1, C-4, D-3
(*d*) A-1, B-5, C-2, D-3

91. Match correct pairs between List I and List II

List I	List II
A. Rivets for air craft body	1. Forging
B. Carburettor body	2. Cold heading
C. Crank shaft	3. Aluminium based alloys
D. Nails	4. Pressure die casting
	5. Investment casting

(*a*) A-3, B-2, C-1, D-5
(*b*) A-3, B-1, C-2, D-4
(*c*) A-3, B-1, C-4, D-3
(*d*) A-3, B-4, C-1, D-2 *[GATE 1996]*

92. Match the casting defects listed in column A with their remedies (causes) given in column B

Column A	Column B
A. Shrinkage cavity	1. Hydrogen in molten metal
B. Run out	2. Too coarse grains
C. Pinhole porosity	3. Improper risering
D. Metal Penetration	4. Faulty mould flask

(*a*) A-3, B-4, C-1, D-2
(*b*) A-3, B-1, C-2, D-4
(*c*) A-3, B-1, C-4, D-3
(*d*) A-1, B-5, C-2, D-3

93. Match correct pairs between List I (process) and List II (product)

List I	List II
A. Investment casting	1. Carburetor body
B. Drop forging	2. Turbine rotor
C. Die casting	3. Cast iron pipe
D. Centrifugal casting	4. Turbine blade
	5. Connecting rods

(*a*) A-4, B-5, C-1, D-3
(*b*) A-3, B-1, C-2, D-4
(*c*) A-3, B-1, C-4, D-3
(*d*) A-1, B-5, C-2, D-3

94. Pattern adopted for those casting in which these are some portion which are structurally weak and are likely to break by force of ramming called
(a) single piece pattern
(b) Loose piece pattern
(c) skelton pattern
(d) follow board pattern.
[*GATE 2006*]

95. Wax pattern in used in
(a) centrifugal casting
(b) shell moulding
(c) die-casting
(d) investment casting.

96. Yellow color on a pattern indicates
(a) surface to be lift unmachined
(b) surface where sand needs to be reinforced
(c) surface to be machined
(d) seats for loose core prints.

97. In sand casting, surface to be left unmachined is marked on pattern by
(a) blue colour
(b) green colour
(c) black colour
(d) yellow colour.

98. Plaster mould casting use patterns made of
(a) wood
(b) wax
(c) granular sand
(d) brass.

99. Main constituent of moulding sand is
(a) alumna
(b) clay
(c) iron oxide
(d) silica.

100. In synthetic sand (_____) is used as a binder
(a) mollases
(b) water
(c) bentonite and water
(d) clay.

101. Which of the following methods are used for obtaining directional solidification for riser design.
1. Suitable placement for chills
2. Suitable placement of chaplets
3. Employing padding

Select the correct answer
(a) 1 and 2
(b) 2 and 3
(c) 1 and 3
(d) 1, 2 and 3
[*IES 1996*]

102. Shrinkage allowance is wooden pattern for cast iron casting is nearly
   (a) 0.1 mm per meter
   (b) 1 mm per meter
   (c) 10 mm per meter
   (d) 40 mm per meter.

103. The shell thickness in shell moulding can be accurately maintained by [*ISE 2005*]
   (a) the type of binder used
   (b) controlling the time during which the pattern is in contact with mould
   (c) controlling the time during which the pattern is heated
   (d) maintaining the temperature of the pattern in the range of 175°C – 380°C.

104. Which one of the following processes produces a casting when pressure force the molten metal into the mould cavity [*IES 2005*]
   (a) die casting
   (b) shell moulding
   (c) continues casting
   (d) investment casting

105. In centrifugal casting the impurities are [*GATE 2002*]
   (a) collected at the centre of casting
   (b) uniformly distributed
   (c) forced towards the outer surface
   (d) trapped near the mean radius of the casting

106. In a centrifugal casting method [*GATE 2007*]
   (a) no core is used
   (b) core is made of sand
   (c) core is made of ferrous metal
   (d) core may be made of any metal

107. Which one of the following pairs is not correctly matched. [*ISE 1997*]
   (a) Aluminium – Pressure die casting
   (b) Jewellery – lost wax process
   (c) Large pipes – centrifugal casting
   (d) Large bells – Loam moulding

108. Process of making hollow casting from permanent mould by a close fitting core is known as
   (a) lost wax casting
   (b) centrifugal casting
   (c) Pressed casting
   (d) slash casting.

109. Misrun is a
   (i) Welding defect
   (ii) Casting defect
   (iii) Metal forming defect
   (iv) Machining defect. [*GATE-2006*]

ANSWERS

1. (*c*)	2. (*c*)	3. (*e*)	4. (*a*)	5. (*c*)
6. (*c*)	7. (*b*)	8. (*d*)	9. (*b*)	10. (*d*)
11. (*a*)	12. (*b*)	13. (*c*)	14. (*a*)	15. (*c*)
16. (False)	17. (*b*)	18. (*b*)	19. (*b*)	20. (*a*)
21. (Agree)	22. (*d*)	23. (Correct)	24. (*c*)	25. (*c*)
26. (*b*)	27. (*b*)	28. (*a*)	29. (*c*)	30. (*a*)
31. (*b*)	32. (*e*)	33. (*c*)	34. (*b*)	35. (Agree)
36. (*c*)	37. (*d*)	38. (Correct)	39. (*c*)	40. (*b*)
41. (*a*)	42. (*d*)	43. (*d*)	44. (*c*)	45. (*b*)
46. (*c*)	47. (Right)	48. (*c*)	49. (*e*)	50. (*c*)
51. (*a*)	52. (*c*)	53. (*b*)	54. (Agree)	55. (*b*)
56. (*e*)	57. (*a*)	58. (*a*)	59. (*a*)	60. (*b*)

61. (*c*)	62. (*c*)	63. (*a*)	64. (*d*)	65. (*b*)
66. (*b*)	67. (*b*)	68. (*b*)	69. (*b*)	70. (*b*)
71. (*a*)	72. (*b*)	73. (*b*)	74. (*d*)	75. (*c*)
76. (*a*)	77. (*b*)	78. (*a*)	79. (*b*)	80. (*b*)
81. (*b*)	82. (*d*)	83. (*b*)	84. (*d*)	85. (*a*)
86. (*c*)	87. (*b*)	88. (*a*)	89. (*b*)	90. (*c*)
91. (*d*)	92. (*a*)	93. (*a*)	94. (*c*)	95. (*d*)
96. (*c*)	97. (*c*)	98. (*d*)	99. (*d*)	100. (*c*)
101. (*c*)	102. (*c*)	103. (*b*)	104. (*a*)	105. (*a*)
106. (*a*)	107. (*a*)	108. (*c*)	109. (*ii*)	

HINTS

Ans. 60: Volume of metal compensated by riser does not constitute cooling during solid state.

$\therefore$ Volume compensated by riser = 7% **Ans.**

Ans. 61: Given metal flow rate = 6.5×10^5 mm³/sec

$$A_1 V_1 = 6.5 \times 10^5 \text{ mm}^3/\text{sec}$$

$$V_1 = \frac{6.5\times10^5}{A_1} = \frac{6.5\times10^5}{650} = 1000 \text{ mm/sec}$$

Applying Bernoulli's theorem at point A and B

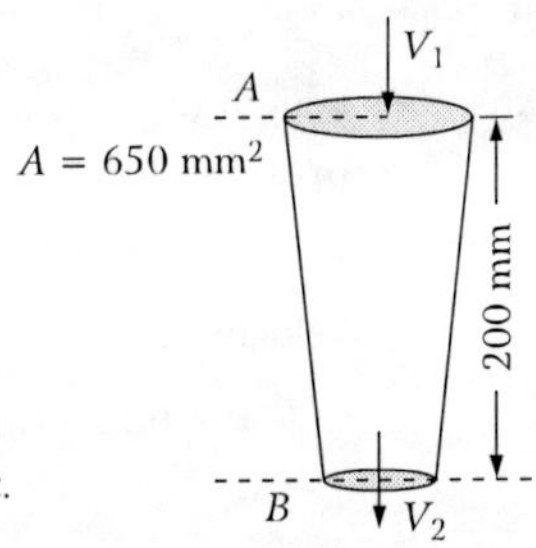

$$h + \frac{V_1^2}{2g} = \frac{V_2^2}{2g}$$

or

$$200 + \frac{(1000)^2}{2\times10^4} = \frac{V_2^2}{2\times10^4}$$

$$V_2 = 2236.068 \text{ mm/sec.}$$

Now at down sprue $Q_2 = A_2 V_2$ or $A_2 = \dfrac{Q_2}{V_2}$

$$A_2 = \frac{6.5\times10^5}{2236.068} = 290.7 \text{ mm}^2 \quad \textbf{Ans.}$$

Ans. 63. Surface of cube = $6l^2$, surface area of sphere = $4\pi r^2$

According to Cine's rule solidification time $\alpha \left(\dfrac{\text{Volume}}{\text{Surface area}}\right)^2$

Given, volume of cube = volume of sphere

$$\frac{t_c}{t_s} = \left(\frac{A_s}{A_c}\right)^2 = \left(\frac{4\pi r^2}{6l^2}\right)^2 = \frac{16\pi^2 r^4}{36 l^4}$$

or

$$\left(\frac{t_c}{t_s}\right) = \left(\frac{4\pi}{6}\right)^2 \left(\frac{r}{l}\right)^4 \quad \textbf{Ans.}$$

Ans. 64. Time taken to fill the mould with top gate

$$t_A = \frac{A \cdot H}{A_g \cdot \sqrt{2gh_m}}$$

where, A = Area of mould

H = Height of mould

A_g = Area of gate

h_m = Gate height

Given, $h_m = H$

$$t_A = \frac{A\sqrt{h_m}}{A_g\sqrt{2g}}$$

Time taken to fill the mould with bottom gate

$$t_B = \frac{2A}{Ag\sqrt{2g}}\left(\sqrt{h_m} - \sqrt{h_m - H}\right)$$

or

$$t_B = \frac{2A}{Ag\sqrt{2g}} \times \sqrt{h_m}$$

$$\frac{t_B}{t_A} = 2$$

or

$$\boxed{t_B = 2t_A} \quad \textbf{Ans.}$$

Ans. 66. Length of sprue h = 20 cm = 0.2 m,

Break of down sprue A = 1 cm^2 = $10^{-4}m^2$

Volume of mould cavity= 1000 cm^3 = $10^{-3}m^{-3}$

Velocity at base of sprue $V = \sqrt{2gh} = \sqrt{2 \times 9.81 \times 0.2} = 1.98$ m/sec

time t is filling the mould cavity = $V = A \times v \times t$

$$10^{-3} = 10^{-4} \times 1 \times 1 \times 1.98 \times t$$

$$t = \frac{10^{-3}}{10^{-4} \times 1.98} = 5.05 \text{ sec} \quad \textbf{Ans.}$$

Ans. 69. Given d = 200 mm or r = 100 mm ≈ 0.1 mm

$K = 0.97 \times 10^6$ S/m^2

$$t \propto \left(\frac{V}{A}\right)^2 \text{ by Chovorinov's rule}$$

$$t = k\left(\frac{V}{A}\right)^2 \text{ for sphere}$$

$$V = \frac{4}{3}\pi r^3 = \frac{4}{3}\pi(0.1)^3$$

$$A = 4\pi r^2 = 4\pi(0.1)^2$$

$$t = 0.97 \times 10^6 \times \frac{0.00419}{0.1256}$$

$$= 1078 \text{ sec Ans.}$$

REVIEW QUESTIONS

1. Define the casting? and Describe the casting process in detail.
2. Compare the casting process with other joining process.
3. Explain the terms "pattern" and "pattern allowances". What are different types of allowances provided when making a pattern? Explain each in brief. *[CCS University 2010]*
4. Describe desirable properties of a moulding sand and state what defects can arise due to absence/lack of these. *[CCS University 2010]*
5. Describe the various kind of patterns in use. What are the allowances provided, when making a pattern? *[UPTU 2010]*
6. What is pattern? Why allowances are provided on a pattern?
7. What are the different types of patterns? Explain single piece pattern. *[UPTU 2006-07]*
8. Explain briefly with neat sketch any three of the following.
   (*a*) Solid pattern (*b*) Split pattern
   (*c*) Skeleton pattern (*d*) Gated pattern
   (*e*) Segmental pattern (*f*) Cope and drag pattern
9. Explain briefly the following pattern allowances
   (*a*) Shrinkage allowance (b) Machine allowances
   (*c*) Distortion allowances (*d*) Rapping/Shaking allowances
10. What do you mean by charging of cupola? Explain different zone and reaction involved in cupola? *[CCS University 2010]*
11. Sketch and briefly describe cupola furnace.
12. Explain in brief die-casting process. *[UPTU 2007]*
13. Write the short notes on: *[UPTU 2009]*
    (*a*) Investment casting (*b*) Centrifugal casting
14. What is the differences between the permanent mould casting and die-casting?
15. Briefly describe with neat sketch, process and application of investment casting process. Also discuss its merits and limitations. *[UPTU 2010]*
16. Write short notes on the following *[UPTU 2010]*
    (*a*) Centrifugal casting (*b*) Casting defects
17. With the help of a neat diagram, explain the working principle of cupola also write its limitations. *[UPTU 2009]*
18. Write short notes on "CO_2 casting" and Investment casting. *[CCS University 2010]*
19. Prove that h/d (height/diameter) ratio of most compact economical riser for side placed riser is one and that for top placed riser is half and that in both cases ratio of volume/area is d/6 (diameter/6). *[UPTU 2002, 2004; CCSU 2010]*
20. What is core? How cores are made and how cores are classified?
21. Describe the Investment casting process also discuss its merits and limitations. *[UPTU 2003, 2004]*

22. What are the common casting defects? What cause them and what measures can be taken to avoid defects from occuring? *[UPTU 2003]*
23. Give the causes and remedies of the following casting defects
    (*a*) Blow holes (*b*) Pin holes/porosity
    (*c*) Rat tail (*d*) Hot tears *[UPTU 2004, 2006]*
24. Name and explain the Non-destrective testing method for casting.
25. Discuss briefly the various methods employed to inspect the casting.
26. What factors are normally responsible for the casting defects?
27. State the advantages and limitations of sand casting.
28. Explain the design consideration of open riser and closed riser. *[UPTU 2008]*
29. Explain in brief the mechanism solidification of casting of pure metal. *[UPTU 2007]*
30. Explain runner and riser as related to casting and compare riser versus runner. *[UPTU 2007]*
31. What are the advantages and disadvantages of casting process?
32. Describe some important feature observed in the cooling curve of pure metal.
33. Write the short note on Runner vs Riser. *[UPTU 2006-07]*
34. What is the Jamming of Cupola? How it can be avoided?
35. Explain the flask less moulding machine?
36. Differentiate between green strength and dry strength of a foundry sand.
37. Write the short notes on
    (*a*) Green sand (*b*) Dry sand
    (*c*) Facing sand (*d*) Parting sand
    (*e*) Floor sand (*f*) Cupola furnace *[UPTU 2004-05]*
38. What are the different materials used for core making? What are oil sand core?
39. Discuss the function of the followings in a mould
    (*a*) Pouring basin (*b*) Sprue
    (*c*) Riser (*d*) Vents
40. Describe the continuous casting.
41. Define aspiration effect in casting. Draw the schematic view of ideal and actual shapes of sprue derive the governing equation to avoid aspiration. *[UPTU 2005-06]*
42. Discuss the role of casting volume and surface area as they relate to the total solidification time and *Chrorinov's rule*.
43. Explain the design consideration of open and closed riser. *[UPTU 2008-09]*
44. Sketch any two gating systems.
45. Explain the salient feature of casting design consideration.
46. Differentiate between a parting gate and a whirl gate with the help of neat sketch.
47. Describe the method of increasing riser efficiency with help of neat sketch.
48. Illustrate and explain any **four** types of gating system.
49. What are different steps in casting? Explain the each in detail.
50. Define sand reclamation.

51. Define the different types of mould.
52. Define loam sand mould and composite mould.
53. What is gate? Explain the various type of gates used in casting with neat sketch.
54. Define the core and explain the basic functions of a core.
55. Define the various cores in detail.
56. Why the riser design is important.
57. Explain the causes and remedies of misrun, hot tear, Rat tail and cold shut.
58. Explain the permanent casting. Explain the advantages and limitations of the same.
59. Define the centrifugal casting in detail. Explain the advantages limitation of the same with applications.

❑❑❑

UNIT-V

UNCONVENTIOANL METAL FORMING PROCESS

PART-I

13 CHAPTER UNCONVENTIONAL METAL FORMING PROCESS

13.1 INTRODUCTION

Unconventional metal forming process is also known as non-traditional forming process. High Energy Rate Forming (HERF) and High Velocity Forming (HVF) Process are known as unconventional metal forming processes. In case of HERF Process, energy required to processing on components released for very short time span, *i.e.*, milliseconds or microseconds on the other hand in HVF workpiece is deformed by the high velocity speed of Ram (or die). In unconventional metal forming process energy is transferred through a medium *i.e.*, water or air. HVF Process is suitable for forging of thin metals while HERF is suitable for computing the metal powder in **powder metal energy, forging, cold welding, cutting, bending**, etc. HVF machines provide very high impact rate at very low operating cost. HERF improves Ultimate tensile strength and produces more uniform strain through out the component. Working principal of these machines are same as drop hammers and impact hammers machines.

Method of Energy release:

1. Chemical—explosive, gas mixture
2. Electrical—magnetic field, spark discharge.

13.2 EXPLOSIVE FORMING

In this process, chemical energy is stored in explosives and finally that energy used to machining the materials. A charge detonated either within pressure or above pressure, transmitting medium generates very high pressure. Same can be used to form parts of large size. Pressure generated basically depends upon the compressibility of energy transmitting medium. Water and air are used for this purpose, air is better medium as compared to water.

Explosive forming is a metal working technique in which an explosive charge is used instead of a punch or press. It can be used on materials for which a press setup would be prohibitively large or require an unreasonably high pressure, and is generally much cheaper than construction a large enough and sufficiently high-pressure press; on the other hand, it is unavoidably a batch process, production one product at a time and with a long setup time.

There are various approaches; one is to place metal plate over a die, with the intervening space evacuated by a vacuum pump, place the whole assembly underwater and detonate a charge at an appropriate height above the plate. For complicated shapes, a segmented die can be used to produce in a single operation a shape that would require many manufacturing steps, or to be manufactured in parts and welded together with concomitant loss of strength at the welds. There is often some degree of work hardening from the explosive-forming process, particularly in mild steel.

Tooling can be made out of fiberglass for short-run applications, out of concrete for large parts at medium pressures, or out of ductile iron for high-pressure work; ideally the tooling should have higher yield strength than the material that is being formed, which is a problem since the technique is usually only considered for material which is itself very hard to work.

Explosive forming was used in the 1960s for aerospace applications, such as the chine plates of the SR-71 reconnaissance plane and various Soviet rocket parts; it continued to be developed in Russia. It proved particularly useful for making high-strength corrugated parts which would otherwise have to be milled out of ingots much larger than the finished product.

Other uses of explosives for manufacturing take advantages of the shaped charge effect, putting the explosive directly in contact with the metal to be worked; this was used for engraving of thick iron plates as early as the 1890s. See also explosively formed projectiles for a variety of military applications of the same kind of technology.

Typical explosives are:

1. Trinitro toluene (TNT) (for high energy)
2. Dynamite (for high energy)
3. Gun Power (for low energy)

Since explosive forming transmits the explosively-generated energy through water, it can simulate a variety of other conventional metal forming techniques. With flat-plate designs, explosive hydro-forming is very comparable to stamp- or press-forming. Where cylindrical workpieces need to be formed, it can simulate spin-forming. Additionally, explosive hydro-forming openly competes with conventional hydro-forming and hot-stretch forming. In fact, where each of these metal-forming techniques come up short, explosive hydro-forming still has more to offer. Explosive hydro-forming can simulate a variety of these forming techniques in a single forming operation.

There are basically two system of explosive forming:

1. Confined system.
2. Unconfined system.

13.2.1 Confined System

This process is suitable for bulging and flaring process for small and circular objects. This process is not very common in practice due to problem of die erosion and die failure as very high pressure required in process. In this process, object (part) is completely enclosed by two or more die parts as explosive is detonated, a shock wave is produced in the energy transmitting medium, which pushes the object into the die process is carried out in the vacuum. Otherwise due to very high temperature object (parts) can be melted or burn even die too. Trapped air also prevent the complete to forming of the part.

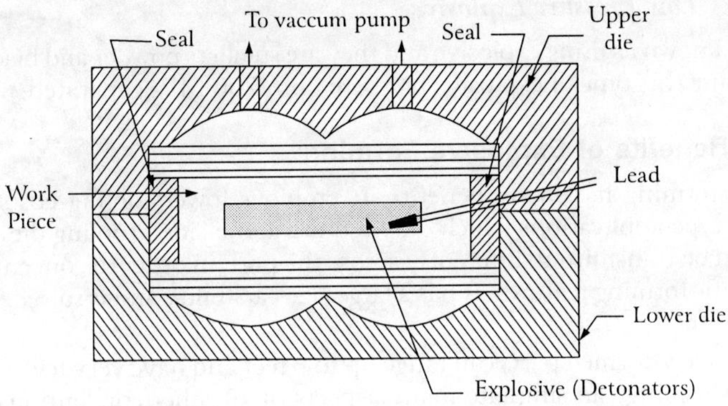

Fig. 13.1 Explosive forming confined system

13.2.2 Unconfined System

This explosive forming system is used to forming the large components (parts) of very thin sheet metal. Even though this process ship plate up to 25 mm thick can be formed too. Object/component is held between the die and entire die and work assembly is kept in a water tank. Explosive is denoted above or within the water. As a result of igniting the explosive waves are produced in the water. Which in turn forces the workpiece into the die. Vacuum is created this process. Unconfined explosive forming process is very much affected by the following factors such as type of explosive used, stand off distance, type of workpiece, medium for energy transmitting.

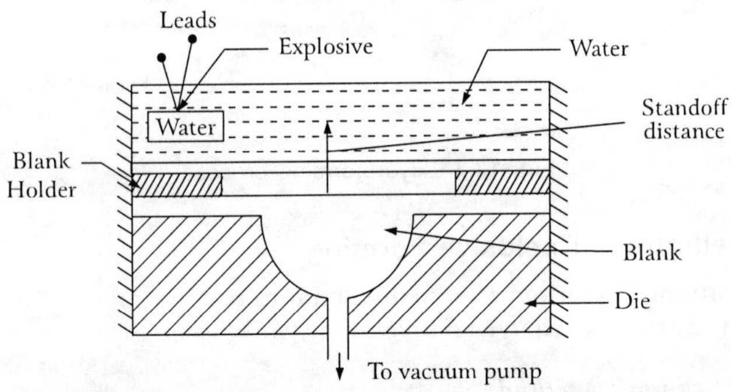

Fig. 13.2 Explosive forming in unconfined system

Commonly water is used as a medium due to the following advantages:

(*a*) Free of cost available.

(*b*) Water is less compressible than air.

(*c*) Noice is produced by explosive which is reduced in water.

(*d*) Less chance of damage of workpiece.

13.2.3 Two Types of Explosives are Used

(*i*) low pressure explosive or firing explosive,

(*ii*) high pressure explosive or detonating explosive.

13.2.3.1 Low Pressure Explosive

They also known as firing explosive and they are smallers powder and black powder. One igniting this type of explosive pressure upto 700 MPa generated.

13.2.4 Benefits of Explosive Forming

Explosive-forming has many benefits. It employs lower tooling costs and uses stamping type applications which only require a one-sided tooling die. Explosive energy can be transmitted differently across the part, in order to concentrate force onto specific forming features. It has a large size capability and is suited to difficult configurations.

Explosively-formed parts can range up to 6 feet and have very few limitations. Explosive-forming can simulate many aspects of all other conventional forming methods, without their respective limitations.

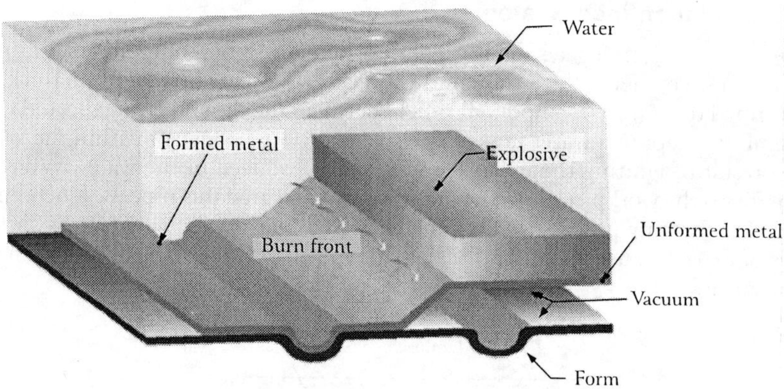

Fig. 13.3 Hydroforming process

13.2.5 Methods of Explosive Forming

Explosive forming operations can be divided into two groups, depending on the position of the explosive charge relative to the workpiece.

13.2.5.1 Standoff Method

In this method, the explosive charge is located at some predetermined distance from the workpiece and the energy is transmitted through an intervening medium like air, oil, or water. Peak pressure at the workpiece may range from a few thousand psi (pounds/inch^2) to several hundred thousand psi depending on the parameters of the operation.

Working:

The system used for standoff operation consists of following parts:

1. An explosive charge
2. An energy transmitted medium

3. A die assembly
4. The workpiece

Figure 13.4 shows an arrangement of Standoff Explosive forming operation. The die assembly is put together on the bottom of the tank. Workpiece is placed on the die and blank holder placed above. A vacuum is then created in the die cavity. The explosive charge is placed in position over the centre of the workpiece. The explosive charge is suspended over the blank at a predetermined distance. The complete assembly is immersed in a tank of water.

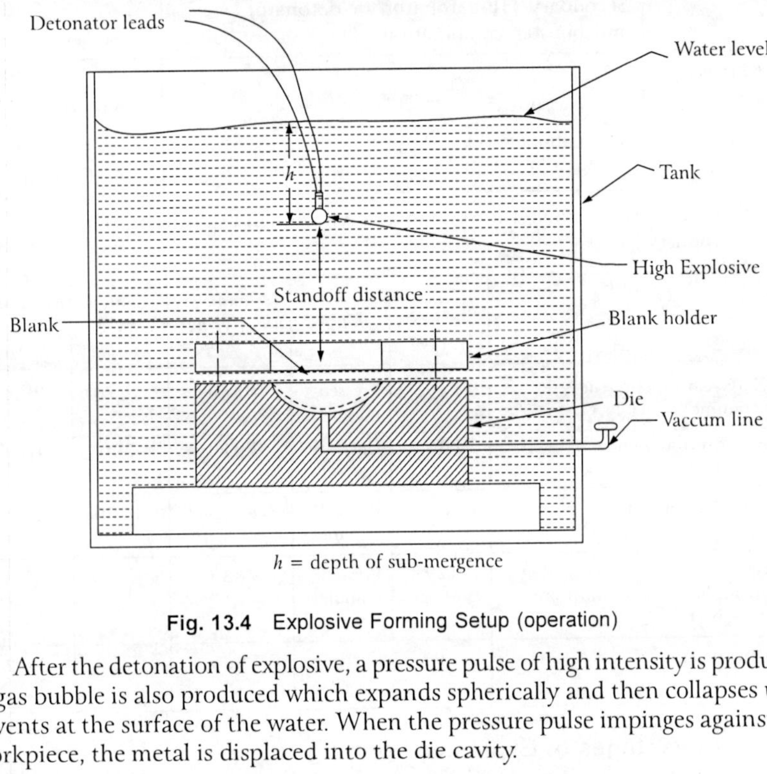

Fig. 13.4 Explosive Forming Setup (operation)

After the detonation of explosive, a pressure pulse of high intensity is produced. A gas bubble is also produced which expands spherically and then collapses until it vents at the surface of the water. When the pressure pulse impinges against the workpiece, the metal is displaced into the die cavity.

13.2.5.2 Contact Method

In this method, the explosive charge is held in direct contact with the workpiece while the detonation is initiated. The detonation produces interface pressures on the surface of the metal up to several million psi (35000 MPa).

13.3 EXPLOSIVES

Explosives are substances that undergo rapid chemical reaction during which heat and large quantities of gaseous products are evolved. Explosives can be solid (TNT-trinitro toluene), liquid (Nitroglycerine), or Gaseous (oxygen and acetylene mixtures).

Explosives are divide into two classes; low explosives in which the ammunition burns rapidly rather than exploding, hence pressure build up is not large, and high explosive which have a high rate of reaction with a large pressure build up. Low explosives are generally used as propellants in guns and in rockets for the propelling of missiles.

TABLE 13.1 Features of Low and High Explosives

Property	*High Explosives*	*Low Explosives*
Method of initiation	Primary HE-ignition, spark, flame, or impact	Ignition
	Secondary HE-detonator, or detonator and booster combination	
Conversion time+	Microseconds	Milliseconds
Pressure	up to about 4,000,000 psi	up to about 40,000 psi

+Time required to convert a working amount of high explosive into high-pressure gaseous products.

Explosive	*Relative power %TNT*	*Form of charge*	*Detonation velocity, m/s*	*Energy kJ/kg*	*Maximum pressure, Gpa*
RDX (Cyclotrimethylene trinitramine), $C_3H_6N_6O_6$	170	Pressed granules	8380	1270	23.4
TNT (Trinitrotoluence, $C_7H_5N_3O_6$)	100	Cast	7010	780	16.5
PETN (Pentaerythritol tetranitrate), $C_5H_8N_{12}O_4$	170	Pressed granules	8290	1300	22.1
Tetryl (Trinitrophenylmethylinitramine, $C_7H_5O_8N_5$)	129	Pressed granules	7835	—	—
Blasting gelatin	99	Cartridge plastic	7985	1220	17.9

13.3.1 Advantages of Explosive Forming

(i) Maintains precise tolerances.
(ii) Eliminates costly welds.
(iii) Controls smoothness of contours.
(iv) Reduces tooling costs.
(v) Less expensive alternative to super-plastic forming.

13.3.2 Facility Requirements

For an explosive forming facility, the general requirements can be broken down without regard for size as follows:

1. **Operations area:**
   (a) either a natural lake or a river, an artificial lake, or a tank in the ground

(b) water supply and pump if not naturally available
(c) vacuum force-pump and pressure gauge
(d) portable electrical power supply *i.e.*, generator etc.

2. **Explosive handling facility:**
(a) storage magazine
(b) shelter and control barricade
(c) Preparation area

3. **Material handling:**
(a) cranes weather mobile or fixed with frame structure
(b) Trollies for transporting die and finished work
(c) Equipment for handling sheet.

4. **Inspection area:**
(a) receiving inspection gauge and instruments
(b) final inspection
(c) reject or rework storage

5. **Inert storage area:**
(a) Inert raw material
(b) inert components
(c) finished part prior to shipment.

6. Die fabrication and maintenance area.

13.4 DIE MATERIALS

Different materials are used for the manufacture of dies for explosive working, for instance high strength tool steels, plastics, concrete. Relatively low strength dies are used for short-run items and for parts where close tolerances are not critical, while for longer runs higher strength die materials are required. Kirksite and plastic faced dies are employed for light forming operations; tool steels, cast steels, and ductile iron for medium requirements.

Material of Die	*Application Area*
Kirksite	Low pressure and few parts
Fiberglass and Kirksite	Low pressure and few parts
Fiberglass and Concrete	Low pressure and large parts
Epoxy and Concrete	Low pressure and large parts
Ductile Iron	High pressure and many parts
Concrete	Medium pressure and large parts

13.5 CHARACTERISTICS

- Very large sheets with relatively complex shapes, although usually axisymmetric.
- Low tooling costs, but high labour cost.

- Suitable for low-quantity production.
- Long cycle times.

13.6 TRANSMISSION MEDIUM

Energy released by the explosive is transmitted through medium like air, water, oil, gelatin, liquid salts. Water is one of the best media for explosive forming since it is available readily, inexpensive and produces excellent results. The transmission medium is important regarding pressure magnitude at the workpiece. Water is more desirable medium than air for producing high peak pressures to the workpiece.

13.7 EXPLOSIVE HYDRO-FORMING

Explosive hydro-forming has been an accepted metal-forming techniques for almost 50 years. It has been used in a wide variety of applications in the automotive, aerospace, and martime industries. SOURIAU PA & E has extensive experience using this technique to form a wide variety of metals. Typical forming projects yield very close tolerances and very high degrees of repeatability.

Explosive hydro-forming can offer significant cost savings on short-run parts because it often only requires a one-sided tooling die. In the explosive hydro-forming process, the water slug applies force evenly over the surface of the workpiece, as it forms into the cavity of the forming die.

Explosive hydro-forming can also efficiently form large parts. Since the power generation comes from an explosive charge, you don't need larger machinery to form large parts, you simply apply a larger charge. Current explosively hydro-formed parts range up to 6 feet.

13.8 ELECTRO-HYDRAULIC FORMING

In electro-hydraulic forming process a high-intensity shock waves generated by discharging stored electrical energy round the electrode, which is dipped in a liquid medium.

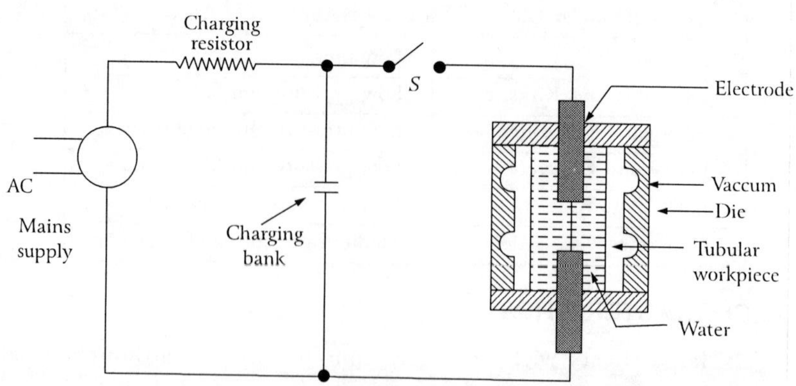

(A)

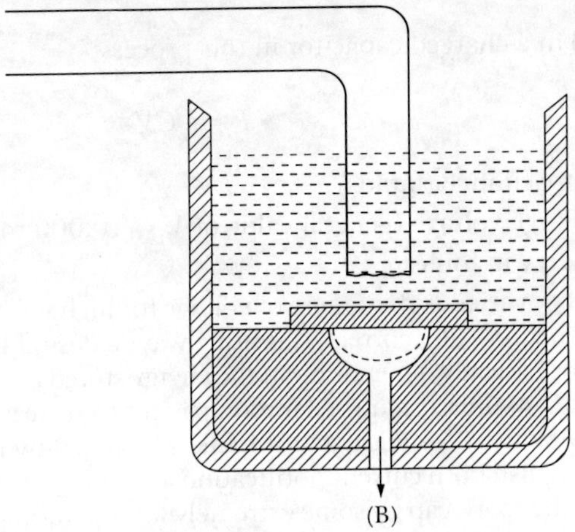

Fig. 13.5 Principle of Electro-hydraulic forming

Figure 13.5 shows the working principle of electro-hydraulic forming system for a tubular part (object). It consist the two electrodes which are connected by a switch to a blank capacitors both electrodes are attached with each other by a thin initiating wire which vaporized during the discharge. In this process part (workpiece) is within the water with electrode as shown in Fig. 13.5. Now capacitor are changed by a power unit via a charging resistor to a suitable voltages and suitable level. When the switch connecting the capacitor to the electrode is closed, stored energy is discharge across the gap at the ends of the electrodes. Electrical energy is converted into shock waves in the water which forms the workpiece into a die.

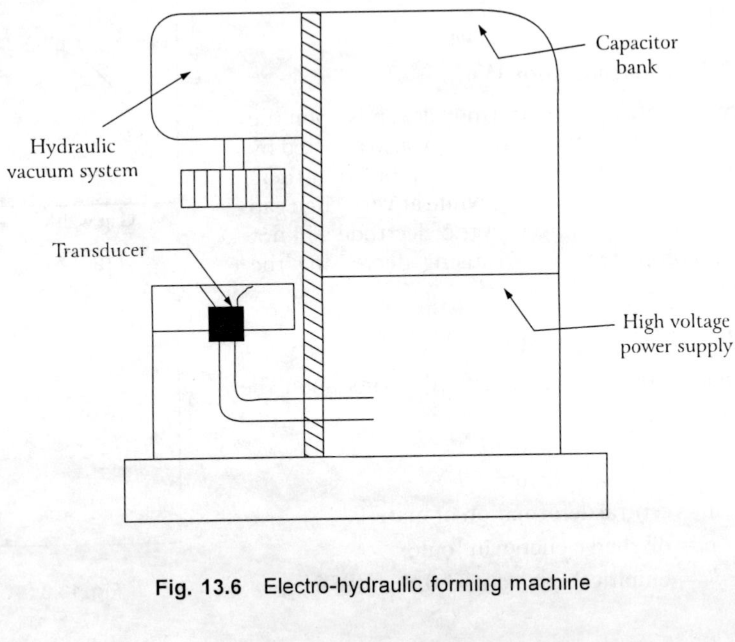

Fig. 13.6 Electro-hydraulic forming machine

Electrical aspects

Energy stroed in a charged capacitor in this process

$$E = \frac{1}{2}CV^2$$

where, C = Capacitance (farads)

V = Charge voltage (general value of V = 10,000 to 30,0000 V)

E = Energy in Joules

in this process energy varies as the square of voltage for high voltage, higher energy is required. It is a theoretically correct, it is not always a desirable course of action and size of capacitor basically depends upon the energy stored in capacitor. Actually, a some what larger container is required for a capacitor to store a specified amount of energy at 20 kV than at 5 kV; capacitor bank is changed by means of a power supply which accomplish both current notification and voltage set up stored energy than dumped into the spark gap by some extremely fast operating triggering device.

13.8.1 Process Variables

Various process variables affect the efficiency of the electro-hydraulic forming process such as:

(*a*) Electrode gap width
(*b*) Charge energy
(*c*) Electrode
(*d*) Die material
(*e*) Stand off distance
(*f*) Initiating wire between electrode
(*g*) Speed of deformation
(*h*) Hydrostatic head

13.8.1.1 Electrode Gap Width

Usually gap between the electrode depends upon the charge voltage. Gap width efficiency also affected by the conductivity of water at zero gap position there will be no explosive discharge while at very large gap the initiating wire connecting the electrode will not vaporise completely. This causing decreasing the efficiency.

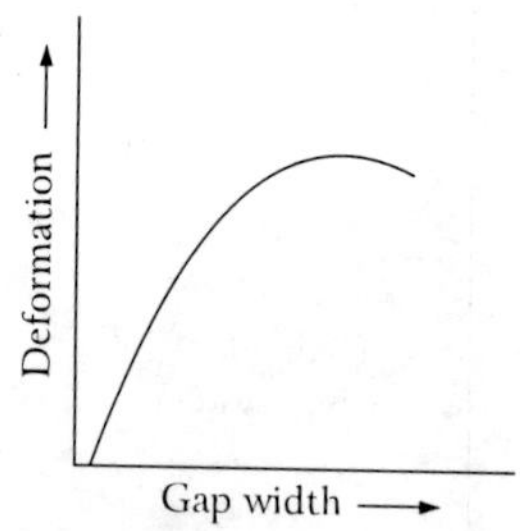

Fig. 13.7

13.8.1.2 Charge Energy

Deformation of material directly depends upon the energy discharge

$$\delta \propto E^K$$

$$\delta = CE^K$$

where, δ = central deformation of material

E = discharge energy in Joule

K = empirical constant = 0.36 – 0.56

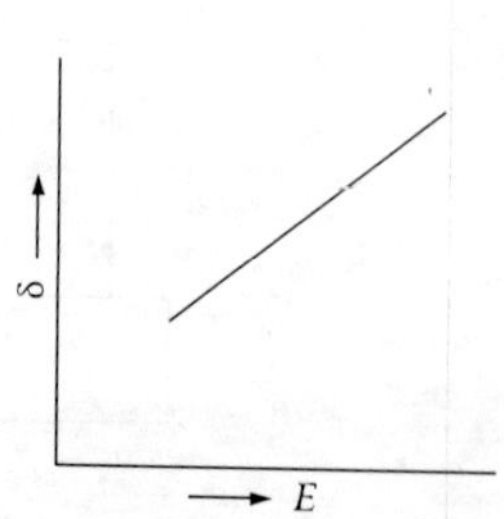

Fig. 13.8

more efficient shock waves can be generated by the low inductance as compared to high inductance and a slow discharge time.

13.8.1.3 Electrode

Electrode material also affect the electro-hydraulic forming process. Commonly material is used for electrode material are:

(*a*) Mild steel
(*b*) Stainless Steel
(*c*) Brass
(*d*) Copper

13.8.1.4 Die Material

Electro-hydraulic forming process is also very much affected by die materials. Die is commonly made by the following materials.

(*a*) Steel
(*b*) Kirksite
(*c*) Epoxy resin

13.8.1.5 Stand-off Distance

Gap between the spark and the workpiece is known as *stand-off distance.* Conical shapes are produced by small stand-off distance where as large diameter tube is formed by the large stand-off distance.

13.8.1.6 Initiating Wire between Electrode

Initiating wire basically used to produce spark ignition. Wire is made by a good conductor's material such as:

(*a*) Copper,
(*b*) Aluminium,
(*c*) Brass,
(*d*) iron, etc.

Diameter of wire should be small and the range of wire diameter is usually from 0.125 mm to 1.5 mm.

13.8.2 Advantages of Electro-hydraulic Forming

(*a*) It is a versatile process and very suitable for forming of tubular or dished shaped part.
(*b*) In process as only one die (female) is used hence resulting the tooling cost.
(*c*) Production rate is more as composed to explosive forming process.
(*d*) Closer dimensional tolerance can be achieved.

13.8.3 Limitations of Electro-hydraulic Forming

(*a*) Only smaller size components can be produce by this process.
(*b*) If the critical impact velocity of put is less than the 32 m/sec, they can not be used.

13.8.4 Application of Process

(*a*) To producing small size components in large quantity.
(*b*) To producing the tubular and dished shaped parts, etc.

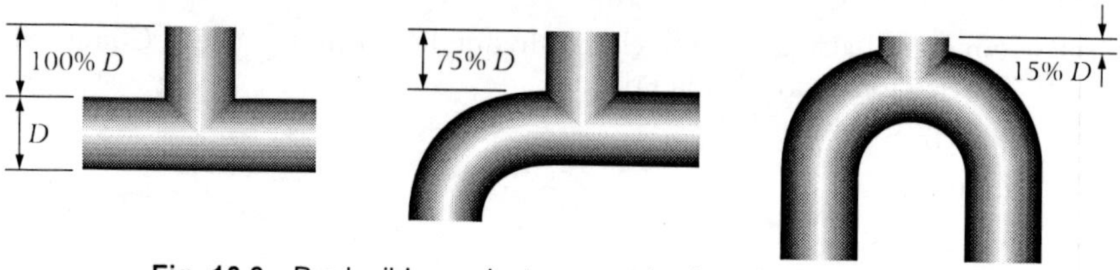

Fig. 13.9 Producible product geometries by tube hydroforming

13.9 ELECTRO-MAGNETIC FORMING (EMF)

It is also know as magnetic pulse forming process. It is the only high velocity forming process. Process is only suitable for a conducting materials. It is the process of electrically forming conducting metals of moderate thickness by the use of a precisely controlled electromagnetic force.

13.9.1 Working Principle

When the current is flow through a conductor a magnetic field is established around the conductor. Strength of magnetic field usually depends upon the current magnitude. In this process, there are two conductors, one is coil and other is workpiece.

Current pulsed from a capacitor bank is passed through a coil which is placed at the both end of workpiece. Current pulse generate high magnetic field around the coil. Their magnetic field induces an eddy current in the workpiece or object. Energy source is bank of capacitors charged to a predetermined voltage. Current is discharged very rapidly through the coil by off a switch. Link between the workpiece and coil and inductive. There is no spark gap or shock wave transmission through water as in electro-hydraulic forming. Process is very suitable for conducting materials like:

(i) Copper
(ii) Aluminium
(iii) Brass
(iv) Silver
(v) Low carbon based steel.

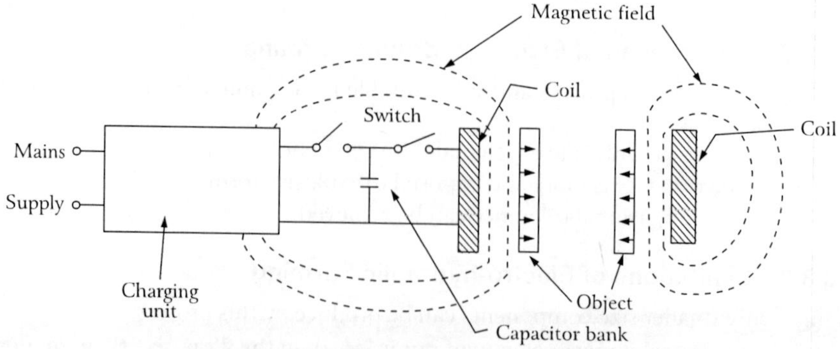

Fig. 13.10 Electromagnatic forming circuit

Efficiency of magnetic pulse forming depends upon the resistivity of the metal being formed. Resistivity should be less than 15 μΩ cm for better result.

13.9.2 Advantages of Electro-magnetic Forming Process

(a) This process can be applied for various types of metals like **silver, gold, copper, mild steel** etc.
(b) Faster rate of production with help of rapid prototyping facilities.
(c) The parameters with which the process depends can be easily controlled using advanced systems.
(d) Environment friendly process means no pollution.
(e) Basic cost for the setup and working is low.
(f) Tools that are required are less expensive.

13.9.3 Disadvantages of Electro-magnetic Forming

(a) Process of deep drawing in forming is difficult and almost impossible.
(b) Components should be symmetrical for the process to be carried out.
(c) It has short duration of the pulse (pressure pulse)
(d) Irregular surface fractures like the holes are problematic since magnetic field will undergo changes in those places.

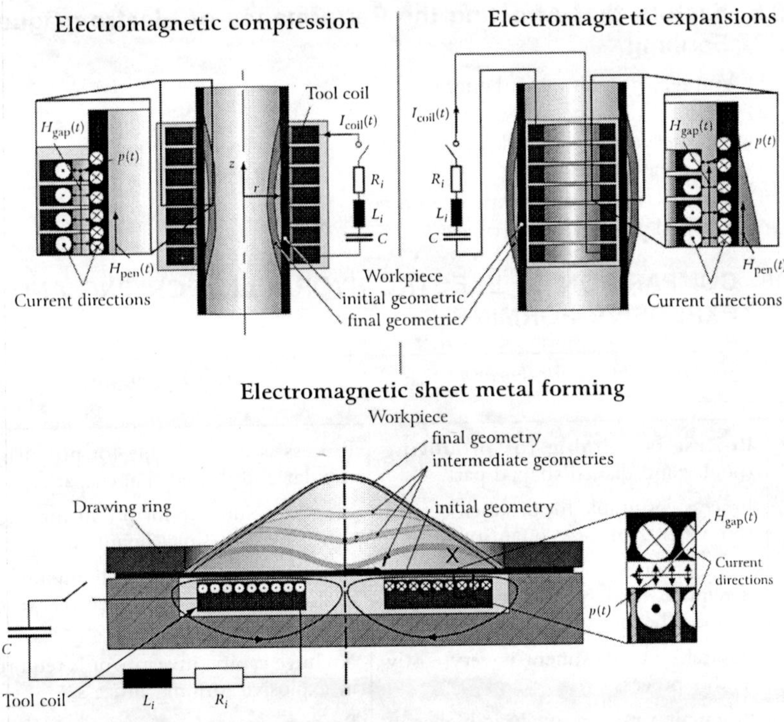

Fig. 13.11 Principle of the electromagnetic forming processes

13.9.4 Electromagnetic Forming has a Number of Advantages and Disadvantages Compared to Conventional Mechanical Forming Techniques

Some of the advantages are:

- Improved formability (the amount of stretch available without tearing).
- Wrinkling can be greatly suppressed.
- Forming can be combined with joining and assembling with dissimilar components including glass, plastic, composites and other metals.
- Close tolerances are possible as **springback** can be significantly reduced.
- Single sided dies are sufficient which can reduce tooling costs.
- Lubricants are reduced or are unnecessary, so forming can be used in **clean room** conditions.
- Mechanical contact with the workpiece is not required, this avoids surface contamination and tooling marks. As a result, a surface finish can be applied to the workpiece before forming.

The principle disadvantages are;

- Non-conductive materials cannot be formed directly, but can be formed using a conductive drive plate.
- The high voltages and currents involved require careful safety considerations.
- Large sheet metal components cannot readily be formed, due to current limitations on the design of very large coils.

13.9.5 Factors that Affecting the Performance of Electro-magnetic Forming

(1) Voltage, frequency and energy
(2) Deformation speed
(3) Characteristics of workpiece material
(4) Coil geometry
(5) Gap between the workpiece and coil
(6) Applied pressure and force.

13.10 COMPARISON OF ELECTRO-HYDRAULIC FORMING AND EXPLOSIVE FORMING

Sl. No.	*Electro-hydraulic Forming*	*Explosive Forming*
1.	Process is suitable for producing tubular and dished shaped part.	Process is also suitable for producing tubular and dished shaped part.
2.	Process is suitable for producing small components in large quantity.	Process is suitable for producing large component in small quantity.
3.	In this process, big size of equipment is requires for large quantity which is very costly too.	Process requires very small amount of explosive charge to produce very large quantity.
4.	Capital cost investment is very costly in this process.	No huge capital investment is requires in explosive forming process.
5.	Production rate normally is higher in this process.	Production rate is not very high as compared to electro-hydraulic forming.

OBJECTIVE QUESTIONS

1. Explosive forming process is suitable for
   (*a*) large metal bodies (*b*) producing metal body
   (*c*) producing high volume of production (*d*) simple parts of soft body.
2. Electro-hydraulic forming process is not suitable beyond
   (*a*) 20 kJ of energy (*b*) 40 kJ of energy
   (*c*) 100 kJ of energy (*d*) none of these.
3. Energy of produced by one gram explosive in explosive forming is
   (*a*) 5 kJ (*b*) 20 kJ
   (*c*) 50 kJ (*d*) 100 kJ.
4. If there is no gap between the electrodes in electro-hydraulic forming then
   (*a*) explosive discharge is minimum (*b*) explosive discharge is maximum
   (*c*) no explosive discharge (*d*) none of these.
5. Workpiece material should be electric conductive in electro-magnetic forming process
   (*a*) low (*b*) medium
   (*c*) high (*d*) none of these.

ANSWERS

1. (*a*)	2. (*c*)	3. (*a*)	4. (*c*)	5. (*c*)

REVIEW QUESTIONS

1. What do you mean by unconventional metal forming process? How do you classify them?
2. Enlist unconventional metal forming processes and briefly describe with neat sketch, working and applications of explosive forming. [*UPTU 2001-02*]
3. What are the different methods by which energy release rates can be obtained? Explain in brief. [*UPTU 2006-07*]
4. What is explosive forming process?
5. Discuss different types of explosive used in explosive forming.
6. Describe Explosive Forming Process Variables.
7. What are the different factors that affecting the explosive forming?
8. Explain with neat sketch die material and die design in explosive forming.
9. Describe mode of energy wave generation in explosive forming.
10. Describe the following explosive forming technique:
    (*a*) Stand-off operation (*b*) Contact operation
11. What are the advantages and disadvantages of explosive forming process?
12. Explain the electro-hydraulic forming operation with the help of a neat sketch.
13. Compare electro-hydraulic forming process and explosive process.
14. Write the short notes on "Electro-hydraulic Forming Process". [*UPTU 2008-09*]
15. Describe the working principle of electro-hydraulic forming process.
16. Describe various electro-hydraulic forming process variable.

17. Discuss following with respect to electro-hydraulic forming process:
    (*a*) Central deflection
    (*b*) Energy conversion mechanism
18. Describe the advantages and disadvantages of electro-hydraulic forming process.
19. What is the electromagnatic forming process? State the principle of electromagnetic forming.
20. Write the applications of electromagnetic forming process and electro-hydraulic forming process.
21. Define the purpose of coil in electro-magnetic forming and explain the various types of coils.
22. State the advantages and disadvantages of electromagnetic forming process.
23. Discuss the characteristic of electromagnetic forming process.
24. Describe the working principle of electromagnetic forming. [*UPTU 2002-03*]
25. Discuss the types of parts that are produced by electromagnetic forming process.

❑❑❑

14 CHAPTER

POWDER METALLURGY

14.1 INTRODUCTION

Powder metallurgy is define as an art of producing the objects/parts/components by heat treatment of compressed metallic powder. In powder metallurgy, metal powder blending and mixing together of same or different metal and compacting this powder mixture in metallic die and later **sintering** or heating the object at a predetermine time to allow the strong bonding between the particle of metal powder to produce a hard, strong and tough product.

Compaction of metal power in die is carried out by cold or hot isostatic pressing. To getting the proper interlocking of powdered metal particles with together and resulting the sufficient strength in green stage of compact.

Few non-metallic materials are also sometimes added to metallic power to providing the sufficient bond and some specific characteristics.

For a single piece of production cost of product is very costly because cost of die is costly. Hence process is very economical for mass production say for more than **10,000** parts at a time. Components produced by powder metallurgy process range small articles such as **gear, cam, ball pens, cutting tools** (inserts) to more porous product like self-lubricating bearing, filters, etc. Powder metallurgy now has become a highly competative technique as compared to other process such as forging and casting. Powder metallurgy process is frequently used for producing the component in automobile industries as well as in aircraft industries. As part

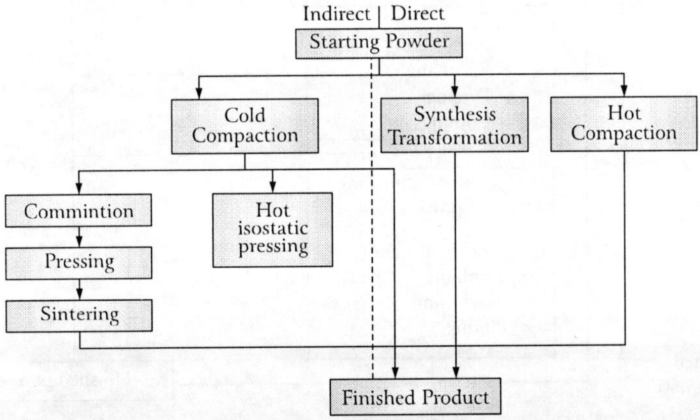

Fig. 14.1 Possible routes in powder Metallurgy (Metal Handbook, 1984)

produced by this process is low in cost and have sufficient strength to bearing the load while component is in operation mode. Mostly the weight of a component produced by powder metallurgy is from **2.5** kg to **50** kg in range.

Powder metallurgy process is very suitable for mixing the two different material like copper (having the **melting point** 1080°C) and tungston (having the melting point 3300°C) and make a product. These two different melting point material cannot be mixed by any other process like casting, etc. revolution in industries has come only by powder metallurgy process.

The most important **advantages** of powder metallurgy are the better dimensional control and high degree of accuracy can be achieved by this process and special characteristics (physical and mechanical properties) too.

Metal powder of **iron, aluminium, nickel, copper, titanium and high melting temperature refractory** is used for producing the products.

14.2 STEPS IN POWDER METALLURGY

Various steps are involved in powder metallurgy process and they are given below:

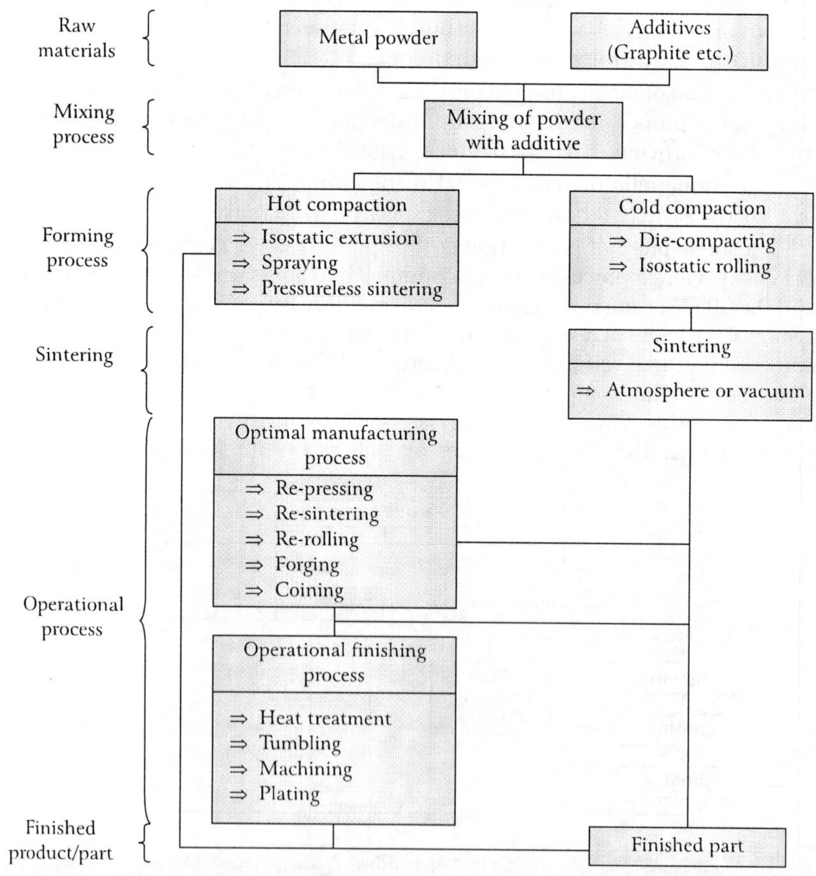

Fig. 14.2 Steps in powder metallurgy process

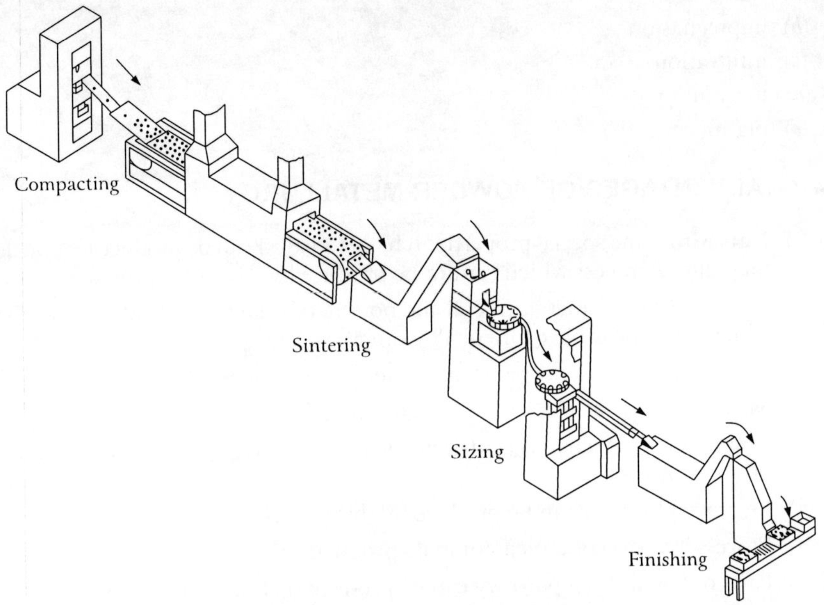

Fig. 14.3 Conventional powder metallurgy process (from Metals Handbook 1984)

14.2.1 Operation in Powder Metallurgy

(*a*) Primary operation

(*b*) Secondary operation

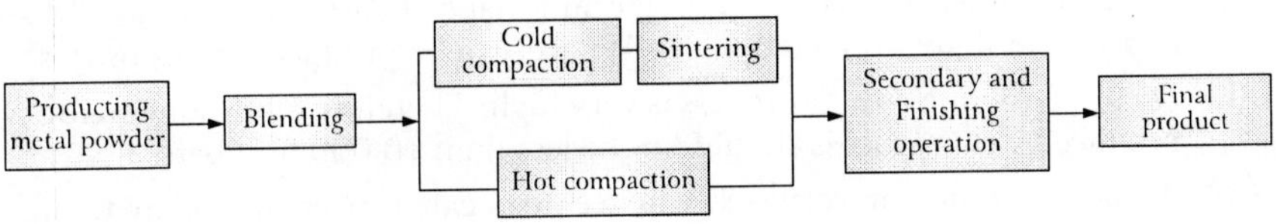

Fig. 14.4 Block diagram of operations performed in powder metallurgy.

14.2.2 Primary Operations

Primary operations include the following steps:

(i) production of metal powder.

(ii) blending (mixing) of ingredient of metal powder.

(iii) compaction of metal powder in between the desired shape of die.

(iv) sintering (heat treatment) of green compact.

14.2.3 Secondary Operations

Usually products made by the powder metallurgy are ready to use but after sintering operation some secondary operations are required to give the final touch of products for market ready as well as to improve the strength, toughness, density, etc. these secondary operations are following:

(*a*) sizing
(*b*) impregnation
(*c*) infiltration
(*d*) machining
(*e*) plating

14.3 ADVANTAGES OF POWDER METALLURGY

(i) Parts with some special properties (characteristics) can be produce by powder metallurgy process which cannot be produce by any other process.
(ii) Same or different metals (metal and non-metal) can be easily mixed together which is not possible (difficult) by other process.
(iii) Parts with the closer dimensional tolerance can be produce easily.
(iv) Machining operations are totally eliminated by powder metallurgy.
(v) Complex parts such as eccentric gears, splinces, irregular shaped hole can be produce by this process.
(vi) Light weight parts can be produce by this process.
(vii) Process is very economical for mass production.
(viii) Parts of any desired porosity can be produce by this process.
(ix) Very thin chance of scrap.
(x) Super hard cutting bits, which are difficult to manufactured by any other methods can be made by powder metallurgy.

14.4 DISADVANTAGES/LIMITATIONS OF POWDER METALLURGY

(i) Principle limitation is related to size and shape of parts, *i.e.*, only small size and simple shape (not complicated) parts can be produce by this process.
(ii) Dies and tooling cost of process is very high. Therefore minimum quantity produce by this process should not be less than 10,000 numbers.
(iii) As metal powder are very costly hence extra care is needed to store them.
(iv) Process is not suitable for all types of metals and alloys.
(v) Chance of corrosion of metal powder during the process hence required extra care. Some metal powder are explosive at room temperature hence not suitable for powder metallurgy.
(vi) A complete dense product is not possible without sintering (heating) the product (after compressing the product).
(vii) During the process compacting of metal powder into die, required a very high pressure which is responsible to wear and tear the die as die cost is very high.
(viii) Parts produce by this process possess **low impact strength** and **low tensile strength** as compared to other methods.
(ix) Sintering of low melting point metal powder (Zn, Sn, Pb) is not so easy.
(x) Since the compacted parts must be ejects from die without fracture hence only limited shapes can be produce by this process.

14.5 APPLICATIONS OF POWDER METALLURGY

Powder metallurgy have the following present applications.

(i) Cemented carbides tips.
(ii) Motor bushes.
(iii) Gear pump rotors.
(iv) Magnetic materials.
(v) Metallic fibers.
(vi) Metal bearing containing porous and graphite.
(vii) Refractory metal composite.
(viii) Friction materials.
(ix) Diamond impregnated tools.
(x) Tungsten filament.
(xi) Cemented carbide.

(*a*) **Porous Product:** Any degree of porosity can be obtained by powder metallurgy in a product, for example, oil impregnated self-lubricating bearing, heat exchanger, sound dampers, etc.

(*b*) **Machine Component:** Various machine components such as **gears, cam, valve, Babbitt bearing, piston rod** and other small products used as engine components.

(*c*) **Cemented Carbide:** Cemented carbide which are used for metal cutting operation are produce by powder metallurgy.

(*d*) **Diamond Impregnated Tools:** These are in the form of inserts made out of iron powder mixture and diamond dust. This types of inserts are basically used for cutting very hard materials.

(*e*) **Refractory Metals:** It is know that ordinary metal cannot withstand at very high temperature therefore to retain their shape and properties at high temperature refractory materials are used, refractory materials basically mixture of Al_2O_3 and SiO_2.

(*f*) **Friction Material:** Due to friction in components wear and tear take place therefore component should be such type that it should not wear easily. A lot of various type friction resistant components are produce by powder metallurgy process.

14.6 PRODUCTION OF METAL POWDER

Metal powder is available in large varieties with different properties metal like **aluminium, copper, titanium, irons, carbon steel, stainless steel, bronze, brass,** etc. are available in the powder form. There are several methods for producing the powder. Even one metal can convert into powder form by more than one method. Size of metal powder varies from 0.1 **μm** to 1000 **μm**. Converting the metal into the powder form basically depends upon desire characteristics of metal powder such as **shape, size, distribution, porosity, chemical purity, etc.**

14.6.1 Different Method for Producing Metal Powder

(*a*) Atomizing.

(*b*) Electrolytic deposition process.

(*c*) Mechanical communication or pulverization.

(*d*) Gaseous reduction.

(*e*) Granulation process.

14.6.1.1 Atomizing Process

In this process, molten metal is forced through a small orifice into a stream of high velocity air, stream or inert gas, this result extremely rapid cooling and disintegration into a very fine powder. Size of metal powder is controlled by orifice (nozzle) design and pressure of gas. Process is very suitable for producing the **Al**, **Zn**, **Pb** and **steel** powder.

Atomization includes—inert gas, atomization, rotating electrode atomization.

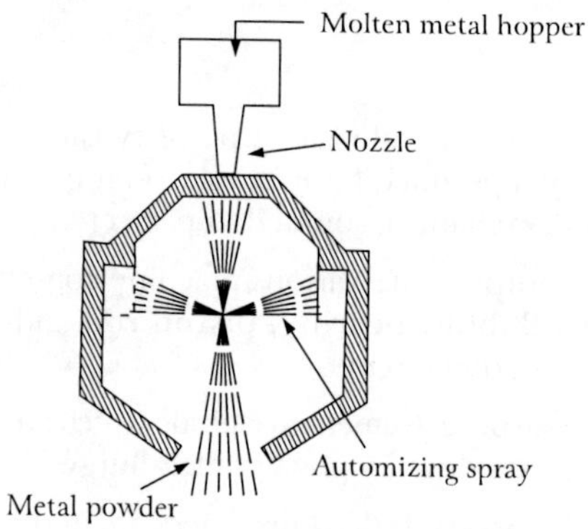

Fig. 14.5 Automization process

14.6.1.2 Electrolysis Process

This process is quite similar as electroplating process. In this process, metal plate that to be converted into powdered is placed in an electrolyte, to work as anode

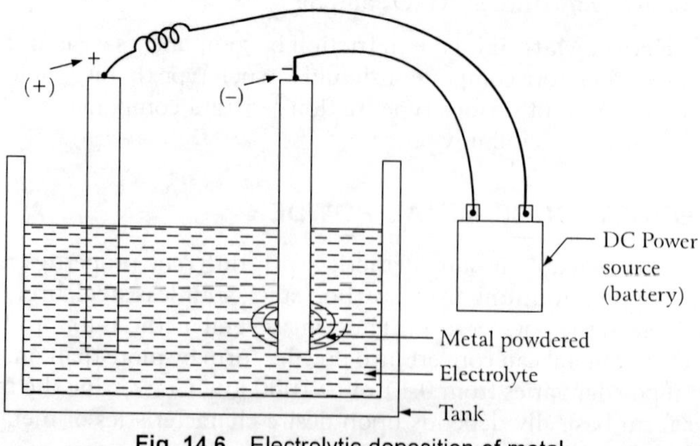

Fig. 14.6 Electrolytic deposition of metal

and other plate is kept in electrolyte to work as cathode. As high amperage current passed in circuit powder is deposited on cathode. Finally cathode plate is taken out from electrolyte solution and scrapped off and scrapped material pulverized to design. Grain size of powder of **Cu, Fe, Ni, Zn, Pb, Sn,** etc. are made by this process.

14.6.1.3 Mechanical Communication or Pulverization

In this method, metals break into very small pieces (two or more pure metal) and then crushed the pieces and finally milling in a ball mill. Under the impact of hard balls, powders repeatedly fracture and weld together by diffusion, forming alloy powder. Process is suitable only for brittle material (very hard) like antimony.

14.6.1.4 Gaseous Reduction

Usually this process consist of grinding the metallic oxide to a finally divided state and then reducing it by hydrogen or CO. This process is applied only for the material which having the melting point above 1100°C. Such as **iron, copper, tungsten, etc.**

14.6.1.5 Granulation Process

In this process, the formation of an oxide film on individual particle when a bath of metal is stirred in contact with air. Process produce the coarse powder with high percentage of oxides.

14.7 BLENDING OF METAL POWDER

Blending of metal powder is defined as the uniform mixing of metal powders, if metal powder is not properly mixed then final **part/component** will not be proper in properties and metallographic structure. Blending of metal powder is carried out for the following purposes:

(*a*) To mix the powder uniformly to obtain uniformity for avoiding any difficulty.

(*b*) To obtain the special mechanical and physical properties as well as characteristic of the powder.

(*c*) To obtain the flow ability property in the metal powder.

(*d*) To avoid the contamination of powders.

(*e*) blending should be done in a controlled condition

(*f*) different type of blending (mixing) equipment is available by the two that blending operation can be performed, i.e., dry or wet condition. Excessive blending should be avoided metal powder like **Al, Mn, Th, Zr, Ti** are explosive owing to their high surface area to volume ratio

(*g*) consequently a great care is required while blending storage, handling of this type of powder.

14.8 COMPACTION (PRESSING) OF METAL POWDER

Next step after blending is metal powder compaction. In compaction process, blended powder are pressed into the die to produce the component of desire density. Figure 14.7 shows the convential method for metal powder compaction by a mechanical or hydraulic press having die and punch. In this process, metal powder is kept in the hollow die and uniform compacting pressure is applied by the punch due to friction between the die wall and metal powder shape of die is achieved by the metal powder.

Some other methods of metal powder are centrifugal compacting, roll pressure compacting, isostatic pressing and metal injection moulding method.

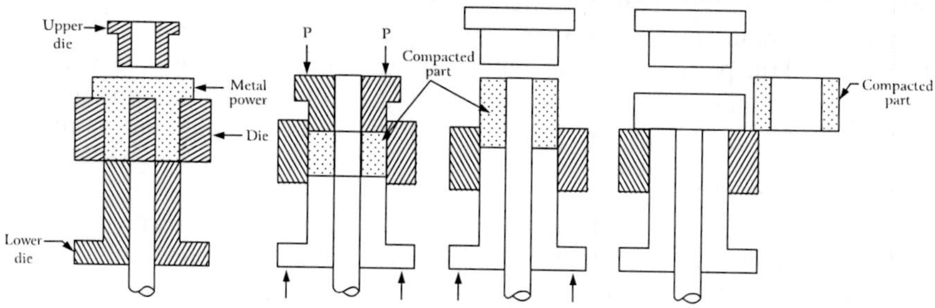

Fig. 14.7 Stages in compaction of metal powder

14.8.1 Dynamic Compaction

Dynamic compaction is used for the bonding and densification of otherwise difficult to from particulate materials, *e.g.*, carbides, nitrides etc.

It include the following:

- Compressed air compaction
- Hydrodynamic compaction
- Explosives
- Gas gun compaction
- Magnetic dynamic compaction

Materials which are compacted by these techniques are:

(i) AIN Aluminium nitride
(ii) SiC Silicon carbide
(iii) BN Boron Nitride
(iv) B_4C Boron carbide
(v) Si_3N_4 Silicon Nitride

The range of compaction times and pressures are 1-100 sec and 10 GPa. The intense pressure pulse during compaction is accompanied by changes which are different from the usual kinetic processes for densification. The shock-wave phenomena are expressed in the Rankin-Hugoniot relationship.

Undesirable gradients of density with accompanying gradients of properties occur which may lead to cracking.

14.9 SINTERING OF GREEN COMPACT

Sintering is defined as a heat treatment process of green compressed compact in a controlled-atmosphere to a temperature below the melting point of metal powder.

Heating done basically to produce the bonding between the individual particles and provide the good strength. Strength of bond usually depends upon the mechanism of diffusion, plastic flow, evaporation of volatile matters, recrystallization grain growth during the sintering.

Important factors which control the sintering:

(i) Temperature

(ii) Time of heating (hr)

(iii) Inside furnace atmosphere

14.9.1 Sintering Process is Carried out in Three Steps

(i) Purge or burn off chamber

(ii) High temperature zone

(iii) Cooling zone

In Sintering process, the Ist step is to increases the adhesion between the metal powder at 150°. Due to the temperature all gases absorbed and in such a way porosity increases. On further heating various changes in the structure encountered. Density of metal increases in Sintering process by applying the pressure. Sintering can be controlled by selection of the metal power and by applying the correct pressure for cold forming. Sintering can be done in a continuous or batch type furnace as per the requirement.

Figure 14.8 shows a continues furnace with three zone, *i.e.*, first zone or burn off chamber, second zone or high temperature zone and third zone or cooling zone.

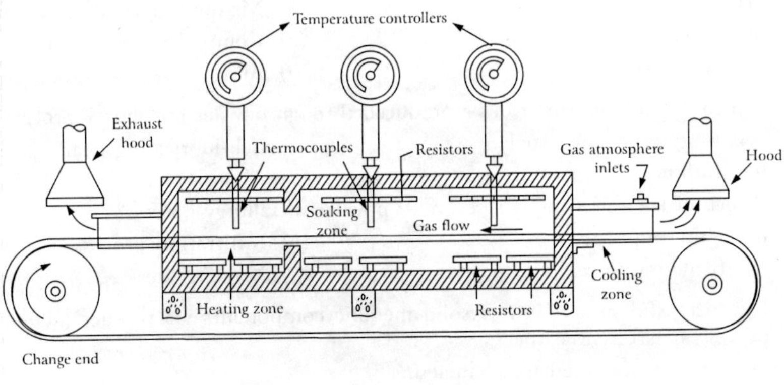

Fig. 14.8 Sintering furnace

14.9.2 Ist Zone/Burn off Zone

In the first zone, all the air gases, lubricants binders are burnt off and evaporated from the green compact. If air of any gas is not properly evaporated from the green compact before high temperature treatment, it can be expand and object will not be properly fabricated. Hence process improves the bonding strength and prevent from any cracking of the green compact.

14.9.3 IInd Zone/High Temperature Zone

In the 2nd zone green compact is heated to the proper sintering temperature and same is kept for 2-24 hours in the furnace in such a way very good atomic diffusion bonding between the green compact particle take place.

S. No.	*Type of Material*	*Sintering temperature*
1.	Iron	1000-1150°C
2.	Copper and brass	750 - 900°C
3.	Tungsten and tantalum	2400°C

14.9.4 IIIrd Zone or Cooling Zone

In the cooling zone sintered part/component gradually cool.

In general sintering process increases in the density of green compact. Various factors that effect the sintering are — **Particle size, time duration for sintering, compacting Pressure and temperature** at which the sintering is carried out.

OBJECTIVE QUESTIONS

1. Sintering is a part of
   (*a*) metal working process (*b*) powder metallurgy process
   (*c*) casting process (d) extrusion process.
2. The following are the main stages in powder metallurgy process
   (*a*) Production of metal powder (*b*) Mixing and blending of powder
   (*c*) Sintering (*d*) Compacting of powder
   (*e*) Secondry process (*f*) All of these.
3. Which of the following can be produced through powder metallurgy process?
   (*a*) Tungsten carbide tools insert (*b*) Self-lubricating bearing
   (*c*) Turbine blade (*d*) All of these.
4. Which of the following is a secondry powder metallurgy process?
   (*a*) Sintering process (*b*) Infiltration process
   (*c*) Blending process (*d*) None of these.
5. For successful production of solid metallic components via powder metallurgy process, it is essential that
   (*a*) no die is required for compacting
   (*b*) they could be closely packed under pressure
   (*c*) these metals should be weldable in solid phase
   (*d*) all of these.
6. Blending process is not required when
   (*a*) different metal powder are used
   (*b*) metal powder of correct particle size and distribution is used
   (*c*) non-metallic particles are added for specific properties
   (*d*) all of these.

7. Which one is a primary powder metallurgy process?
(*a*) Coining
(*b*) Plating
(*c*) Machining
(*d*) Powder production
(*e*) None of these.

8. Metal powders can be produced by:
(*a*) machining
(*b*) mechanical pulverization
(*c*) shooting
(*d*) all of these.

9. Which process of the following are used for production of metal powders?
(*a*) Electrolystic process
(*b*) Water atomization
(*c*) Reduction process
(*d*) Rotating electrode process
(*e*) Centrifugal atomization
(*f*) All of these.

10. The main characteristic of metal powder which make them suitable for powder metallurgy process
(*a*) Flowability
(*b*) Particle size, shape and distribution
(*c*) Sintering ability
(*d*) Compressibility and porosity
(*e*) All of these.

11. Self-lubricating properties to a sintered parts are provided by
(*a*) coining
(*b*) plating
(*c*) impregnation
(*d*) infiltration.

12. Strength of a green compact is increased by
(*a*) pre-sintering
(*b*) compacting
(*c*) impregnating
(*d*) all of these.

13. Strength, density and hardness of sintered parts can be increased by:
(*a*) sizing
(*b*) blending
(*c*) infiltration
(*d*) all of these.

14. Machining of sintered parts is done to
(*a*) cutting thread, groove and under cut
(*b*) producing small holes
(*c*) finish with close tolerance
(*d*) all of these.

15. To enable their easy machining some extremely hard material like tungsten carbide are subjected to:
(*a*) sintering
(*b*) infiltration
(*c*) presintering
(*d*) all of these.

16. Which of the following parts can be produced through powder metallurgy process?
(*a*) Cutting tools and die
(*b*) Porous metal bearing
(*c*) Tungsten filament wire for bulb
(*d*) All of these.

17. Process of converting loose metal powder into a green compact of required shape and size by using punches and die is known as
(*a*) Briquetting
(*b*) Coining
(*c*) Sizing compacted
(*d*) Impregnating.

18. Sintering of briquetted parts provided
    (*a*) dense structure
    (*b*) increased strength
    (*c*) good bonding of particles
    (*d*) alloying of constituents
    (*e*) all of above.

19. Powder metallurgy process is best suitable for
    (*a*) mass production
    (*b*) batch production
    (*c*) one-off production
    (*d*) none of these.

20. In hot isostatic process, which is powder metallurgy process, which one of the following is achived?
    (*a*) Simultaneous impregnating and infiltrating
    (*b*) Simultaneous compacting and sintering
    (*c*) Simultaneous compacting and sizing
    (*d*) None of these.

ANSWERS

1. (*b*)	2. (*f*)	3. (*d*)	4. (*b*)	5. (c)
6. (*b*)	7. (*d*)	8. (*d*)	9. (*f*)	10. (*e*)
11. (*c*)	12. (*a*)	13. (*c*)	14. (*d*)	15. (*c*)
16. (*d*)	17. (*a*)	18. (*e*)	19. (*a*)	20. (*c*)

REVIEW QUESTIONS

1. What do you mean by the term 'Powder Metallurgy'? Mention four unique application of their method.
2. How do you define the "Powder Metallurgy Process"?
3. What are the basic stages in "Powder Metallurgy Process"?
4. How powder metallurgy is different from other manufacturing process? Give two examples of the products manufactured by powder metallurgy process also give advantages of their process. [*UPTU 2004-05*]
5. What is a Green Compact? What special case should be taken in removing it from the die?
6. Write the short notes on:
    (*a*) Manufacture of component by powder metallurgy process
    (*b*) Pre-sintering [*UPTU 2003-04*]
7. Describe briefly the main operation involved in making a powder metallurgy product.
8. Mention the advantages/limitations and applications of powder metallurgy. Describe each step involved in powder metallurgy. [*UPTU 2001-02*]
9. What do you understand by blending of metal powder? And what is its purpose?
10. Explain the process of making briquettes. What are the purpose of compacting metal powder die?
11. Define flowability and compressibility in relation to metal powder.
12. What are the secondary powder metallurgy processes? Describe infiltration, impregnation, sizing and coining operations.

13. Describe the following techniques used in powder metallurgy process:
    (*a*) Injection moulding (*b*) Isostatic processing
    (*c*) Hot isostatic processing
14. Write short notes on:
    (*a*) Osprey process (*b*) Ceracon process
    (*c*) Centrifugal compacting
15. Describe the advantages and limitations of powder metallurgy.
16. What are the problem associated with transfer of compacting pressure on particles of metal powder?
17. What is the purpose of sintering? Explain how it is done?
18. What are the advantages of isostatic compaction?
19. Explain the basic steps of powder metallurgy process. Explain in brief "Sintering". Why sintering is done in a controlled manner? [*UPTU 2006-07, 2008-09*]
20. How the tungsten filament wire is produced?
21. What is the purpose of impregnating powder metallurgy product?
22. What are the materials used for electrical contacts and resistance welding electrode?

❑❑❑

15

CHAPTER

JIGS AND FIXTURES

15.1 INTRODUCTION

Basic object of mass production is to reduces the per unit cost of product of interchangeability to facilitate cost assembling.

Jigs and fixtures are production tools used to accurately produce duplicate and interchangeable item (parts). They are so designed of manufactured that a large number of parts can easily machined or assembled identically. Jigs and fixtures also ensure the interchangeability of parts. They also eliminate the special set up (which is very costly) for each part.

15.2 JIGS AND FIXTURES

Jigs and fixtures are production tools used to accurately manufacture duplicate and interchangeable parts. They are specially designed so that large numbers of components can be machined or assembled identically, and to ensure interchangeability of components.

The economical production of engineering components is greatly facilitated by the provision of jigs and fixtures. The use of a jig or fixture makes a fairly simple operation out of one which would otherwise require a lot of skill and time.

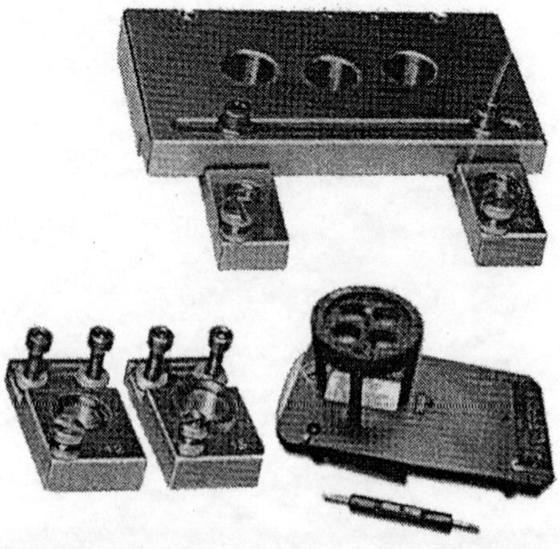

Fig. 15.1 Jigs and fixtures

Both jigs and fixtures position components accurately and hold components rigid and prevent movement during working in order to impart greater productivity and part accuracy. Jigs and fixtures **hold or grip** a workpiece in the predetermined manner of firmness and location, to perform a machining operation on the workpiece during manufacturing operation.

A jig or fixture is designed and built to hold, support and locate every component (part) to ensure that each is drilled or machined within the specified limits.

The correct relationship and alignment between the tool and the workpiece is maintained. Jigs and fixtures may be large (**airplane** fuselages are built on picture frame fixtures) or very small (as in watch making). Their use is limited only by job requirements and the imagination of the designer.

The jigs and fixtures must be accurately made and the material used must be able to withstand wear and the operational (cutting) forces experienced during metal cutting.

15.2.1 Jigs

Jig is a device which is used to hold and locate a part for a specific operation in such a way that they guide one or more cutting tools or in a simple word it could be said that Jig is a device which holds as well as guide the cutting tools. Usually Jigs are **not fixed** on a machine.

They are basically used for **drilling** and **reaming** purpose. A simple drill Jig is shown in Fig. 15.2. Object (workpiece) that to be drilled is hold in drilling Jig. Commonly Jigs are fitted with hardened steel bushing. They are fitted in such a way that they are not fixed, they are free to moved to permit the proper resistering of the work and tool.

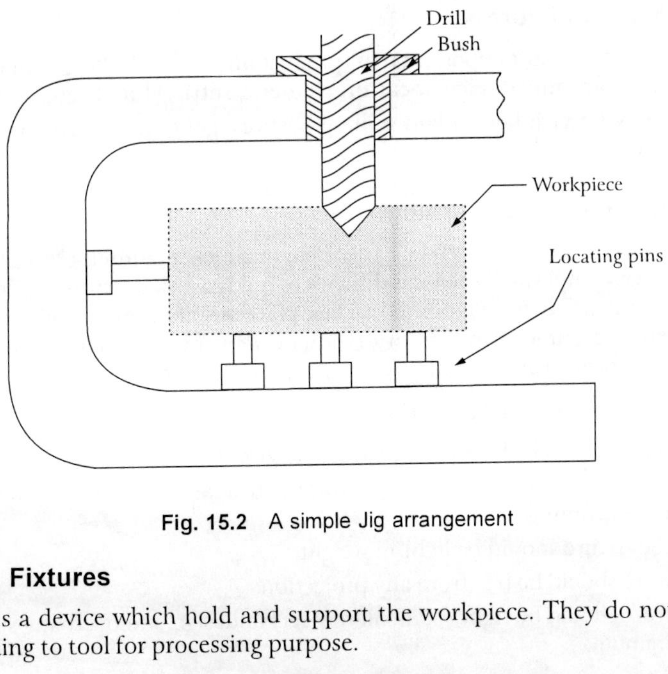

Fig. 15.2 A simple Jig arrangement

15.2.2 Fixtures

Fixture is a device which hold and support the workpiece. They do not provide any guiding to tool for processing purpose.

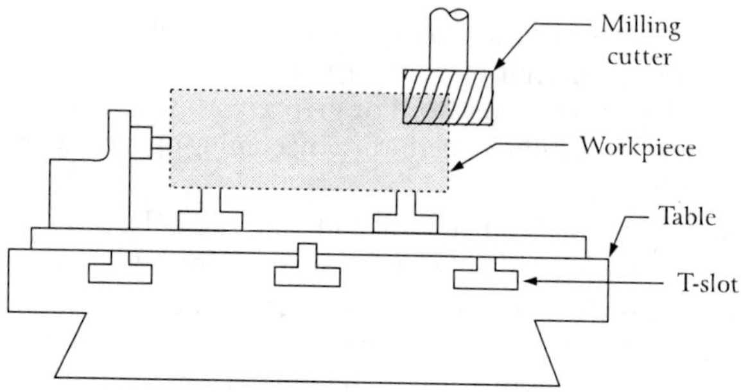

Fig. 15.3 A simple fixture arrangement

15.3 TYPES OF FIXTURES

Fixtures are usually classified by the types of machining operations to performed. Common fixtures are given below:

(i) Milling fixture
(ii) Lathe fixture (Turning)
(iii) Welding fixture
(iv) Grinding fixture
(v) Indexing fixture
(vi) Broaching fixture
(vii) Boring fixture
(viii) Tapping fixture

15.3.1 Milling Fixture

Milling is one of the most important mass production process. Milling fixture consists **base, T-bolt, clamping device, location element, setting block**, etc.

They may be vice Jaws, plain milling fixtures, gang milling fixture, straight milling fixture.

15.3.2 Lathe Fixture (Turning)

Simplest types of workpiece can be easily hold in chuck, mandrels etc. but if shape of workpiece is complicated then it is difficult to hold a work piece, in this condition, a fixture is needed. They are mounted on face plate or spindle of machine by the use of lathe fixture, accuracy, Productivity, rigidity can be increases. On the other hand vibration, scrap, etc. can be reduces by the used for fixtures.

Design Consideration of Lathe Fixture:

(a) fixture should be balanced, to avoid by vibration.
(b) clamp should be positioned on surfaces, on areas, which are rigid before and after machining.
(c) lathe fixture should be light in weight.
(d) fixture should be free from any projection.
(e) fixture should be right (ok in all respect) in design and overhang should be minimum.

(f) parts to be fixtured should, when ever possible, be gripped with their largest diameter.
(g) the part should be positioned in the fixtures so that most of machine operation can be performed in the first fixturing.

15.3.3 Welding Fixture

In welding process proper holding of workpiece is required, if workpiece is not properly clamped by a fixture more chance of defective (distorted part) welding. Purpose of welding fixture is to hold the parts to be welded in a proper manner to avoid defective welding.

Main object of a welding fixtures are:

(i) To hold the parts in a convenient position for welding
(ii) To provide proper clearance for filler metal
(iii) To reduce the distortion

15.3.3.1 Factors that affect the welding fixture design

(i) cost of fixture material
(ii) type of material to be welded
(iii) types of welding operation performed
(iv) complexity of weld
(v) size and shape of part to be welded.

15.3.3.2 Common Welding fixtures are

(i) Gas welding fixture
(ii) Electric arc welding fixture
(iii) Resistance welding fixture
(iv) Thermit welding fixture etc.

15.3.4 Grinding Fixture

In grinding process proper clamping of workpiece is required otherwise high degree of accuracy and tolerances can not be obtained. In surface grinding magnetic chuck (table) are used to hold the plain workpiece. For cylindrical workpiece grinding chucks are used.

15.3.5 Indexing Fixture

Indexing fixtures are used in milling machine while indexing for gear cutting.

15.3.6 Broaching Fixture

Broaching fixtures are used when internal surface of any hole required accurate finishing.

Functions of a broaching fixture:

(a) holding the workpiece firmly
(b) locating the workpiece accurately
(c) guide the tool in respect of Job.
(d) index the workpiece between the cut.

15.4 COMPARISON BETWEEN THE JIGS AND FIXTURE

Sl. No.	*Jig*	*Fixture*
1.	From the construction point of view Jigs are lighter in weight.	Fixtures are generally heavier in construction.
2.	Jigs hold and position the workpiece as well as it guide the tool.	Fixture hold the work and position the workpiece but they do not guide the tool.
3.	They are quicker in handling and clamping with the table.	They are rigidly bolted on the machine table.
4.	Jigs portionally used in drilling, reaming taping and counter bearing operations.	Fixture are commonly employed for holding work in milling, grinding, planning/turning operations.
5.	Common types of Jigs are **Drill Jigs, Template Jigs, Plate Jigs, Box Jigs,** etc.	Common type of fixtures are **plate fixture, angle fixture, indexing fixture** and **milling fixture**.

15.5 ADVANTAGES OF JIGS AND FIXTURES

(i) It saves operation labour by minimizing the handling operations.

(ii) It increases the interchangeability.

(iii) It increases the productivity by increasing the speed, feed, depth of cut as object is firmly gripped.

(iv) It increases the machine tools versality.

(v) By the application of it, no skill labour is required to perform the operation. As tool guiding element ensure the correct positioning of the tool with respect of workpiece.

(vi) It reduces the cost quantity control of machine parts.

(vii) It increases the machining accuracy. Ensure uniform quality in manufacture, therefore, all similar parts are interchangeable.

(viii) It reduces the scrap quantity of material.

(ix) Through use of jigs and fixtures heavy and complex shapes parts can be hold rigidly on machine table.

(x) By the use of jigs and fixture reduces the accident and improve the safety environment.

15.6 PRINCIPLES OF LOCATION

A body in space have twelve degree of freedom means body is free to move in any direction location system should prevent linear and rotational motion of the work-piece along and around the X, Y and Z axes for required precision and accuracy.

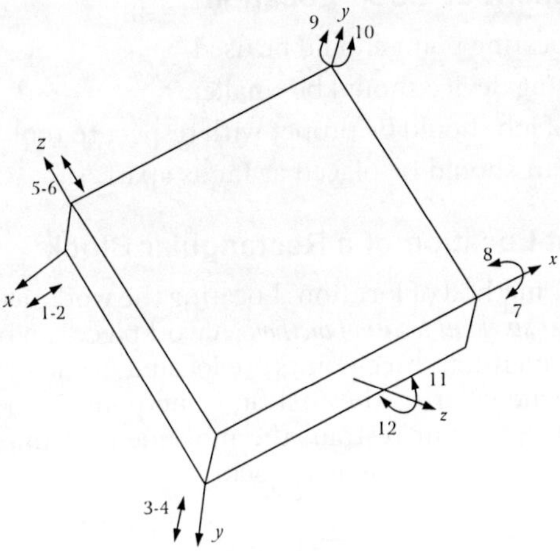

Fig. 15.4(a) 12-degree of freedom

15.6.1 Degree of Freedom (D.O.F.)

A workpiece or object in space free to move has 12 d.o.f

(i) Along 3 axes in opposite direction = Six (6) degree of freedom

(ii) rotate along these axes clockwise and anticlockwise = 6 d.o.f.

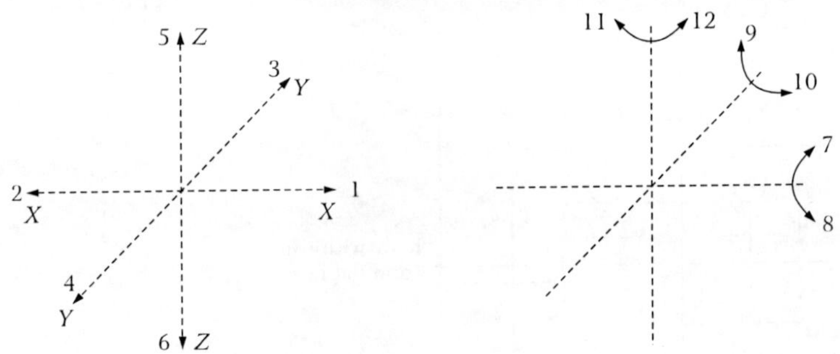

(a) 6-translatory (b) 6-rotary

Fig. 15.4(b)

15.6.2 Twelve Degree of Freedom

(*a*) 6 movement of translation—axial movement in either direction along the axes *XX*, *YY* and *ZZ*.

(*b*) 6 rotational movement, i.e., rotation of body about *XX*, *YY*, *ZZ* axes in either clockwise or anticlockwise direction. Each direction of movement is considered one degree of freedom.

It should be ensure for designing location facilities that arrangement should be for positive restraints.

15.6.3 Requirement of Good Location

(i) Minimum locating point should be used.
(ii) Size of locating device should be smaller.
(iii) Alignment of job should be proper with respect to tool.
(iv) Locating point should be placed as far as apart.

15.6.4 Six Point Location of a Rectangular Block [UPTU 2009-10]

It is known as 3-2-1 method of location. Locating the workpiece/component from six point is known as *six point location method.* A workpiece can be located by means of six point. In this method, three points are located in the first plane, two in the second plane, perpendicular to the first and one point is in the third plane as shown in Fig. 15.5. Six point restrain the movement of uneven object guiding principal is same as that of rectangular block.

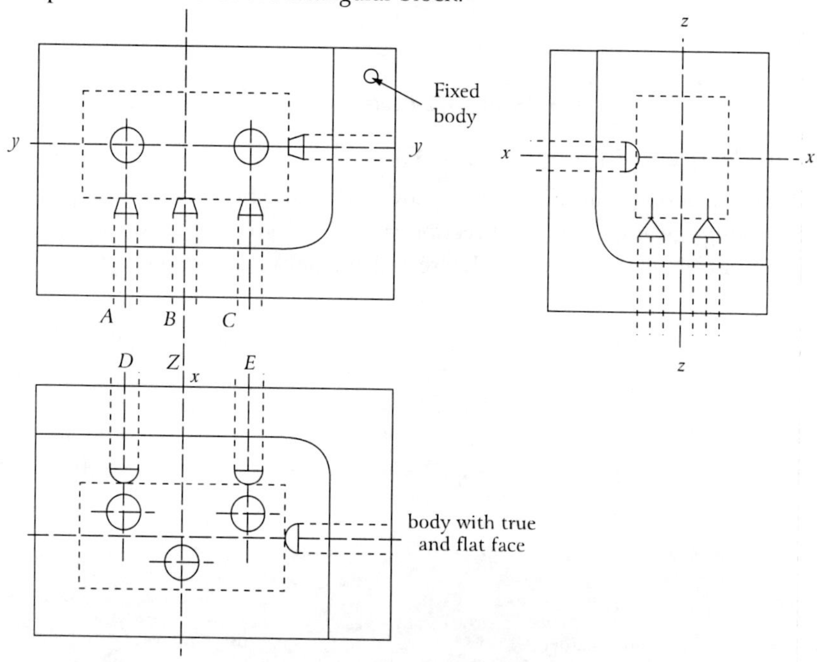

Fig. 15.5 3-2-1 principle of location

15.7 GENERAL CLASSIFICATION

The terms "jigs" and "fixtures" are rather loosely used by shopmen. While this is necessarily so in some cases, in most instances it is more correct to apply the term jig to a device which holds the work and automatically locates the cutting tool so that each piece produced is a duplicate of all the others. **Fixtures**, on the other hand, do not automatically locate the cutting tool. While fixtures may be used to produce duplicates, this result is usually gained by means of a cutting tool locating jig separated from the fixture itself. **Fixtures** are essentially **work-holding devices**.

Objects of These Tools

While several effects are gained by using jigs and fixtures, they all reduce to one thing namely production. For example, by the proper use of jigs and fixtures, production is made more uniform, giving interchangeability of parts. If jigs and fixtures are properly used, production is attended by a reduction of labour cost, both when the machine parts are being produced and when the parts are assembled to produce the completed machine.

In one case, the jig is both a rapid production tool and an interchangeability tool. In other case, the jig is merely a convenient tool for getting rapid production.

As a Work Holder

It is usual in the design of jigs to next consider how the piece shall be held in the prospective jig. The points or surfaces upon the piece which are those best suited for location points and surfaces are decided upon. If the piece has been previously machined, the surface machined usually offers the best location to work. On the other hand, the surfaces of the stock are rough, as in an ordinary casting, the selection of the locating surfaces or surface is usually a more difficult one. Usually some surface or hole will be essentially more important than all the remaining surfaces or holes. In such a case, the jig designer uses location points which will position the important hole or surface, afterward considering the points of lesser importance.

A flat surface, if it has previously been machined, is usually located against a flat surface, if not previously machined, a flat surface should be given line or point contact.

15.8 DRILL JIGS

A jig or a fixture is designed to fulfill the following requirements:

(*a*) Design of jigs/fixtures should be such that they can used on a number of machine tools.

(*b*) Machine tool should not required further any modification to holding the object.

(*c*) Design should not create bottlenecks in the production area.

A drill jig is basically used to produce the hole in a large number of components/workpiece on the basics of inter-changeable. In practicle life a large varies of drilling jigs are used.

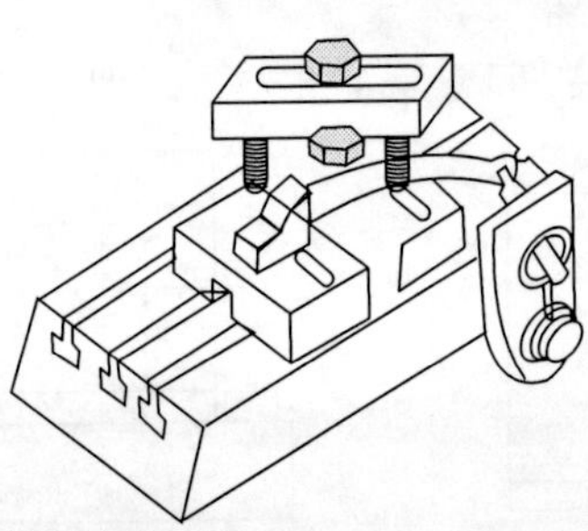

Fig. 15.6(*a*) A standard unit fixture

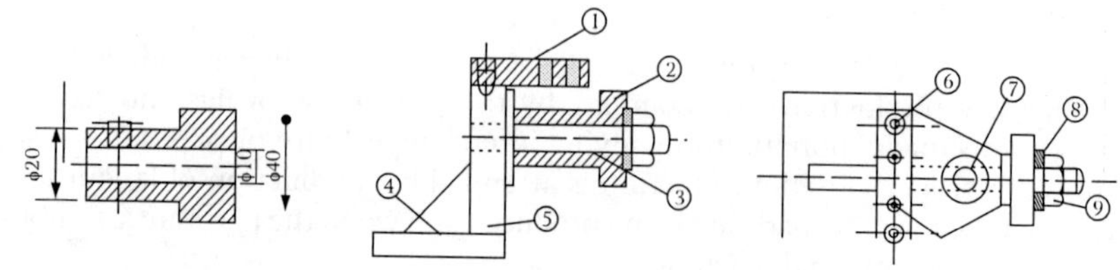

Fig. 15.6(*b*) Drilling Jig with components: 1. Jig plate, 2. Workpiece, 3. Pin, 4. Web, 5. Body, 6. Allen Screw, 7. Bush, 8. C-washer, 9. Nut

15.8.1 Different Type of Drilling Jigs

(*a*) Template Jig
(*b*) Channel Type Jig
(*c*) Plate Type Jig
(*d*) Diameter Jig
(*e*) Box Type Jig
(*f*) Leaf Jig
(*g*) Indexing Jig
(*h*) Open Type Jig
(*i*) Vice Type Jig

While a study of the illustrations in this book will show the student that jigs are an important factor in all production machines, perhaps in no other machine is their importance so complete as ' in the drilling of holes'. For this reason a drill jig will be used to illustrate jig construction. Figure 15.7 shows the different Jigs.

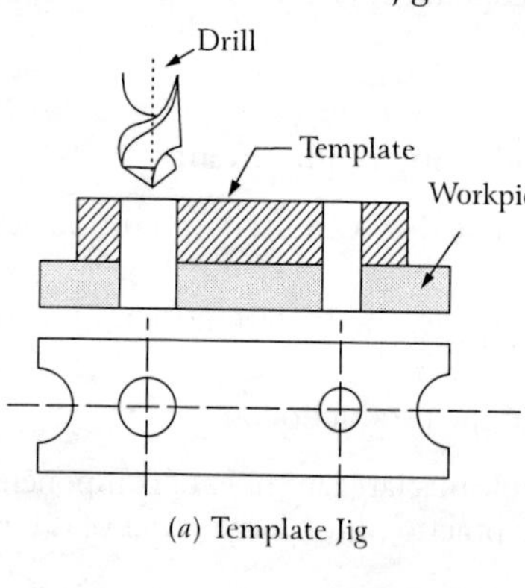

(*a*) Template Jig

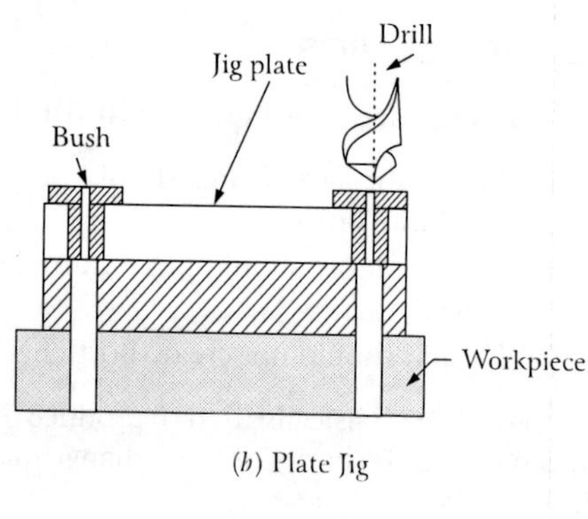

(*b*) Plate Jig

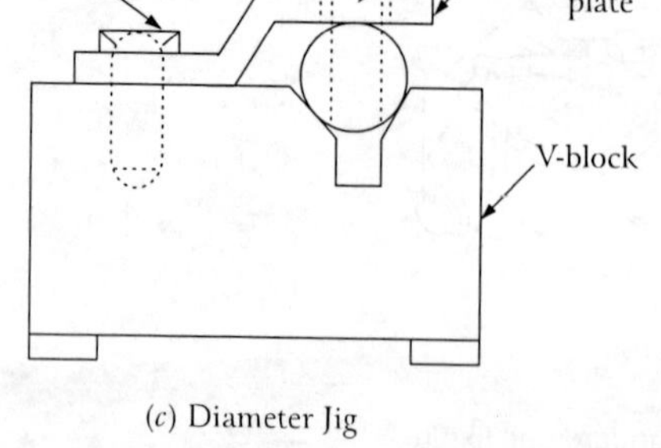

(*c*) Diameter Jig

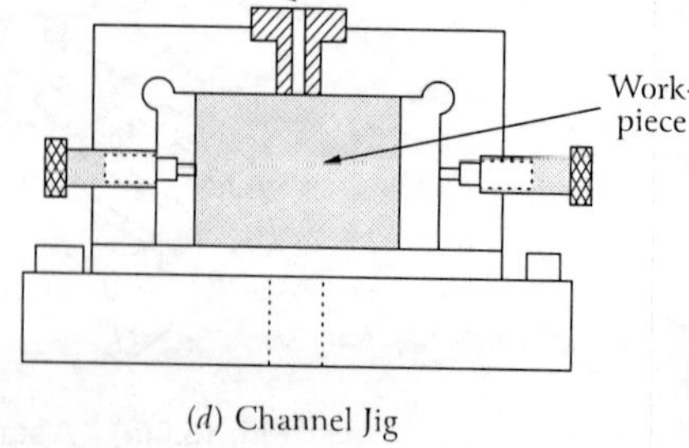

(*d*) Channel Jig

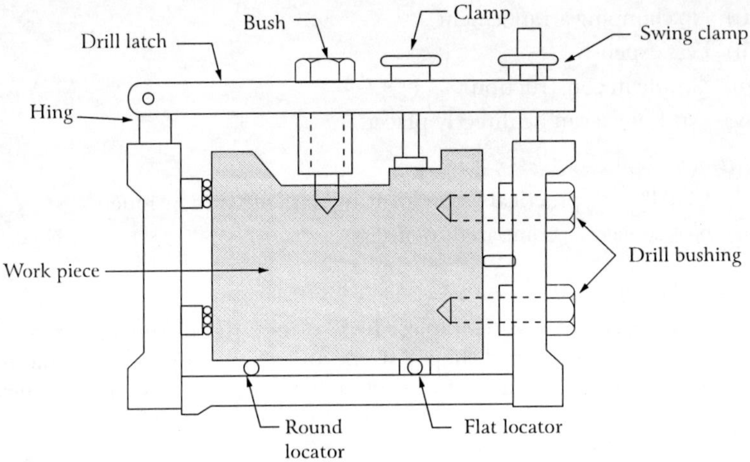

(*e*) Box Jig

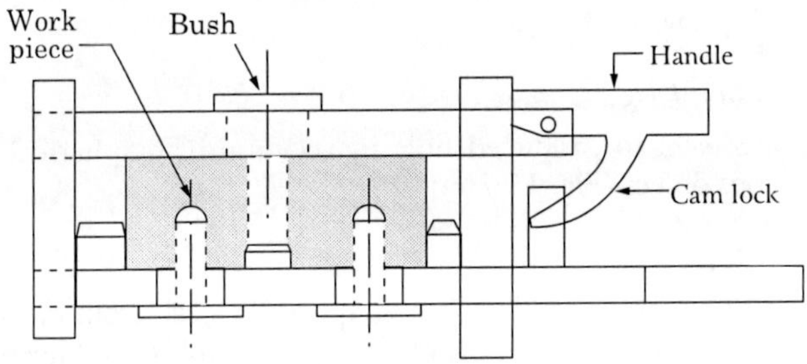

(*f*) Leaf Jig

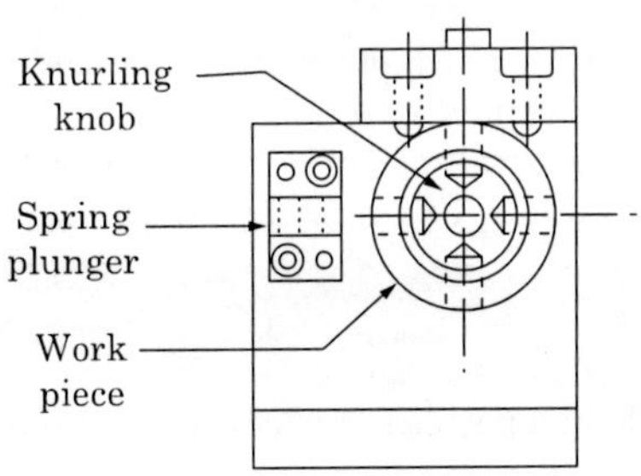

(*g*) Index Jig

Fig. 15.7 Different type of Jigs

15.8.1.1 Template Jig

This is one of the simplest and common type Jig, used for drilling operation. It is used for mass production and for high degree of accuracy. It is not clamped only workpiece is gripped by it. It may or may not be consist the bush. Figure 15.7(a) shows the template Jig.

Advantages:

(i) No clamping arrangement.
(ii) Less expensive.
(iii) Simple in construction.
(iv) Workpiece can be directly placed.

Limitations:

(i) Usually not practical when locating data are dimensioned.
(ii) Not so easy as compared to others.

15.8.1.2 Channel Jig

Consisting a **bush** and it is a simplest form of box Jig. In this jig arrangement workpiece is kept in a channel shaped frame and gripped by a screw. Channel Jig is used only for simplest shapes of workpiece Fig. 15.7(*d*) shows the channel Jig.

15.8.1.3 Plate Jig

Plate Jig usually consist **plate**, locating pins and **bushes** and a C-shape clamping device is needed. They are used to produced hole in a long strip. Clamping element is not a part of Jig (Fig. 15.7(*b*)).

15.8.1.4 Diamond Jig

Diamond Jig are used to produced hole in a cylindrical, spherical shape or a diametrical shapes object (Fig. 15.7(*c*)).

15.8.1.5 Box Jig

As the name deplict it is made like a box shape and closed from mostly all sides. This type of Jig used for part of a awkward shape (complex) and need machining more than one plane (Fig. 15.7(*e*)).

15.8.1.6 Leaf Jig

Leaf Jig consist a hinged cover/leaf which can sewing open to insert the workpiece and closed by supporting dowel pins for proper alignment of workpiece. Drill bushs are located in the leaf. Main advantage of leaf Jig that small and simple part can easily handled and main limitation of leaf Jig that more chance of wear or distortion of Pivot Pins.

15.8.1.7 Indexing Jig

They are used to drill hole in a pattern they consist and indexing mechanism.

15.9 LOCATING WORK IN DRILL JIGS

Figure 15.8(a) shows the use of pins or studs used as side-locating points in simple jig work, and Fig. shows how V's are similarly used on curved surfaces. While these are simple examples, they illustrate a principle which can easily be applied to more complicated cases.

The use of locating pads is shown in Fig. 15.8(a) and Fig. 15.8(b) shows how an inserted pin may be used for supporting a plane surface. Where pins are used for location points, Fig 15.8 shows the sides against the pieces which are located usually flatted somewhat to bring surface contact rather than line contact. Hardening the pins will also prevent excessive wear.

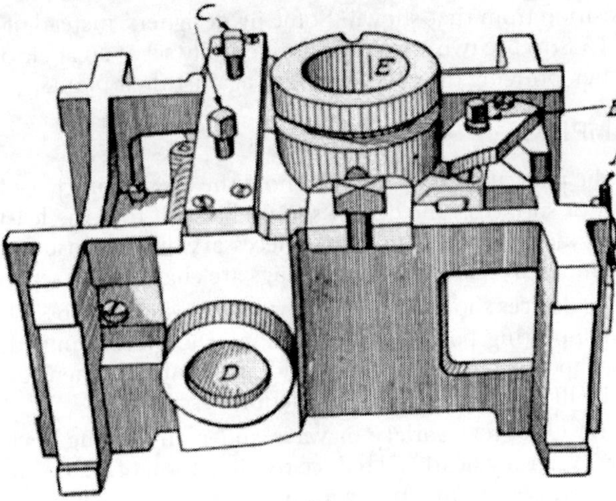

(*a*)

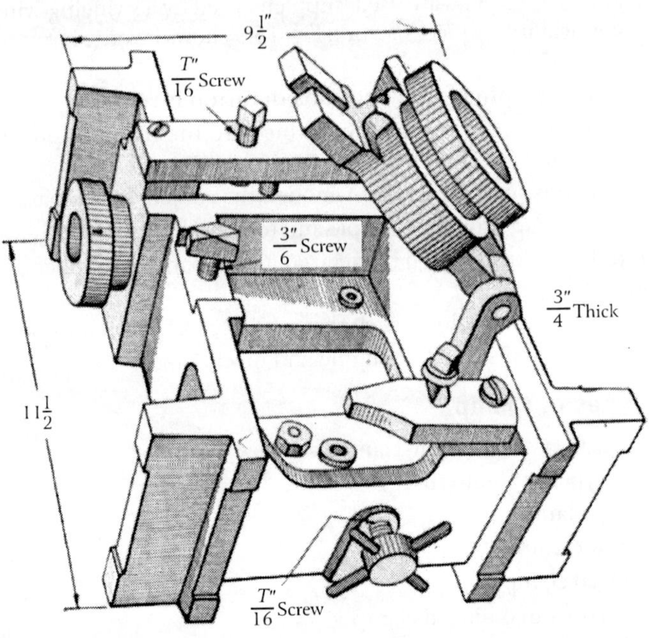

(*b*)

Fig. 15.8 Two views of closed box drill jig courtesy of "American Machinist"

15.10 LOCATING POINTS WITH ADJUSTMENTS

In some cases, it is well to have locating surfaces or points adjustable. In Fig. 15.8(a) the inserted pin, if threaded into base B could, for example, be raised to some other position from that shown. Some jig designers, instead of the V-block shown in Fig. 15.8(b) use two set screws horizontally set at an angle of 45 degrees with one another, bringing the curved surface against their points.

15.11 CLAMPING

This refers to the particular devices which hold the piece being jigged against the location points or surfaces. The design should be such that the least number of clamping devices may be used, so that no unnecessary time is consumed in charging the jig, as this limits production unless the jigs are charged as a separate job.

All clamping devices should exert their pressure, wherever possible, directly in line with the supporting points. If this is done the piece clamped will not be sprung out of shape. As an aid in understanding the already mentioned points, a simple jig will be illustrated and its construction described.

This is done in a great variety of ways and many of the devices are very ingenious. However, they nearly all reduce to some form of clamp either straight, bent, or forked, pressed against the work by either a set screw, a cap screw, or a cam. In Fig. 15.9 it will be noted that the clamping is done by a strap similar to that shown in Fig. 15.9 and the piece is pushed into position by knurled head set screws. In Fig. 15.9 set screws are used, supplemented by a swinging-wing clamp at the side pressing against the piece of work. Fig. 15.9 shows favorite forms of clamping devices.

For effective clamping following consideration is required:

1. To avoiding distortion of the supporting part, the clamping pressure should be applied only on supporting part.
2. Clamping pressure should be so enough that object can be easily hold.
3. Design of clamp should be simple and fool proof.
4. Clamp should be designed to facilitate the maximum number of operation in one setting.
5. Design of clamp should be such that it can be easily lifted while unloading.
6. Clamp operation should be simple and mechanized.

15.11.1 Types of Clamp

For high production quick acting clamps are used.

Following types of clamps used:

(i) Strap clamp
(ii) Screw clamp
(iii) Hinged clamp
(iv) Cam operated hinged clamp
(v) C-Clamp
(vi) Heel clamp
(vii) Toggle clamp
(viii) Quick acting Cam operated clamp.

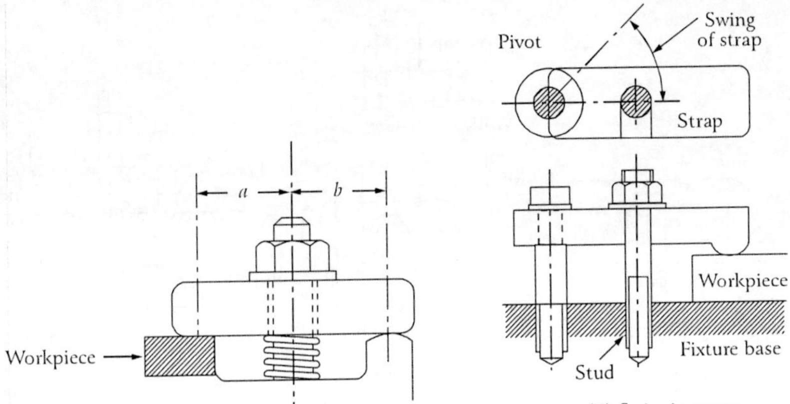

(i) Simple bridge type

(ii) Swinging type

(a) Strap clamps

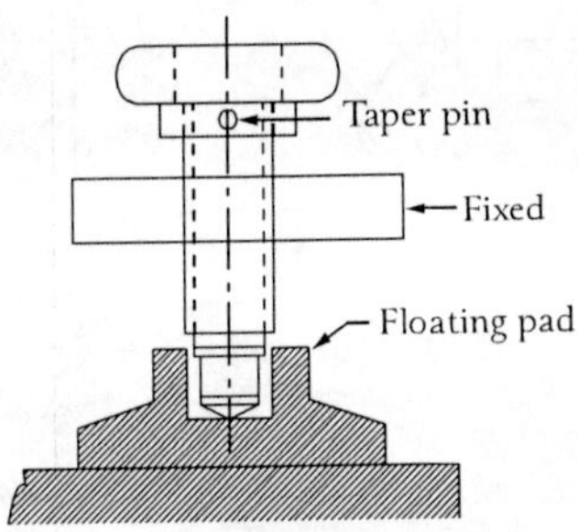

(b) Screw clamp

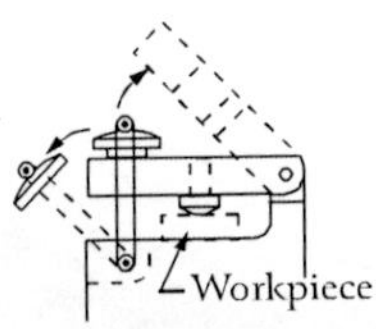

(c) Hinged clamp

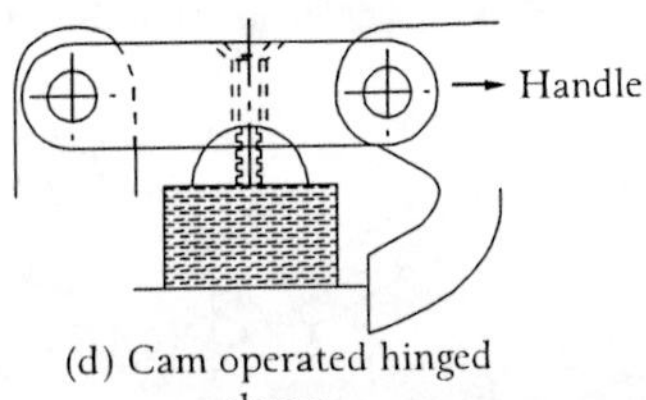

(d) Cam operated hinged clamp.

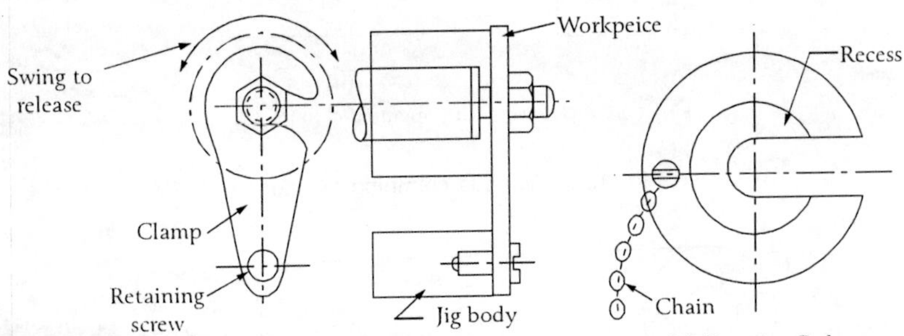

(i) Free-C-clamp

(ii) Capative C-clamp

(*e*) C-clamps

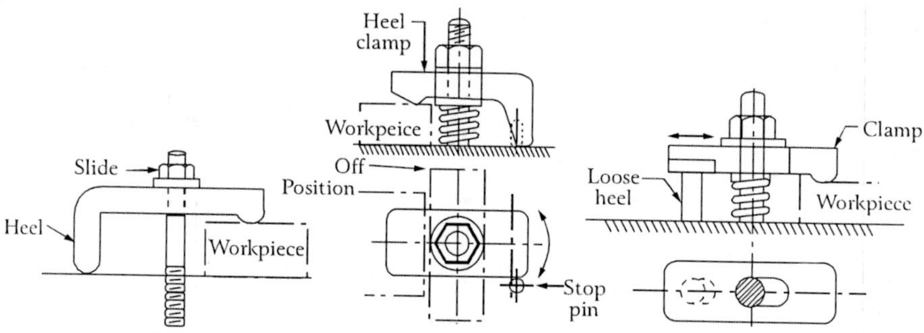

(i) Dog (ii) Solid heel (iii) Loose-guided heel

(*f*) Heel clamps

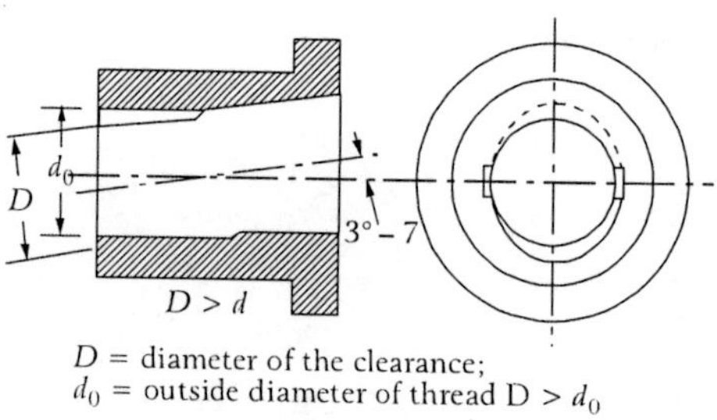

(g) quick acting nut

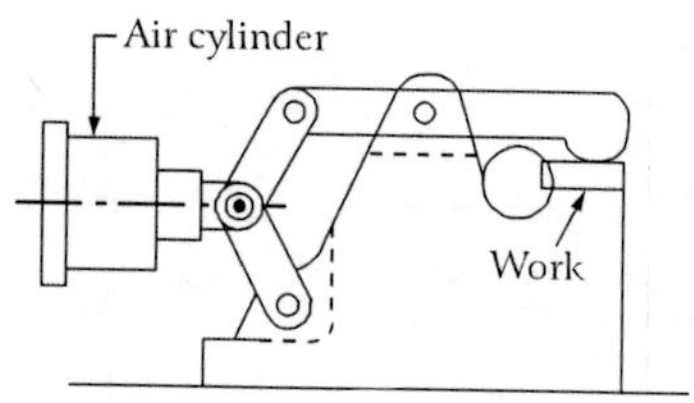

(h) Toggle clamp

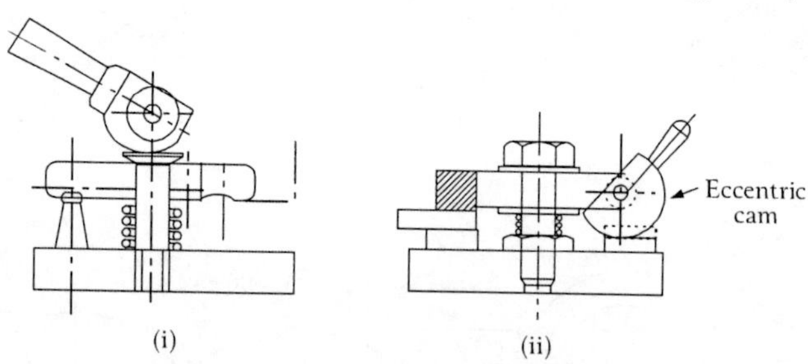

(*i*) Quick acting cam operated clamps

Fig. 15.9 Various clamping devices

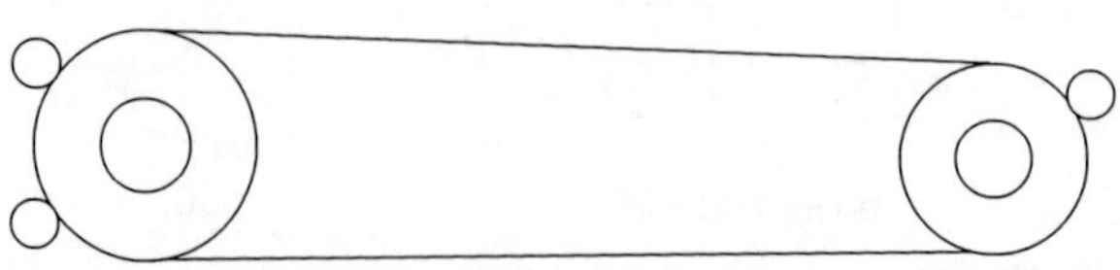

Fig. 15.10 Method of using locating stubs or pins

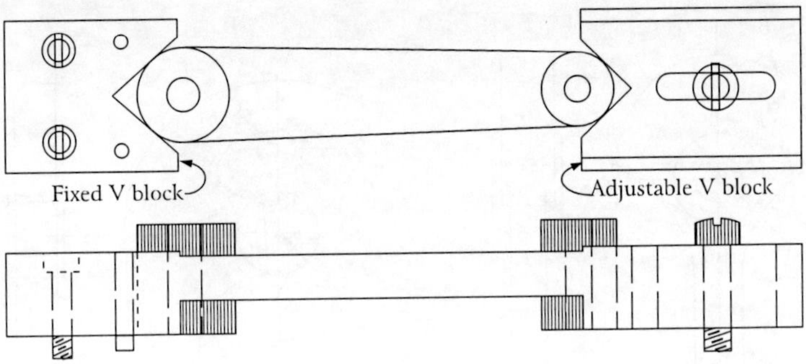

Fig. 15.11 Diagrams showing fixed and adjustable V's

Jig Body

While steel may be used for the body or frame of a jig, it is a usual thing to use cast iron. If cast iron is used the jig can be more or less completely worked out in the pattern, and possibilities of alteration in design may show as desirable. When it is realized that many shops use jigs weighing hundreds of pounds in their production work, it is clearly seen why cast iron is largely used for jig bodies:

Bearing Points

Only in the smaller sizes do drill jigs rest upon a surface of any considerable area. It will be noted, by reference to that supporting points, termed feet, are provided on those sides of the jig which are to rest upon the work table: The height of the feet must be sufficient to clear all bushings, holding screws, or other projecting parts. Also their bearing area must be sufficiently large to prevent their slipping into the bolt.

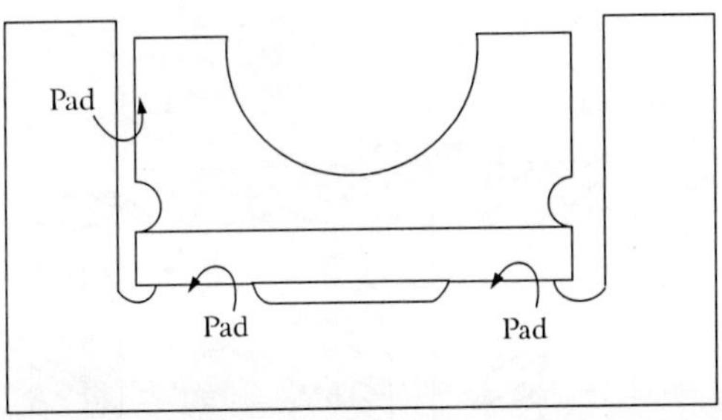

Fig. 15.12 Diagram showing use of locating pads

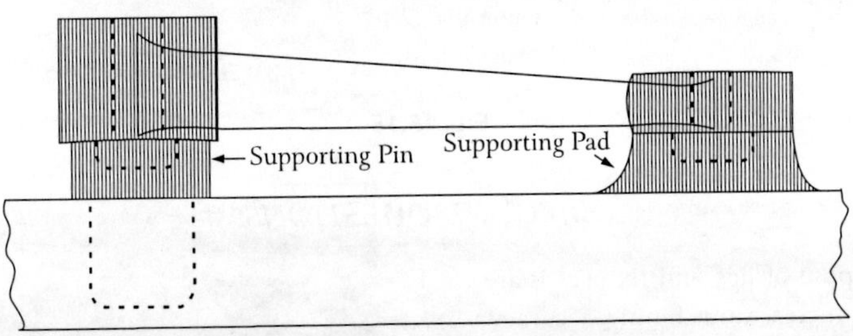

Fig. 15.13 Sketch showing adjustable locating points

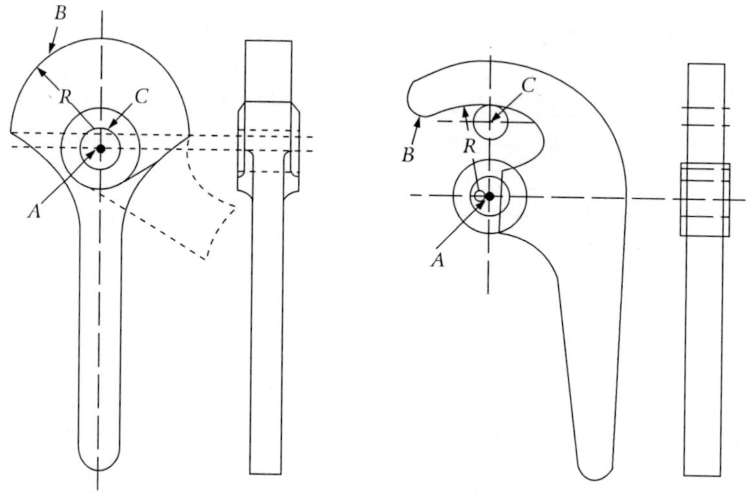

C-clamps

Fig. 15.14 Diagrams showing Cams of eccentrics used for clamping coutesy of "Mechinery". New York City.

Drill Bushes

Drill bushes may be the following:

(i) plain bush

(ii) flanged bush

(iii) screwed bush

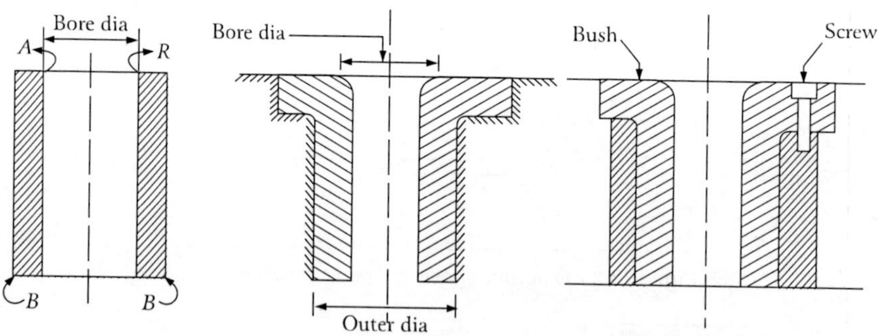

(*a*) Plain bush (*b*) Flanged bush (*c*) Screwed bush

Fig. 15.15

OBJECTIVE QUESTIONS

1. Purpose of jigs and fixtures is/are to

 (*a*) increase machining accuracy

 (*b*) facilitate interchangeability

(*c*) increases productivity
(*d*) decreases expenditure on quality control
(*e*) all of the above.

2. Jigs are heavier than fixture in construction:
(*a*) yes (*b*) no (*c*) no idea.

3. A jig is define as a device
(*a*) is used to check the accuracy of workpiece
(*b*) holds and locate a workpiece guides and controls one or more cutting tool
(*c*) holds and locate a workpiece during an inspection or for a manufacturing operation
(*d*) all of the above.

4. Fixture is used
(*a*) to check the accuracy
(*b*) for holding and guiding the tool in drilling, reaming or tapping operations
(*c*) for holding the work in milling, grinding, planning or turning operations
(*d*) none of the above.

5. Jigs are used
(*a*) to check the accuracy of workpiece
(*b*) for holding the work in milling, grinding, planning or turning of operations
(*c*) for holding and guiding the tool in drilling, reaming, or tapping operations
(*d*) none of the above.

6. Fixture is known as a device which:
(*a*) to check the accuracy of workpiece
(*b*) holds and locate a workpiece and guide and control one or more cutting tool
(*c*) holds and locate a workpiece during inspection or for manufacturing operation
(*d*) hold and position the workpiece but do not guide the tool.

7. Jigs and fixtures are used in industry to
(*a*) reduces the time of machining
(*b*) enable employment of class-skilled operator
(*c*) increases the rate of production
(*d*) increases the machining accuracy.

8. Six-point location of a rectangular block uses the following method of location
(*a*) 1-1-4 method (*b*) 3-2-1 method
(*c*) 2-2-2 method (*d*) any one of above.

9. A jig is commonly used to the following type of machine
(*a*) shaper (*b*) drilling machine
(*c*) lathe machine (*d*) milling machine

10. Commonly used material for making bushes is
(*a*) hardened steel (*b*) aluminium
(*c*) wrought iron (*d*) copper.

11. Jigs and fixutre are used when
(*a*) mass production of parts is required
(*b*) high degree of accuracy is required
(*c*) a few component are to be produced
(*d*) different type of component are to be produced.

12. Material commonly used for making jigs and fixture are:
(*a*) aluminium (*b*) brass
(*c*) wrough iron (*d*) steel.

13. The floating position of the holding fixture in a rotary transfer device is used to
    (*a*) improve the accuracy
    (*b*) improve upon the acceleration and acceleration characteristic
    (*c*) reduce the tendency to over index
    (*d*) reduce the cycle time.

14. Choose the correct statement
    (*a*) A fixture is used to guide the tool as well as to locate and clamp the workpiece.
    (*b*) No arrangement to guide the tool is provided in jig.
    (*c*) A jig is used to guide the tool as well as to locate and clamp the workpiece.
    (*d*) Jigs are used in a machine to locate and clamp the workpiece and guide the tool. [*GATE 1999*]

15. Consider the following statements regarding fixture employed for holding workpiece during machine
    1. The location is based on 3-2-1 principle.
    2. The number refer to pins employed in three mutually perpendicular planes to arrest all the degree of freedom.
    3. Fixture also provide tool guidance, of these statements.

    (*a*) 1, 2 and 3 are correct (*b*) 2 and 3 are correct
    (*c*) 1 and 3 are correct (*d*) 1 and 2 are correct [*IES 1995*]

16. In the selection of a suitable material for the manufacturing of a drill bush, the critical property to be considered is
    (*a*) Toughness (*b*) Wear resistance
    (*c*) Elasticity (*d*) Rigidity

17. Cylindrical Job should always be clamped in a .... for drilling
    (*a*) V-block (*b*) three jaw
    (*c*) collect chuck (*d*) Vice

18. Clamping should always be arranged:
    (*a*) Directly above the points supporting the tool
    (*b*) Directly below the point supporting the work
    (*c*) Directly above the point supporting the work
    (d) None of the above

19. Most of the locating and clamping devices are made up of:
    (*a*) Aluminium (*b*) Mild steel
    (*c*) Cast Iron (*d*) Steel

20. A workpiece is space free to move in any direction can have
    (*a*) 12 degree of freedom (*b*) 9 degree of freedom
    (*c*) 6 degree of freedom (*d*) 3 degree of freedom

21. Simplest type of Jig of the following is:
    (*a*) Box Jig (b) Template Jig
    (*c*) Leaf Jig (d) Plate Jig.

22. Choose the correct statement
    (*a*) A fixture is used to guide the tool as well as to locate and clamp the workpiece
    (*b*) No arrangement to guide the tool is provided in Jig
    (*c*) A Jig is used go guide the tool as well as to locate and clamp the workpiece
    (*d*) Jigs are used in CNC machine to locate and clamp the work piece and guide the tool.

ANSWERS

1. (*e*)	2. (*b*)	3. (*b*)	4. (*c*)	5. (*c*)
6. (*d*)	7. (*c*)	8. (*b*)	9. (*b*)	10. (*d*)

11. (*a*)	12. (*c*)	13. (*b*)	14. (*c*)	15. (*a*)
16. (*b*)	17. (*a*)	18. (*c*)	19. (*d*)	20. (*a*)
21. (*b*)	22. (*c*)			

REVIEW QUESTIONS

1. What are the advantages of using jigs and fixtures?
2. What do you understand by "Degree of Freedom"? Explain.
3. What are jigs and fixtures? Explain the same.
4. .... is a body which holds and positions the workpiece and guide the cutting tool during the machining operation.
5. A .... should be securly fastened to the table of the machine upon which the work is done.
6. The term .... refers to establish a definite location relationship between the workpiece and the cutting tool is jig or fixture.
7. A .... is a device that holds the workpiece firmly against the locators provided.
8. A channel jig is the simplest form of................ jig.
9. Define the following terms—*(a)* jig *(b)* fixture.
10. What is the difference between jig and fixture? *[UPTU 2001-02, 2002-03, 2003-04]*
11. What is a clamp?
12. What are the functions of a clamping device?
13. Explain briefly the principles of jig and fixture design.
14. Describe the 3-2-1 principle of location. *[UPTU 2003-04, 2009-10]*
15. Describe briefly the principle of location.
16. What is the meaning of "locating and clamping" of a component?
17. What are the common method of locating circular surface? What is V-angle for locating circular surface and also gives shortcoming of V-location? *[UPTU 2004-05]*
18. What are the economical aspect of jigs and fixture differentiate between drill Jigs and a fixture? *[UPTU 2004-05]*
19. Explain the following fixtures
    (*a*) Milling fixture
    (*b*) Grinding fixture
    (*c*) Turning fixture
    (*d*) Indexing fixture.
20. Explain the following jigs
    (*a*) Template Jig
    (*b*) Diameter Jig
    (*c*) Channel Jig
    (*d*) Plate Jig
    (*e*) Ring Jig
    (*f*) Universal Jig.
21. What are the basic requirement of a jig?
22. List different types of drilling bush.
23. Write the basic criteria considering in designing of jigs and fixtures.

❑❑❑

16

CHAPTER

MANUFACTURING OF PLASTIC COMPONENTS

16.1 INTRODUCTION OF PLASTICS

Plastic basically defined as a non-metallic material which can be moulted into any shape. They are organic materials or in other words it can be said that, plastics are natural or synthetic resine or their compounds which can be casted, moulded as films or coatings.

Plastic usually consists carbon chemically combined with hydrogen, oxygen and other non-metallic materials. Natural material include **wood**, **coal**, **natural rubber**, **animal fibres of food**. Where as synthetics includes **solvent**, **adhesive rubber**, **lubricants**, **cutting oils**, etc.

They are used on a daily basis throughout the world. The word plastic is a common term that is used for many materials of a synthetic or semi-synthetic nature. The term was derived from the **Greek plastikos**, which means "fit for molding." Plastics are a wide variety of combinations of properties when viewed as a whole. They are used for shellac, cellulose, rubber, and asphalt. We also synthetically manufacture items such as clothing, packaging, automobiles, electronics, aircrafts, medical supplies, and recreational items. The list could go on and on and it is obvious that much of what have today would not be possible without plastics.

One way plastics changed the world was in cost. It was so much cheaper to manufacture than other materials and the various ways it could be used was staggering. For instance, the use of polymers, which are substances with a higher molecule mass and which have a large number of repeating units, is common today. There are synthetic polymers, which are produced on a large scale and have many properties and uses. **And** there are naturally occurring polymers, which include starches, cellulose, proteins, and latex. Polymers are molecules (monomers) that join together like a chain with one or more monomers. The polymers are changed depending on the incorporation of these monomers. If the atoms in the monomers are combined with the polymer, it is called an addition polymer. When some of the atoms of the monomers are released into small molecules, as in liquid, then the polymer is called a condensation polymer. A double bond between carbon atoms is most common in addition polymers.

In the early part of the twentieth century, a big boom occurred in polymer chemistry when polymer materials such as **nylon** and **Kevlar** came on the scene. Much of the work done with polymers focuses improvement while using existing technologies, but chemists do have opportunities ahead. There is a need for the development of new applications for polymers, always looking for less expensive materials that can replace what is used now. Chemists have to be more aware of what the market yearns for, such as products with a green emphasis, polymers that break down or are environmentally friendly. Concerns such as these have brought new activity to the science arena and there are always new discoveries to be made.

Although **hard** plastics like **Bakelite** were used as early as 1909, it wasn't really big until later. In the early 1920's, German chemist Hermann Staudinger first challenged the doubts of other chemists, on the existence of molecules having molecular weights greater than a thousand. He proposed that there were macromolecules composed of at least 10,000 atoms. While his studies focused on rubber, it soon was recognized that polymeric macromolecules made up many important natural materials. This was followed by the creation of synthetic analogs with a variety of properties such as fiber materials, flexible films, paints and tough light solids. Plastics molecules are also "polar" in nature, which means that they act like magnets. Just as a compass arrow points to magnetic north, when in the presence of a field or voltage these molecules will align themselves with each other. Because of this discovery, modern society was forever changed.

16.2 COMMON PLASTICS AND THEIR APPLICATIONS

Due to their relatively low cost, ease of manufacture, versatility, and imperviousness to water, plastics are used in an enormous and expanding range of products, from paper clips to space ships. They have already displaced many traditional materials, such as wood, stone, horn and bone, leather, paper, metal, glass, and ceramic, in most of their former uses.

Plastics are different types which are given below:

- A chair made with a polypropylene seat.
- **Polyester (PES)** – Fibers, textiles.
- **Polyethylene terephthalate (PET)** – Carbonated drinks bottles, peanut butter jars, plastic film, microwavable packaging.
- **Polyethylene (PE)** – Wide range of inexpensive uses including supermarket bags, plastic bottles.
- **High-density polyethylene (HOPE)** – Detergent bottles and milk jugs.
- **Polyvinyl chloride (PVC)** – Plumbing pipes and guttering, shower curtains, window frames, flooring.
- **Polyvinylidene chloride (PVDC) (Saran)** – Food packaging.
- **Low-density polyethylene (LDPE)** – Outdoor furniture, siding, floor tiles, shower curtains, clamshell packaging.
- **Polypropylene (PP)** – Bottle caps, drinking straws, yogurt containers, appliances, car fenders (bumpers), plastic pressure pipe systems.
- **Polystyrene (PS)** – Packaging foam/"pea nut", food containers, plastic tableware, disposable cups, plates, cutlery, CD and cassette boxes.
- **High impact polystyrene (HIPS)** – Refrigerator liners, food packaging, vending cups.
- **Polyamides (PA) (Nylons)** – Fibers, toothbrush bristles, fishing line, under-the-hood car engine moldings.
- **Acrylonitrile butadiene styrene (ABS)** – Electronic equipment cases (e.g., computer monitors, printers, keyboards), drainage pipe.

- **Polycarbonate (PC)** – Compact discs, eye glasses riot shields, security windows, traffic lights, lenses.
- **Polycarbonate/Acrylonitrile Butadiene Styrene (PC/ABS)** – A blend of PC and ABS that creates a stronger plastic. Used in car interior and exterior parts, and mobile phone bodies.
- **Polyurethanes (PU)** – Cushioning foams, thermal insulation foams, surface coatings, printing rollers (Currently 6th or 7th most commonly used plastic material, for instance the most commonly used plastic found in cars).

16.3 SPECIAL PURPOSE PLASTICS

Like other materials some special purpose plastics are there and they are given below:

- **Melamine formaldehyde (MF)** – One of the aminoplasts, and used as a multi-colourable alternative to phenolics, for instance in moldings (*e.g.*, break-resistance alternatives to ceramic cups, plates and bowls for children) and the decorated top surface layer the paper laminates (*e.g.*, Formica).
- **Plastarch material** – Biodegradable and heat resistant, thermoplastic composed of modified corn starch.
- **Phenolics (PF) or (phenol formaldehydes)** – High modulus, relatively heat resistant, and excellent fire resistant polymer. Used for insulating parts in electrical fixtures, paper laminated products (*e.g.*, Formica), thermally insulation foams. It is a thermosetting plastic, with the familiar trade name Bakelite, that can be molded by heat and pressure when mixed with a filler-like wood flour or can be cast in its unfilled liquid form or cast as foam (*e.g.*, Oasis). Problems include the probability of moldings naturally being dark colours (red, green, brown), and as thermoset it is difficult to recycle.
- **Polyetheretherketone (PEEK)** – Strong, chemical-and heat-resistant thermoplastic, biocompatibility allows for use in medical implant applications aerospace moldings. One of the most expensive commercial polymers.
- **Polyetherimide (PEI) (Ultem)** – A high temperature, chemically stable polymer that does not crystallize. Polylactic acid (PLA)- A biodegradable, thermoplastic found converted into a variety of aliphatic polyesters derived from lactic acid which in turn can be made by fermentation of various agricultural products such as corn starch, once made from dairy products.
- **Polymethyl methacrylate (PMMA)** – Contact lenses, glazing (best known in this form by its various trade names around the world; *e.g.*, Perspex, Orogact, Plexiglas), aglets, fluorescent light diffusers, rear light covers for vehicles. It forms the basis artistic and commercial acrylic paints when suspended in water with the use of other agents.
- **Polytetrafluoroethylene (PTFE)** – Heat-resistant, low-friction coatings, used in things like non-stick surfaces for frying pans, plumber's tape and water slides. It is more commonly known as **Teflon**.
- **Urea-formaldehyde (UF)** – One of the amino plastic and used as a multi-colourable alternative to phenolics. Used as a wood adhesive (for plywood, chip board, hardboard) and electrical switch housings.

16.4 PROPERTIES OF PLASTICS

- Plastics are anti-corrosion resistant, they do not rusted with water contacts.
- Plastics are light in weight.
- Plastic have low strength as compared to metals and non-metals.
- Plastics have good damping capacity.
- Surface finish produced by a plastic parts is very good.
- Plastics are bad conductor of electricity.
- Plastic parts can be produced in any colour even it can produced as clear as transparent glass.
- Any machining operation can be easily performed on plastic material or in other words workability of plastic is very good.
- Plastic possesses very good thermal and electrical insulation properties.
- Plastic do not easily react with chemical.

16.5 CLASSIFICATION OF PLASTICS

Plastics are basically two types:

(i) Thermoplastic, and
(ii) Thermosetting plastic.

16.5.1 Thermoplastic

This type of plastics are softer under heat and pressure application. They are softer at elevated temperature but becomes harder on cooling. They can be heated and cooled at many times; but cannot be heated above its decomposition temperature or in simple words we can say that **these** plastic can be remelted.

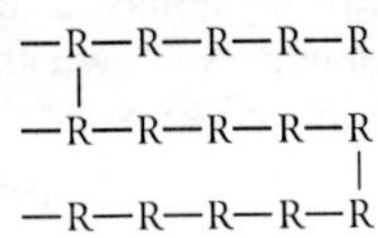

Thermoplastic

16.5.1.1 Characteristics of Thermoplastic

- They consist the long molecular chain and chains are held close to each other by secondary weak forces of type **van der Walls forces**.
- They possesses high plasticity which increases as temperature increases.
- They are very soft on heat and affected by many solvent.
- They are highly plastic and can be easily moulded in any shape by application of heat.
- They do not go any chemical change on heating.
- They possesses very low melting point and do not go so hard as thermosetting plastic.
- They can be reused as they have resale value.

16.5.1.2 Some Common Thermoplastics are:

(i) Furan resin
(ii) Acetal
(iii) ABS

(iv) Polybutylene – PB
(v) Polyethylene or Polythene $(C_2H_4)_n$
(vi) Polythene Cross Linked
(vii) Poly propylene (PP)
(viii) Polyvinyl chloride (PVC)
(ix) Chlorinated Polyvinyl Chloride
(x) Poly vinylidene Fluoride
(xi) Teflon or polytetrafloro ethylene (PTFE)
(xii) Cellulose nitrate $C_6H_{10-n}\ O_5\ (NO_3)_n^-$
(xiii) Bitumen
(xiv) Silicones
(xv) Fluoro-carbons
(xvi) Polystyrene – CH_2 – CH (C_6H_5)

16.5.2 Thermosetting Plastic

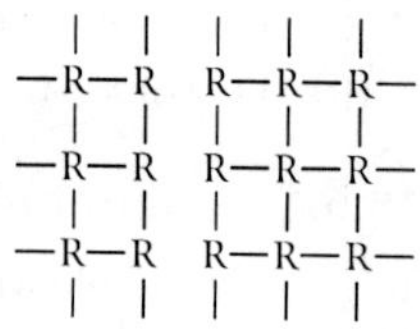

Thermosetting

This type of plastic are also know as **thermosets or heat-setting plastics.** They are formed into the shape at the temperature of 130°C - 180°C and pressure applied results in a permanent hard product or in very simple words we can say a plastic that cannot be remelted by the application of heat is known as ***thermosetting plastic.*** This type of plastic cannot be reused. These are hard and dense as their forming under heat and pressure. These plastics may distort under load.

16.5.2.1 Characteristics of thermosetting plastic

- This type of plastic cannot be remelted as once they have set and hardened.
- They consist the resines formed by condensation.
- Thermosettings plastic resines possess 3-dimensional molecular structure and have very high molecular weight.
- They are hard, tough, non-swelling and brittle due to cross linking.
- They possess very good tensile strength.
- They are capable to withstand high temperature.
- Thermosetting plastics are generally used to make heat resistant products.
- Thermosetting plastic can be heated up to 260°C without melting.

16.5.2.2 Some Common Thermosetting Plastics

(i) Polysters (unsaturated),
(ii) Phenols formaldehyde (PF)—"Bakalite",
(iii) Urea,
(iv) Epoxics,
(v) Melamine.

16.5.3 Thermosetting Plastic Materials

16.5.3.1 Urea Formaldehyde

It is produced by the condensation of urea and aqueous formaldehyde [urea is produced by mining of liquid carbon dioxide (CO_2) and liquid amonia (NH_3) under the application of pressure and heat.

Properties:

- Urea formaldehyde is very hard, brittle and rigid.
- It have low impact strength.
- Excellent electrical properties.
- Excellent resistance to organic solvent.

Uses (Applications)

- Plywood bonding adhesive.
- Paperback adhesive such as belt sending.
- Domestic electrical appliances like switch covers, plug, socket bases.
- Surfacing material on wood.

Product Manufacturing Method:

- Compression.
- Transfer moulding.

16.5.3.2 Phenol Formaldehyde (Bakalite)

Phenol formaldehyde is the most popular and widely used resine. It is obtained by the reaction of phenol with formaldehyde. Chemical formula of phenol is C_6H_5OH and formaldehyde is (HCOH).

Properties:

- Bakalite possesses very excellent heat and electrical resistance properties.
- They are very hard, rigid and scratching resistance.
- They can resist heat up to 150°C.
- They possesses good water resistance properties.
- Excellent electrical insulating properties.

Uses (Applications)

Laminated phenolic plastic is available in different shape and size as a raw materials for manufacturing of:

- Gears,
- Handles,
- Marine coupling,
- Transmission thrust washer,
- Telephone hand sets,
- Knobs, etc.

16.5.3.3 *Epoxy*

Epoxy, also known as **polyepoxide**, is a thermosetting polymer formed from reaction of an epoxide "resin" with polyamine "hardener". Epoxy has a wide range of applications, including fiber-reinforced plastic materials and general purpose adhesives.

Chemistry of Epoxy

Structure of unmodified epoxy prepolymer resin, *n* denotes the number of polymerized subunits and is in the range from 0 to about 25.

Structure of TETA, a typical hardener. The amine (NH) groups react with the epoxide groups of the resin during polymerization.

Epoxy is a copolymer; that is, it is formed from two different chemicals. These are referred to as the "resin" or "compound" and the "hardener" or "activator". The resin consists of monomers or short chain polymers with an epoxide group at either end. Most common epoxy resins are produced from a reaction between epichlorohydrin and bisphenol-A, though the latter may be replaced by similar chemicals. The hardener consists of polyamine monomers, for example triethylenetetramine (TETA). When these compounds are mixed together, the amine groups react with the epoxide groups to form a covalent bond. Each NH group can react with an epoxide group, so that the resulting polymer is heavily crosslinked, and is thus rigid and strong.

The process of polymerization is called "curing", and can be controlled through temperature, choice of resin and hardener compounds, and the ratio of said compounds; the process can take minutes to hours. Some formulations benefit from heating during the cure period, whereas others simply require time and ambient temperatures.

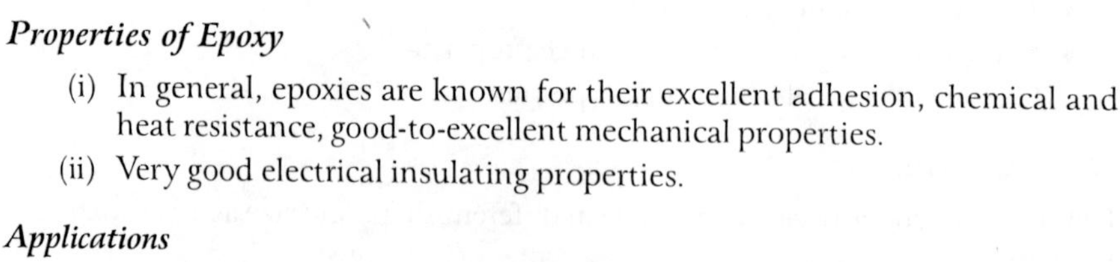

Properties of Epoxy

(i) In general, epoxies are known for their excellent adhesion, chemical and heat resistance, good-to-excellent mechanical properties.

(ii) Very good electrical insulating properties.

Applications

- Coating, adhesive and composite material
- High electrical resistance for electronic application.

16.5.3.4 *Polyester*

Polyester is a category of polymers which contain the ester functional group in their main chain. Although there are many polyesters, the term "polyester" as a specific material most commonly refers to polyethylene terephthalate (PET).

Polyesters include naturally occurring chemicals, such as in the cutin of plant cuticles, as well as synthetics through step-growth polymerization such as polycarbonate and polybutyrate. Natural polyesters and a few synthetic ones are biodegradable, but most synthetic polyesters are not.

Depending on the chemical structure, polyester can be a thermoplastic or thermoset; however, the most common **polyesters** are **thermoplastics**.

Applications

(i) Fabrics woven or knitted from polyester thread or yarn are used extensively in apparel and home furnishings, from shirts and pants to jackets and hats, bed sheets, blankets, upholstered furniture and computer mouse mats.

(ii) Industrial polyester fibers, yarns and ropes are used in tyre reinforcements, fabrics for conveyor belts, safety belts, coated fabrics and plastic reinforcements with high-energy absorption.

(iii) Polyester fiber is used as cushioning and insulating material in pillows, comforters and upholstery padding.

(iv) Polyesters are also used to make bottles, films, tarpaulin, canoes, liquid crystal displays, holograms, filters, dielectric film for capacitors, film insulation for wire and insulating tapes.

(v) Polyesters are widely used as a finish on high-quality wood products such as guitars, pianos and vehicle/yacht interiors.

(vi) Thixotropic properties of spray-applicable polyesters make them ideal for use on open-grain timbers, as they can quickly fill wood grain, with a high-build film thickness per coat. Cured polyesters can be sanded and polished to a high-gloss, durable finish.

Composition of the main chain	*Number of repeating units*	*Examples of polyesters*	*Examples of manufacturing methods*
Aliphatic	Homopolymer	Polyglycolide or Polyglycolic acid (PGA)	Polycondensation of glycolic acid
		Polylactic acid (PLA)	Ring-opening polymerization of lactide
		Polycaprolactone (PCL)	Ring-opening polymerization of caprolactone
	Copolymer	Polyethylene adipate (PEA)	
		Polyhydroxyalkanoate (PHA)	
Semi-aromatic	Copolymer	Polyethlene terephtalate (PET)	Polycondensation of terephthalic acid with ethylene glycol
		Polybutylene (PBT) terephthalate	Polycondensation of terephthalic acid with 1,4-butanediol
		Polytrimethylene terephthalate (PIT)	Polycondensation of terephthalic acid with 1,3-propanediol
		Polyethylene naphthalate (PEN)	Polycondensation of at least one naphthalene dicarboxylic acid with ethylene glycol
Aromatic	Copolymer	Vectran	Polycondensation of 4-hydroxybenzoic acid and 6-hydroxynaphthalene-2- carboxylic acid

16.6 DIFFERENCE BETWEEN THE THERMOSETTING PLASTIC AND THERMOPLASTIC

Sl. No.	*Thermosetting Plastic*	*Thermoplastic*
1.	They are more stronger and hard as compared to thermoplastic.	They are not so stronger and hard.
2.	Thermosetting plastic cannot be remelted (reused).	Thermoplastic can be remelted again and again, *i.e.*, they can be reused.
3.	They have 3-D network of primary covalent bond by the cross linking between the bonds.	They are linear polymers and having no cross linking and branching.
4.	Thermosetting plastic once hardened and set they cannot soften by the application of heat.	There is no such type limitation.
5.	They cannot recycle.	They can be recycle.
6.	Object casted by thermosetting platic can be used at very high temperature *i.e.*, upto 260°C without damage.	Objected casted by thermoplastic cannot be used at high temperature.
7.	They are generally supplied in monomeric.	They are usually supplied as granular form.
8.	Thermosetting plastics are used for making **camera body**, **TV cabinet telephone receiver**, **electric plug**, etc.	Thermoplastic generally used for making the **soft toys**, **photographic films**, **combs**, **pen body**, insulating tape, hoses, etc.

16.7 MATERIAL OF PLASTIC

Furan resine: **Furan** is a heterocyclic organic compound, consisting of a five-membered aromatic ring with four carbon atoms and one oxygen. The class of compounds containing such rings are also referred to as furans.

Furan is a colourless, flammable, highly volatile liquid with a boiling point close to room temperature. It is toxic and may be carcinogenic. Furan is used as a starting point to other specialty chemicals.

16.7.1 Chemistry

Furan is aromatic because one of the lone pairs of electrons on the oxygen atom is delocalized into the ring, creating a $4n+2$ aromatic system (see Huckel's rule) similar to benzene. Because of the aromaticity, the molecule is flat and lacks discrete double bonds. The other lone pair of electrons of the oxygen atom extends in the plane of the flat ring system. The "sp" hybridization is to allow one of the lone pairs of oxygen to reside in a p orbital and thus allow it to interact within the *pi* system.

Due to its aromaticity, furan's behavior is quite dissimilar to that of the more typical heterocyclic ethers such as tetrahydrofuran.

It is considerably more reactive than benzene in electrophilic substitution reactions, due to the electron-donating effects of the oxygen heteroatom. Examination of the resonance contributors shows the increased electron density of the ring, leading to increased rates of electrophilic substitution.

(i) *Acetal*

An acetal is a molecule with two single-bonded oxygen atoms attached to the same carbon atom. Traditional usages distinguish **ketals** from acetals (whereas a ketal has two carbon-bonded R groups and is formally derived from a ketone, an acetal has one or both carbon-bonded R groups as a hydrogen and is formally derived from an aldehyde). Current IUPAC terminology classifies ketals as a subset of acetals.

(ii) *ABS - Acrylonitrile Butadiene Styrene*

Characteristic of ABS

- strong and rigid
- resistant to a variety of basies and acids
- some solvents and chlorinated hydrocarbons may damage the material
- maximum usable temperature 160°F (71°C)
- common uses in DEV - Drainage, Waste and Vent - pipes

(iii) *PB - Polybutylene*

- flexible pipe
- used for pressurized water systems
- usable for hot and cold water
- only compression and banded type joints used

(iv) *PE - Polyethylene*

Polyethylene was the first type of plastic to sell more than 1 billion pounds a year in the United States. Today, most plastic bottles are made of polyethylene, as well as parts of domestic appliances and supermarket bags.

Properties of Polyethylene

They posseses the following properties.

(i) More flexiable over a temperature range.
(ii) They are transparent odourless and light in weight
(iii) they are relatively low in cost
(iv) they do not react with any chemical
(v) they are corrosion resistant
(vi) they are bad conductor of electricity
(vii) high resistivity and dielectric strength.

Types of Polyethylene

On condensation of polymerisation, polyethylene can be made by two types:

(i) High density polythlene (HDPF)
(ii) Low density polythlene (LDPE)

1. ***High density Polyethlene:*** They are obtained by low pressure polymerisation process. They are softening at 140°C and their specific gravity is 0.96.

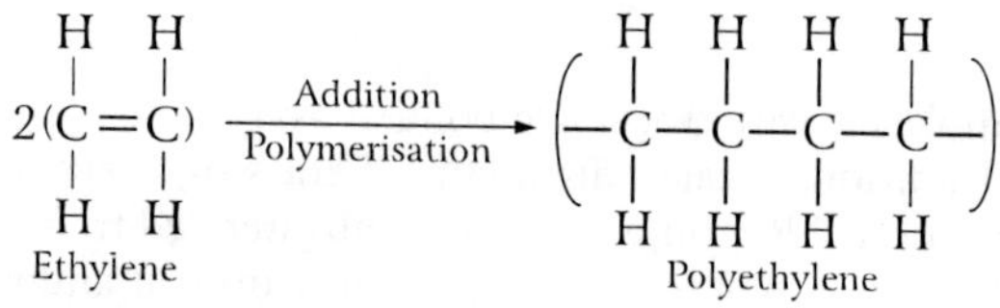

Applications

- flexible pipe
- used for pressurized water systems - sprinkler
- pipe and water tank storage.
- fan and blower casing.
- moisture proofing
- packing purpose
- ice-cube trays.

(v) *PEX - Polyethylene Cross Linked*

- flexible pipe
- used for pressurized water systems - sprinkler.

(vi) *PJP- Polypropylene*

Polypropylene is a soft and versatile type of plastic used in various types of containers as well as plastic seats and ropes

Characteristics of PJP

- temperature up to 180°F (82°C)
- highly resistant to acids, bases and many solvents
- usable in laboratory plumbing
- light weight.

(vii) *PVC - Polyvinyl Chloride*

Better known as PVC or vinyl, polyvinyl chloride is used for pipes, fittings, window and door frames as well as **cables, hoses, flooring** and **even clothing**.

Characteristic of PVC

- resistant to a variety of acids and bases
- may be damaged by some solvents and chlorinated hydrocarbons
- maximum usable temperature 140°F (60°C)
- usable for water, gas and drainage systems
- not useable in hot water systems
- strong and rigid

(viii) *CPVC - Chlorinated Polyvinyl Chloride*

- similar to PVC - but designed for water up to 180°F (82°C)

PVDF - Polyvinylidene Fluoride

- strong and very tough material
- resistant to abrasion, acids, basics, solvents and much more

- usable to 280°F(138°C)
- usable in laboratory plumbing

(ix) *Silicone resin*

Silicone resins are a type of silicone material which is formed by branched, cage-like oligosiloxanes with the general formula of $R_nSiX_mO_y$, where R is a non-reactive substituent, usually Me or Ph, and X is a functional group H, OH, Cl or OR. These groups are further condensed in many applications, to give highly crosslinked, insoluble polysiloxane networks.

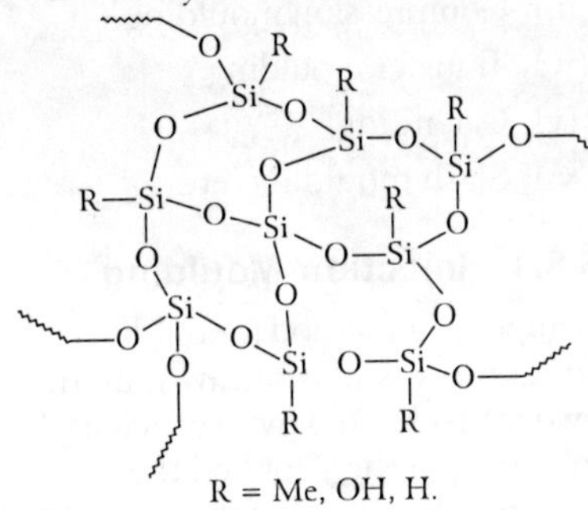

R = Me, OH, H.

When R is methyl, the four possible functional siloxane monomeric units are described as follows:

- "M" stands for Me_3SiO,
- "D" for Me_2SiO_2,
- "T" for $MeSiO_3$ and
- "Q" for SiO_4.

Properties of Silicon Resin

- Silicone resin are mixtures of solvent (toluene or xylene) with combined hydrolyzates of methyl chlorosilanes and phenyl chlorosilanes contain some amount of hydroxyl groups. The hydroxyl can further condensed into a hard film under heating. Solid content is 50~70% on sale.
- Outstanding insulation property, high-and-low-temperature resistance, corona and electric arc resistance, water-proof.
- Outstanding resistance to weather-aging, ozone and chemicals.
- With silicone resin, the H-grade motor, other electric appliances would be size-small, weight-light, last-long and could be used under harsh environment.
- When over loaded, the motor or other electric appliance would not go out of order and would not bum heavily. The gasses giving off are not poisonous. The remains after burning is silicon dioxide.

Applications of Silicon Resin

Brand	*Main Applications*
941	Painted glass fabric lamination, etc.
955	Transformer and motor-coil dipping painting for heat-resistance and for moisture-resistance of electron elements
1053	Motor and transformer-coil dipping heat-resistance painting material
1151	Painting material of wire sheathing of glass fabric
1152	Varnished glass fabric for insulation of motor groove and for wire sheathing
1153	Soft mica-sheet, soft mica-belt, soft glass fabric, glass fabric tube, moisture-proof condenser, dipping and paper treatment

16.8 METHODS OF PLASTIC PROCESSING

Various methods are employed to processing the plastic. Some of them is given below:

(i) Injection moulding,
(ii) Extrusion,
(iii) Compression moulding,
(iv) Transfer moulding,
(v) Calending,
(vi) Slash moulding, etc.

16.8.1 Injection Moulding

An important method for producing parts **of thermoplastic** is **injection moulding**. In this process plastic raw material is available in the form of granular powder. Powder first heated by a sufficient device (heater, attached in machine) to drive off moisture and then applied the pressure (by hand) to feeding the liquid plastic into a very narrow orifice of the die (mould cavity). Pressure applied for a pre determine time ones it is ensure that mould cavity is completely full by liquid plastic and release the pressure. Temperature of the material rises between 150°C to 300°C. After the plastic has been solidified and sufficiently hardend in the mould cavity, mould opened and the produces part knocked out. Speed of this process is very fast and process is very suitable for mass production; it can produce from 12 to 16 thousand parts in a shift. Process is very suitable for producing the **cups, housing, toot handles, toys, knobs, plumbing fitting, telephone receivers**, etc.

Various capacity's injection machine is available such as 12,000 mm^3 – 2.2×10^6 mm^3 and pressure range various from 100 MPa – 150 MPa.

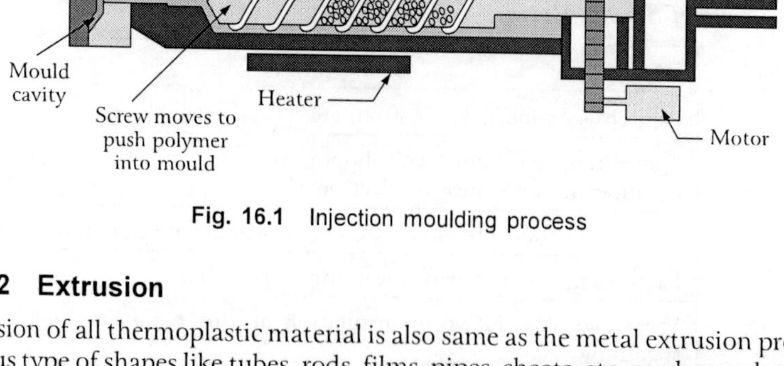

Fig. 16.1 Injection moulding process

16.8.2 Extrusion

Extrusion of all thermoplastic material is also same as the metal extrusion process. Various type of shapes like tubes, rods, films, pipes, sheets, etc. can be produced by extrusion process of plastic. In extrusion of plastic process, plastic is kept in the container and passes through die of required cross-section by applying of pressures. Plastic is feed through hopper in the container. Heater arrangement is also provided in the machine. Earlier extrusion machine was ram type. This process is not suitable for **thermosetting plastic.** Main advantage of this machine, simple in construction and controlled pressure is possible.

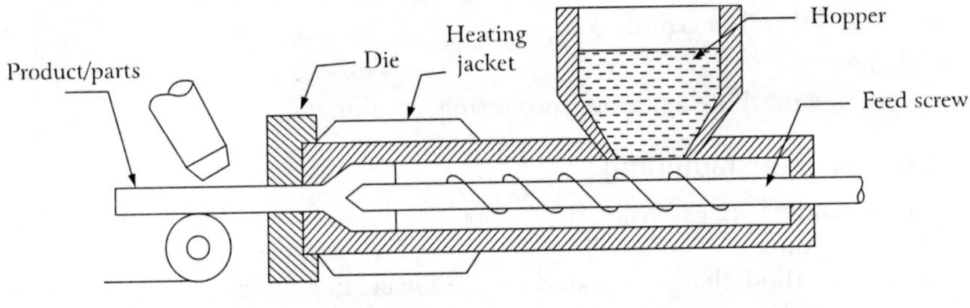

Fig. 16.2 Extrusion moulding process

16.8.3 Compression Moulding

Compression moulding is some what like the closed die forging process, most of thermosetting plastics and few thermoplastics plastic can be moulded by (120 – 250°C) this process. In this process plastic (in the form of particles) is kept in a heated mould and compressed same at a suitable pressure and temperature. Mould is kept closed for sufficient time to allow the chemical change. When the plastic is completely trapped between the male and female die, it is referred as "**positive mould**".

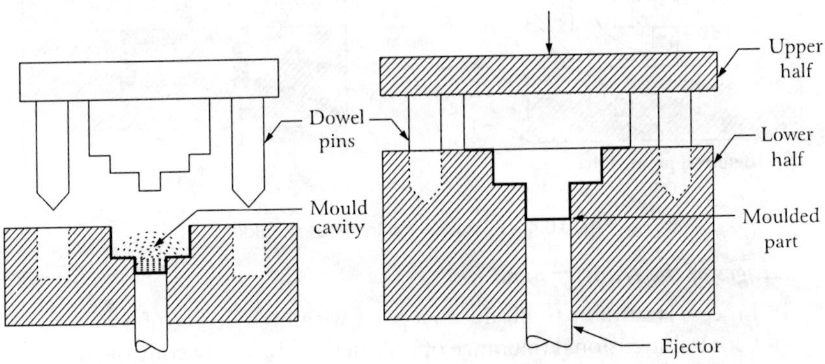

Fig. 16.3 Principle of compression moulding

Advantages of compression moulding

1. Simple in construction
2. Initial cost of set up is low
3. Operation setting is easy
4. Parts produced by this process have not so much stress since material is not forced via gates.

Limitations of Process

1. Usually process is suitable for thermosetting plastics
2. Production rate is not so fast
3. Dimensional accuracy of parts produced by this process is not so good.

Applications

- Electrical and electronic composts,
- handles,
- washing machine agitators and hosing, container caps, etc.

16.8.4 Transfer Moulding

Transfer moulding is a extension of compression moulding and quite similar to injunction moulding.

In this method charge is placed in the lower die, charge is heated upto its plastic stage and then compressive force is applied when the curing is completed. Mould cavity is the replica of shape that to be produced by this process for applying pressure, a plunger/ram is used and pressure range from 30–150 MN/m^2 is associated with this process.

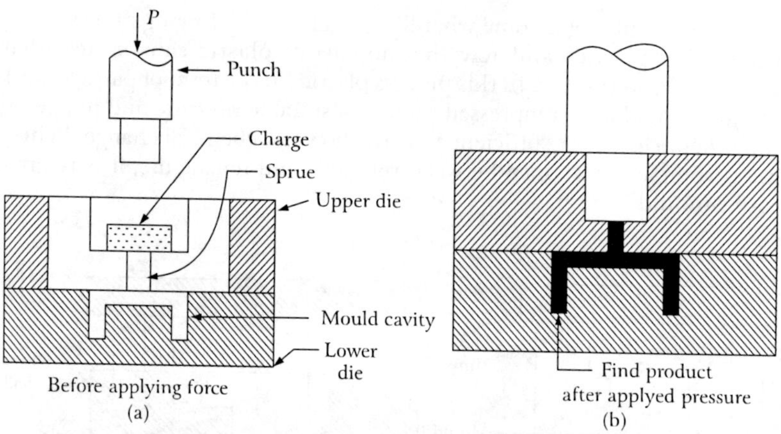

Fig. 16.4 Principle for transfer moulding

Advantages of the transfer moulding

(i) Process is suitable to produced a parts with high degree of accuracy.
(ii) Closer dimensional tolerance of product can be easily controlled.
(iii) Process is suitable for mass production.
(iv) Medium tooling cost.
(v) Complex shapes of parts can be produced by this method.
(vi) Operation is quite simple, parts is not directly subjected to compressive force.

Limitation of transfer moulding

(i) Method is only suitable for small size of parts
(ii) Only thermosetting plastics can be used in this method.

16.9 WELDING OF PLASTICS

Welding of plastic is very important, as plastic is frequently used in various item (part) and on applying the load plastic component can break easily. Therefore the joining of plastic components/parts is very essential. As we know that thermosetting plastics cannot be welded by conventional method as they have no effect on heating thermosetting plastics are available as semi-finish products. Same thermosetting plastic can be welded by mechanically or cemented. On the other hand thermoplastic can be heated easily and they are welded easily too.

In broad manner plastic can be welded by two process:

Welding of Plastic

Heat used from external source such as stream of hot gases, hot tool. In this process heat is transferred into workpiece by means of conduction, convection and partly radiation.	Heat is generated by an external source within the workpiece of conversion of some energy, like friction, current, chemical, reaction, infrared light, ultra-sonic sound, etc.

16.9.1 Most Commonly Plastic Welding Process

(i) Friction welding.
(ii) Hot-gas welding.
(iii) Ultrasonic welding.
(iv) Infrared welding.
(v) High frequency induction welding.
(vi) Nuclear welding.

16.9.1.1 Friction Welding

Plastic can be easily welded by friction welding method. In this method two plastic parts are held in the friction welding machine (just like the lathe machine). One of plastic part rotating with very high speed and comes in contact to other plastic parts. Mechanical energy is directly converted into heat energy. Due to high temperature friction generated and parts welded.

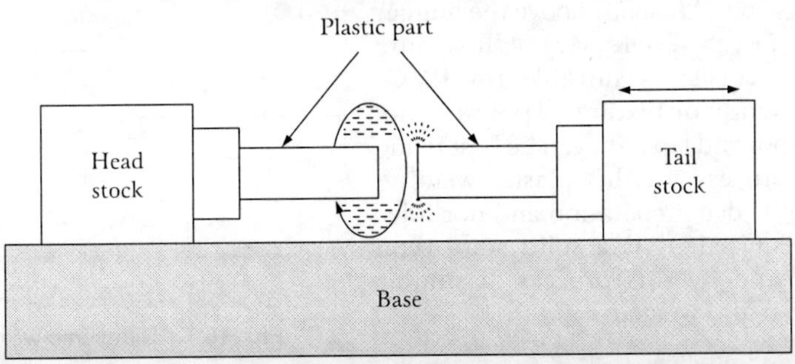

Fig. 16.5 Friction welding machine

Friction welding have the following limitations:

(i) Flash may produce when pressure is applied on the workpiece to joining purpose.

(ii) One of the piece must be a body of revolution and its section at the joint must be a circle or an annuals.

(iii) Only butt joint can be formed.

(iv) Only hollow workpiece can be joint.

16.9.1.2 Hot Gas Welding

By this process, the plastic parts are heated with a hot gas flame of welding torch. Flame may be generated by burning of C_2H_2, H_2, CO, N_2 gases with O_2.

In this process, edge of plastic part is heated by the flame and groove is filled with the help of a filler rod. As the filler rod is forced down by hand. It welds to the softened edges and forms a joint. This plastic welding technique is somewhat like the gas welding process of metal.

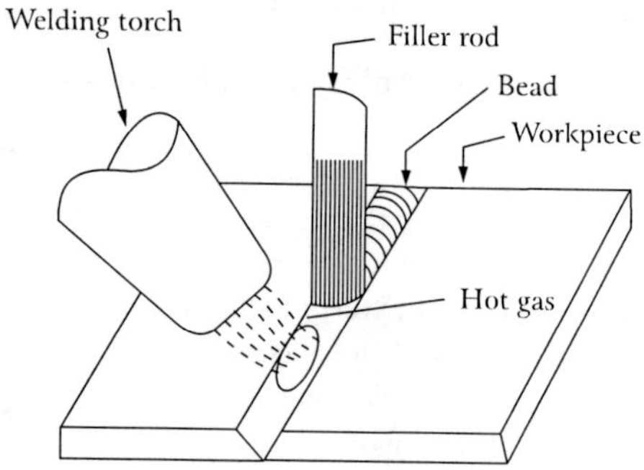

Fig. 16.6 Hot gas welding

Main limitations of this process are:

- Weld bead strength is very low.
- Welding speed rate is slow.
- Only thin sheets of plastic can be welded.
- Reduced plasticity in the weld and near-weld area.
- Danger on over heating.

16.9.1.3 Ultrasonic Welding

Working principal of ultrasonic welding for plastic is same as ultrasonic welding of metal. In this process the coalescence is produced by the local application of high frequency vibratory energy as the parts are held together under pressure. High frequency vibration is generate by the high frequency sound (usually above the human hearing *i.e.*, > 20 K Mz). Ultrasonic welding process is suitable for **PVC, acrylics, synthetic textile** and **polystyrene.** Only T-spot and lap joint can be best made by this process. In this plastic welding process no edge preparation and nor filler rod is required. It is a solid state type welding and by this process dissimilar plastic can also welded.

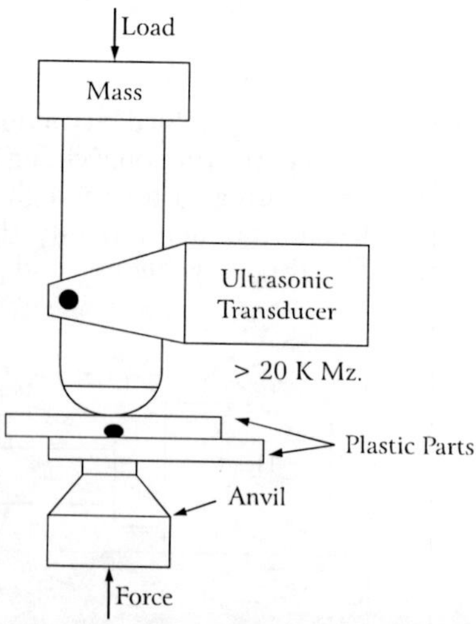

Fig. 16.7 Ultrasonic welding

16.9.1.4 *Infrared Welding*

In this process, heat is produce to joining the plastic by the means of infrared height like—sylite glower. Process is very suitable for joining the polythylene films. A chrome steel is carried out on a black backing plate of a formed plastic, sponge rubber or thick rubberized fabric. Resistance of the back up plate held firmly against the object (plastic part) supplies necessary pressure of welding.

OBJECTIVE QUESTIONS

1. Thermoplastic
   (*a*) once set cannot melted again (*b*) can be soften by heat
   (c) degrade and decompose by heat (*d*) are cross linked material.

2. Thermosetting plastic
   (*a*) softened on heating and hardened on cooling for many times
   (*b*) moulded by heating and cooling
   (*c*) formed by condensation polymerization
   (*d*) none of the above.

3. Plastic bottle manufactured by *[JWM 2010; GATE 2010]*
   (*a*) calendering process (*b*) injection moulding process
   (*c*) transfer moulding process (*d*) blow moulding.

4. Long plastic tubes and rods are manufactured by:
   (*a*) injection moulding (*b*) compression moulding
   (*c*) blow moulding (*d*) extrusion.

5. Thermosetting plastic are those material which
   (*a*) do not become hard by the application of heat and pressure and no chemical change take place
   (*b*) used as a friction lining for brakes and clutches
   (*c*) Take the shape under the application of heat and pressure and results in a permanent hard product
   (*d*) flexible and can withstand considerable near under suitable working condition.

6. Plastics can be joined of meaning of
   (*a*) welding (*b*) adhesive
   (*c*) fastness (*d*) all of the above.

7. Bakelite is a
   (*a*) non-plastic material (*b*) thermoplastic
   (*c*) thermosetting plastic (*d*) none of these.

8. Which one of the following is a thermoplastic?
   (*a*) epoxides (*b*) cellophone
   (*c*) silicon (*d*) all of these.

9. Polyster is a
   (*a*) thermoplastic (*b*) non-plastic material
   (*c*) thermosetting plastic (*d*) none of these.

10. Injection moulding is commonly used for processing
    (*a*) thermoplastic (*b*) thermosetting plastic
    (*c*) non-plastic (*d*) none of these.

11. Which one is synthetic elastomer?
   (*a*) Vulconised rubber (*b*) Natural rubber
   (*c*) Neoprene (*d*) All of above.

12. Most of the thermosetting plastic can be moulded by
   (*a*) slush casting (*b*) compression moulding
   (*c*) injection moulding (*d*) none of above.

13. Transfer moulding is a variation of
   (*a*) compression moulding (*b*) extrusion process
   (*c*) injection moulding (*d*) none of these.

14. Thin plastic sheets and felons are commonly produced by
   (*a*) calending process (*b*) compression moulding
   (*c*) transfer moulding (*d*) all of above.

15. Plastic rods, tubes, pipes, rapes, sheets, etc. are generally produced by
   (*a*) injection moulding (*b*) extrusion moulding
   (*c*) compression molding (*d*) none of these.

16. Narrow neck plastic bottle are produced from thermoplastic by *[JWS 2010]*
   (*a*) transfer moulding
   (*b*) blow moulding
   (*c*) compression moulding
   (*d*) injection moulding.

17. Flexible toys, artificial flowers, etc. are produced (thermoplastic) by
   (*a*) slush moulding
   (*b*) extension moulding
   (*c*) transfers moulding
   (*d*) none of these.

18. Which of the following are thermoplastics
   (*a*) Polystrene (*b*) Poly propylene
   (*c*) PVC (*d*) All of above.

19. PET, LDPE, PP and PS are the trade name of plastic used in
   (*a*) packing industries (*b*) durable goods industries
   (*c*) producing of enclosures (*d*) all of above.

20. PETE or PET plastic are commonly for making the
   (*a*) carpet, films and engg carbons (*b*) fiber fill for winter coats
   (*c*) liquid soap and soft drink bottle (*d*) all of above.

21. Cellulose nitrate is a
   (*a*) thermosetting plastic (*b*) thermoplastic
   (*c*) non-plastic (*d*) ceramic.

22. Polystrene is a
   (*a*) non-plastic (b) thermoplastic
   (*c*) thermosetting plastic (*d*) none of these.

23. Which of the following are used as laminating material in low pressure laminating of plastics?

(*a*) Paper
(*b*) Nylon
(*c*) Cotton
(*d*) Hemp
(*e*) Asbestos
(*f*) All of these.

24. Twist drill used for drilling holes in plastic should contain:
(*a*) 0° rake
(*b*) 5° rake
(*c*) 10° rake
(*d*) 23° rake.

25. Plastic material is used for component which requires characteristic as
(*a*) low density, machinability and large plastic deformation
(*b*) low density, machinability and high strength
(*c*) machinability, high strength and large plastic deformation
(*d*) high strength, large plastic deformation and low density. [*IES 2010*]

26. Which one of the following material is used for car tyres as a standard material?
(*a*) Nitrile rubber
(*b*) Styrene-butadiene rubber
(*c*) Butyl rubber
(*d*) Any of the above defending upon the need. [*IES 1997*]

27. Tensile strength of plastic lies in the range
(*a*) 1–100 N/m^2
(*b*) 100–300 N/m^2
(*c*) 400–800 N/m^2
(*d*) 1000–1500 N/m^2

28. Structure of a polymer is
(*a*) long chain
(*b*) cubic
(*c*) rhombic
(*d*) closed pack hexagonal [*IES 2008*]

ANSWERS

1. (*b*)	2. (*c*)	3. (*d*)	4. (*d*)	5. (*c*)
6. (*d*)	7. (*c*)	8. (*b*)	9. (*c*)	10. (*a*)
11. (*c*)	12. (*b*)	13. (*a*)	14. (*a*)	15. (*b*)
16. (*b*)	17. (*a*)	18. (*d*)	19. (*a*)	20. (*d*)
21. (*b*)	22. (*b*)	23. (*d*)	24. (*a*)	25. (*a*)
26. (*d*)	27. (*a*)	28. (*a*)		

REVIEW QUESTIONS

1. What are plastics? How are they classified?
2. What are the advantages of plastic over metals?
3. Differentiate between the thermoplastic and thermosetting plastic. [*UPTU 2003-04*]
4. Describe the important features of themoplastics. Give the examples of this type of plastic.
5. Enlist six important thermoplastic materials.
6. What are the characteristics of plastic?
7. What are the laminated plastic? Give their important properties and applications (uses).
8. Give the name of various additive used for plastics.

9. Write a short note on structure of plastic.
10. What are "fibre glass reinforced plastic." State their properties and applications (uses).
11. Explain briefly following method of joined thermoplastic
    (*a*) friction welding
    (*b*) ultrasonic welding
    (*c*) induction welding
    (*d*) fusion bonding
    (*e*) vibration welding
    (*f*) solvent bonding.
12. Write the short notes on 'machining of plastic'.
13. What are the latest trends in plastic or what is the future of plastic?
14. Write the short notes on the following:
    (*a*) Epoxies
    (*b*) Phenolics
    (*c*) Polyesters
    (d) Aminos.
15. Explain the important characteristics/properties and applications of the following plastics:
    (*a*) Nylons
    (*b*) Acetals
    (*c*) Fluoro carbons
    (*d*) Polyamides
    (*e*) Acrylics
    (*f*) ABS.
16. What is injection moulding? Describe its advantages.
17. Write the short notes on:
    (*a*) Welding of plastics,
    (*b*) Resines and Adhesives
    [*UPTU 2008-09*]
18. Write the past, present and future of plastic, plastics products and their uses (applications). [*UPTU 2001-02*]
19. Explain the extrusion of plastic. [*UPTU 2005-06*]
20. Explain the role of resine and adhesive to be used in plastic industries. Name few resins and adhesives. [*UPTU 2005-06*]

❑❑❑

EXPERIMENTS

Say minimum 8 experiments out of following (or such experiment)

1. Design of pattern for a desired casting (containing hole).
2. Pattern making for a given desired shape.
3. Sand testing (at least one such as grain fineness number determination).
4. Injection moulding with plastics.
5. Forging, hand forging processes.
6. Forging - power hammer study and operations.
7. Tube bending with the use of sand on a tube bending m/c.
8. Press work experiment such as blanking/piercing, washer, making etc.
9. Wire drawing/extrusion on soft material.
10. Rolling-experiment.
11. Bending and spring back operation.
12. Powder metallurgy experiment.
13. Jigs and Fixture experiment.
14. Any other suitable experiment of manufacturing science/process/technique.

E

EXPERIMENT-1

OBJECTIVE

To make a mould of given pattern and produce a casting.

TOOLS USED

Molding sand, cope and drag, molten metal, gating system, peen hammer, sprue pins, swab trowels, shovel, vent wire, strike off bar, draw spike, gate cutter etc.

THEORY

MOULDING SAND

Silica sand found in many natural deposits is suited for moulding purpose because it can withstand high temperature without breakdown. Sand is low in cost, has long life, and is available in a wide range of grain sizes and shapes. The disadvantages are that sand has a high expansion rate when subjected to heat and has a tendency to fuse with the molten metal.

Types of Moulding Sand

Foundry sand can be grouped as:

1. **Natural Sand:** It contains sufficient amount of binding clay and hence can be used directly.
2. **Synthetic Sand:** These are clay free high silica sand. Suitable binders are added to them to make usable for foundry work.

Moulding sand can be further classified.
According to its composition and use, into the following categories:

(i) **Green Sand:** Foundry sand containing moisture is known as green sand. It is a mixture of silica sand with 20 % to 30 % clay and water from 6-8 %. Moulds prepared from green sand do not require any baking before pouring the molten metal into them.

(ii) **Dry Sand:** It is the sand which is free from moisture and possesses greater strength than green sand.

(iii) **Facing Sand:** These forms the face of the mould *i.e.*, rammed around the pattern surface. It is the fresh prepared and well tempered foundry sand.

(iv) **Parting Sand:** It is sprinkled on the pattern and the parting surfaces of the mould so that the sand mass of one flask does not stick to that of the other or to the pattern. Dry silica sand and 'burnt' sand are used for this purpose.

(v) **Floor or baking sand:** It is the sand which left on the floor after the casting has been removed from the mould. It is used to back up the facing sand and to fill the rest of the flask.

(vi) **Core Sand:** It is used for making cores and has high silica content.

(vii) **Oil Sand:** It is the silica sand using oil binders.

Types of Moulds

Depending on the materials used, moulds are classified as follows:

1. **Green-sand moulds:** These are mixture of sand, clay, and water. "Green" means mould contains moisture at the time of pouring.
2. **Dry sand moulds:** Fairly coarse moulding sand mixed with a binding material is used. Flasks are of metal, since moulds must be oven baked before being used. It is free from gas troubles due to moisture.
3. **Loam moulds:** It is first built up with bricks or large iron parts are then plastered over with a thick **loam mortar**, the shape of the cavity being obtained with sweep or skeleton patterns. The mould is then allowed to dry throughly.
4. **CO_2 Moulds:** Clean sand is mixed with sodium silicate and the mixture is rammed around a pattern. When CO_2 gas is pressure-fed into the mould, the sand mixture hardens. Very smooth and intricate castings are obtained. It is used for core making.
5. **Metal Moulds:** These are used mainly in the die-casting of low-melting temperature alloys. Metal moulds are accurate with smooth finish and eliminates much machine work.
6. **Special Moulds:** Plastics, cement, plaster, paper, wood and rubber are all mould materials used to fit particular applications.

PROCEDURE

Procedure for making a mould

(i) Firstly, sand is rammed by rammer.

(ii) The water is sprinkled over the sand and suitable quantity of bentonites, dextrin are mixed with sand for bonding strength.

(iii) The prepared sand is poured in the mould box and rammed keeping the pattern at bottom of the mould box.

(iv) After this mould box is rolled over and pattern is withdrawn.

(v) The mould surface is dried for a certain time.

(vi) The molten metal is poured to make desired casting.

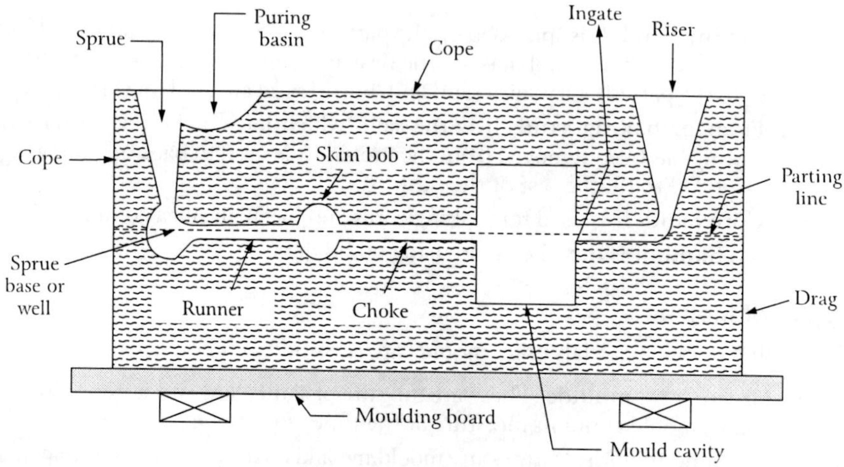

Fig.1 Schematic illustration of sand moulding

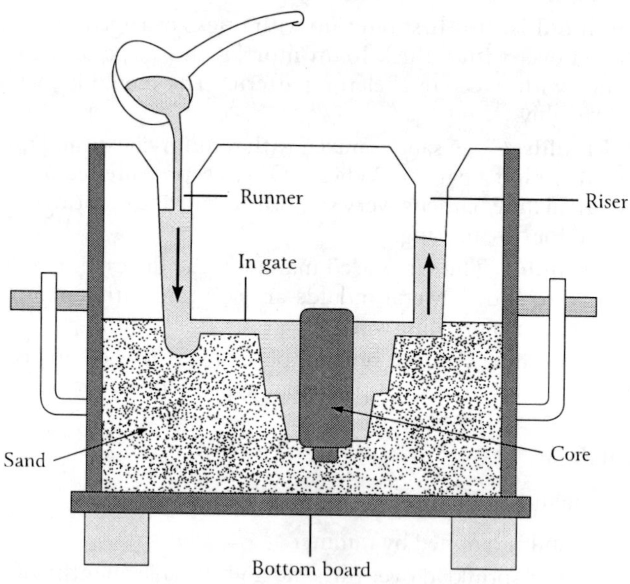

Fig. 2 Sand-casting process layout

PRECAUTIONS

- Ramming should be done properly otherwise mould may break up during rolling over.
- Sand mixing should be done with the help of hand by wearing hand gloves.
- The mould prepared must be strong enough to hold the weight of the metal.
- To keep a safe distance from furnace during melting.
- Molten metal should be poured in the mould gradually to avoid splashing.
- Proper safety measure should be adopted.

E

EXPERIMENT-2

OBJECTIVE

To make a wooden pattern of specific dimensions for desired casting.

APPARATUS OR TOOLS USED

Chisel, Mallet, iron jack planer, Rip saw, Scriber, Marking block, Engineering scale, Try square, wood working bench vice.

THEORY

It is defined as a model of anything to constructed that it may be used for forming an impression called mould in damp sand or other suitable materials.

Various types of pattern are used in foundry practice.

1. Single piece pattern
2. Split pattern
3. Match plate pattern
4. Cope and drag pattern
5. Gated pattern
6. Sweep pattern
7. Skeleton pattern
8. Segmental pattern
9. Loose piece pattern

Pattern Allowance

Pattern allowances are feature as it affect the dimensional characteristics of the casting. Thus when patterns are certain allowances must be given on the size specified in the finished component drawing so that a casting can be produced with particular specification, selection of correct pattern allowance greatly helps in reduction in machining cost and avoid in rejection.

The various pattern allowances are as follows:

(a) Draft Allowance
(b) Shrinkage Allowance
(c) Machining Allowance
(d) Camber Allowance
(e) Distortion Allowance

PROCEDURE

1. The given wood piece pattern are smoothen with jack planer and then flat piece is cut to the required dimension by rip saw.
2. Layout of required dimensions is marked on the smoothen workpiece as per desired shape with the help of smellier scale and try square.
3. The cut wooden workpiece is held in bench vice and required design is made with the help of chisel and light wooden mallet and rip saw.

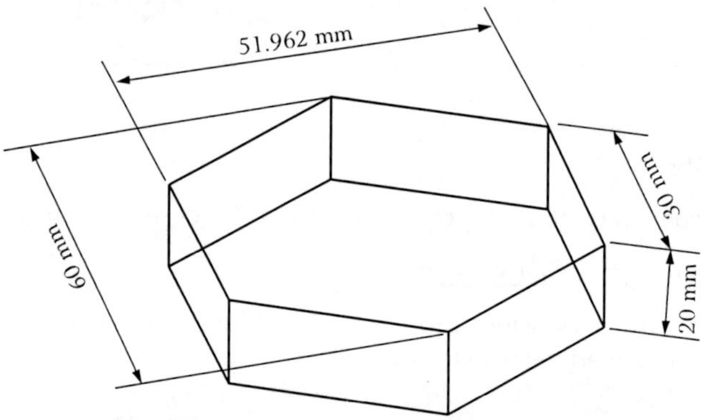

Fig. 1 The Hexagonal pattern for casting

PRECAUTIONS

1. Flat piece should be held properly in vice.
2. Concentration should be put during striking.
3. Keep a safe distance from worktable.
4. Always use Aperon for safety purpose.
5. Hold/grip all the tools firmly.
6. Follow always Do's and Don't mention in the lab.

E

EXPERIMENT-3

OBJECT

To prepare a blank and punch of given size on an aluminium sheet.

MATERIALS USED

Hand press, aluminium sheet, etc.

THEORY:

BLANKING

This is the operation carried out on processes and consist of cutting the outside contour of a stamping. Production of sheet metal blanks of flat shapes requires a single action press equipped with tools comprising a punch. A corresponding die, stripper to keep the sheet and for spacing successive cuts. The operation of cutting inside the contour i.e., holes and slots is called piercing.

FINE BLANKING

This is a press working technique. It utilizes special press and precession tools and dies to produce parts in final assembly shapes. Parts produced have clearly sheared surface over the entire material thickness. In conventional methods one-third thickness is sheared and balance is fractured which has to be shaved, milled etc. The dimensional accuracy obtained is also high.

PUNCHING

This is also press operation and consist of cutting holes of various shapes in sheet. This operation is generally continuous and at a time two or more holes may be punched.

TRIMMING

It consists of removal of cutting away of excess material left around the parting lines in the previous operation.

NOTCHING

This is the operation of removing small amount of metal from the edges of workpiece by the notching action usually obtained with the help of nibbler. It is a shearing process in notching process material removed only outside edge of the workpiece (object).

LANCING

It is a special form of piercing operation in which the entire contour is not cut, the blanked material remains attached with the sheet.

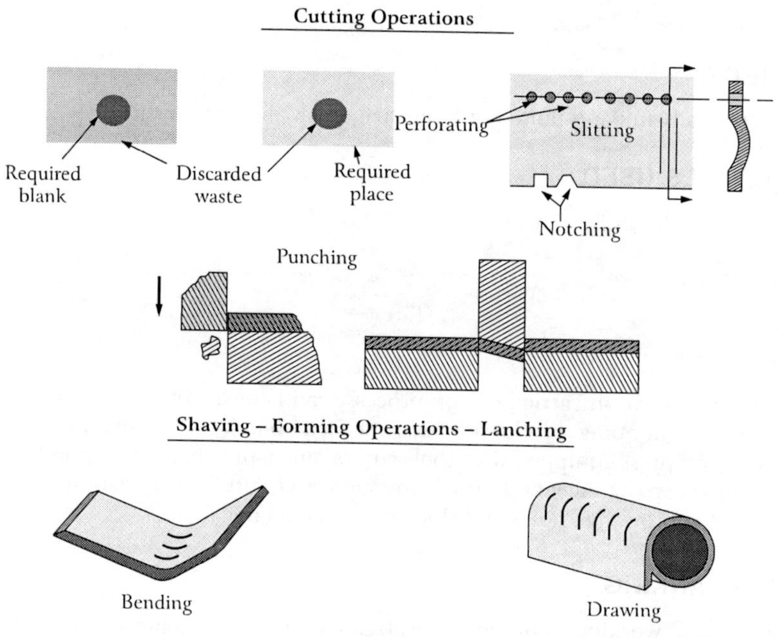

Fig. 1 Forming and cutting operations

OPERATIONS

1. At first required size of the aluminium sheet is cut.
2. It is clamped to the fixture or die securely.
3. After that punch is moved down by turning wheel to complete the operation.

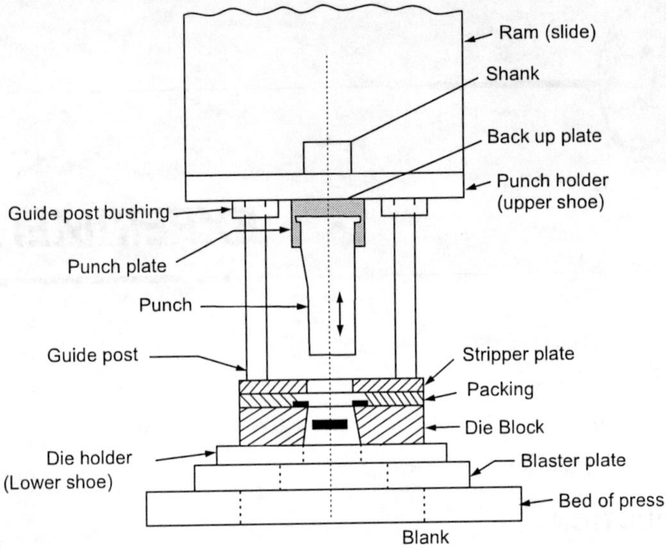

Fig. 2 A simple cutting die

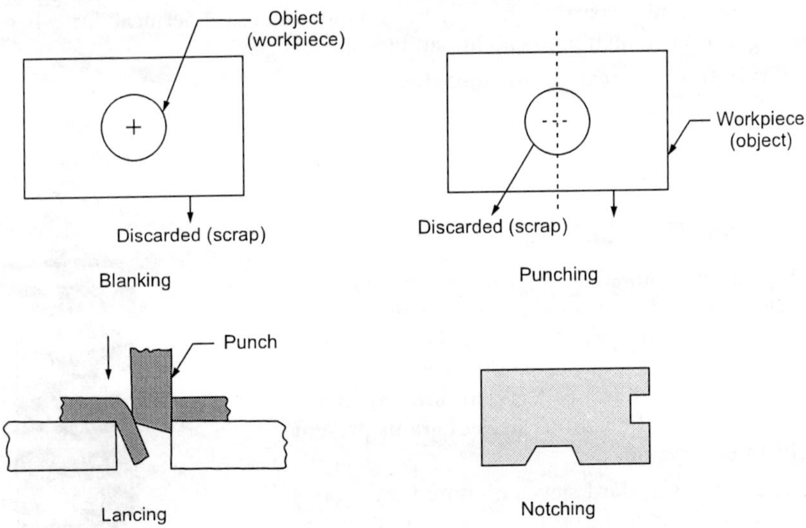

Fig. 3 Sheet metal operations

PRECAUTIONS

1. Always keep a safe distance from press.
2. To secure tightly the aluminium sheet with the fixture.
3. Not do touch any part of the device except the wheel used for moving up or down the punch.
4. Do not operate the machine without any prior permission of operator.

E

EXPERIMENT-4

OBJECT

To determine the grain fineness number of a given sand sample.

EQUIPMENTS USED: Sand Sieve machine

INTRODUCTION

Granular particles of various sizes and shapes provide variable interstices (space between grains) and hence are directly responsible for permeability and compactness of the sand. Granular particles have higher strength but lower permeability, whereas round grains have high permeability and lower strength.

Grain fineness test determines the:

(*a*) grain size,
(*b*) distribution and
(*c*) grain fineness.

EXPERIMENTAL SET UP

The apparatus required for determining grain fineness consists of a number of standard sieves mounted one above the other, on a power-driven shaker. The shaker vibrates the sieves and the sand placed on the top sieve gets screened and collects on different sieves depending upon the various sizes of grains present in the molding sand.

There are ten standard sieves mounted one above the other and under the bottom most sieve is placed a pan. The top sieve is the coarsest and the bottom-most sieve is the finest of all the sieves. The in-between sieves are placed in order of fineness from top to bottom.

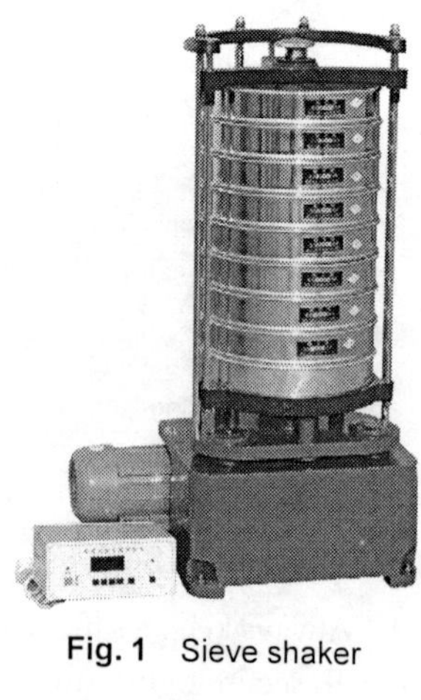

Fig. 1 Sieve shaker

EXPERIMENTAL PROCEDURE

To carry out this test, a sample of dry sand, free from clay is placed on the top most sieve bearing U.S. series. A set of standard testing sieves having U.S. Bureau of

Standard Meshes 12, 20, 30, 40, 50, 70, 100, 140, 200 and 270 are mounted on a sieve shaker. The above sample is vibrated for a definite period of time. After this, weight of the sand retained on each sieve can be obtained.

To obtain the grain-fineness, weight of the sand retained by each sieve is multiplied by 2, which gives the percentage of weight retained by each sieve. This percentage is again multiplied by a multiplying factor.

The A.F.S. (American Foundry Men's Society) grain fineness number will be

$$= \frac{\text{Total sum of product}}{\text{Total sum of the \% of sand retained by different sieves and pan}}$$

CALCULATION

Sieve Number (ASTM)	*Calculation for A.F.S. Grain Fineness Number*				
	Sieve size (microns)	*Sand-retained*		*Multiplying Factor*	*Product*
		Weight	*% Wt. Retained (2 × Wt)*		
12	—	—	—	5	—
20	850	—	—	10	—
30	600	—	—	20	—
40	425	—	—	30	—
50	300	—	—	40	—
70	212	—	—	50	—
100	150	—	—	70	—
140	106	—	—	100	—
200	75	—	—	140	—
270	53	—	—	200	—
Pan...	—	—	—	300	—
Total			—		—

PRECAUTIONS

1. Place the sieve according to their required sequence.
2. Use appropriate amount of sand during experiment.
3. Carry out experiment safely.

E

EXPERIMENT-5

OBJECT

To make a 90° band of given specimen of tube.

EQUIPMENT USED: Tube bending machine.

THEORY

Bending is the cold working process involving plastic deformation in which the total surface area remain constant. Bending of pipe tube rolled or extruded shape involves practices and theories of simple supported beam. The cantilever and become and many of the problems of forming and deep drawing in bending flow of metal occurs in plastic range and there is permanent change in shape thinning of outer wall of a bent tube of other hollow section and corresponding thickness of the inner wall of the bend are noticeable effect of plastic flow.

(a) Before Tube bending

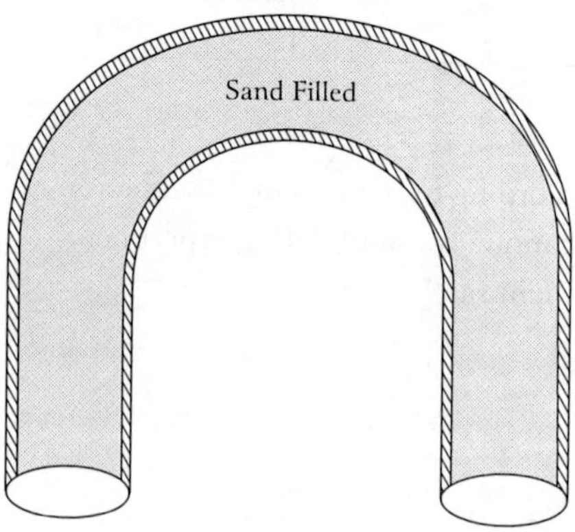

(b) Tube After bending

Fig. 1

Some common type of bending operations:

1. Stretch Bending

The process consists of gripping a sheet of material or a rolled or extruded section at each end in suitable jaws and starching it over a die mode to the required contour until complete forming has been achieved by extending and thinning the metal over a profile.

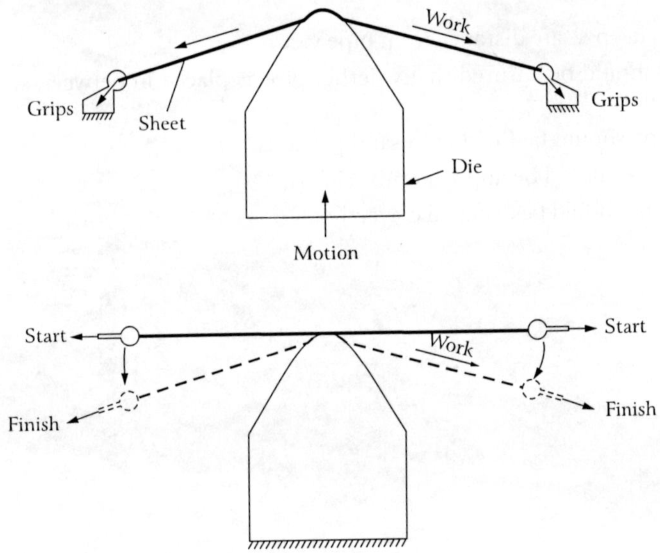

Fig. 2 Stretch bending

2. Roll Bending

This operation is carried out at a series of roll stations and a flat sheet of strip.

Transformed into desired section progressively in stages. The metal gauges or section thickness remains constant and only bending operation take place in each stage.

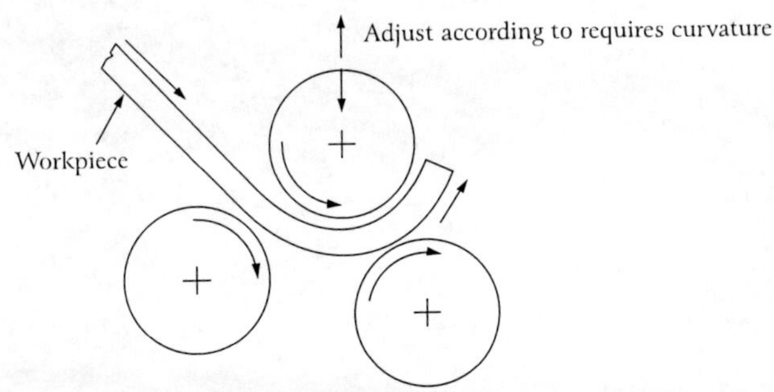

Fig. 3 Roll bending

3. Draw Bending

This operation is used for bending hollow section. Machine used for this operation has rotating die to which the piece to be bent is clamped and is drawn against a roller.

A mandrel is used to eliminate the following effect of tubular section.

PRECAUTION

(i) To keep a safe distance from pipe vice

(ii) It should be ensured that whether pipe is placed in between support and roller.

(iii) Pipe should be filled with sand properly.

(iv) Force should be apply on tube slowly.

(v) Tube should be fastened correctly.

E

EXPERIMENT-6

OBJECTIVE

To Perform Wire Drawing operation on Wire Drawing Machine.

APPARATUS USED

Wire Drawing Machine

THEORY

A wire by definition is circular with small diameter so that it is flexible. The process of wire drawing is to obtain wires from bigger diameters through a die. Wire drawing is always a cold working process.

PROCEDURE

The end of the rod or wire is to be further reduced, is made in to a point shape and inserted through the die opening. This end is then gripped on the other side with gripper, which would then pull the wire through the die.

The steps followed for wire drawing are:

1. Before wire is to be drawn, the stock needs to be prepared for wire drawing.
2. The material should be sufficiently ductile since the tensile force pulls it.
3. For easy entrance of wire in to die, the end of the stock is made pointed to facilitate the entry.

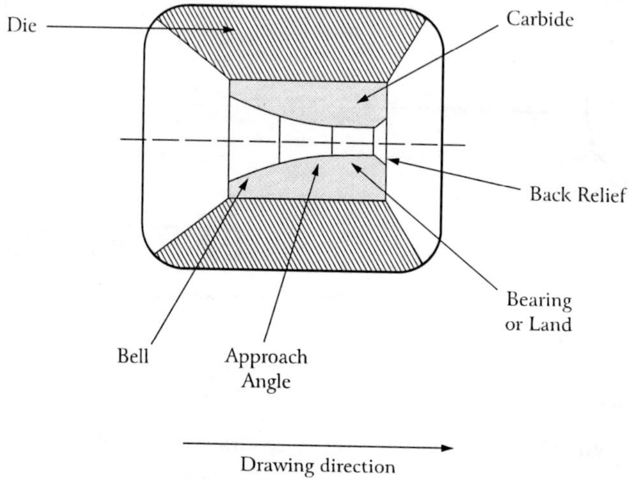

Fig. 1 Wire Drawing Die

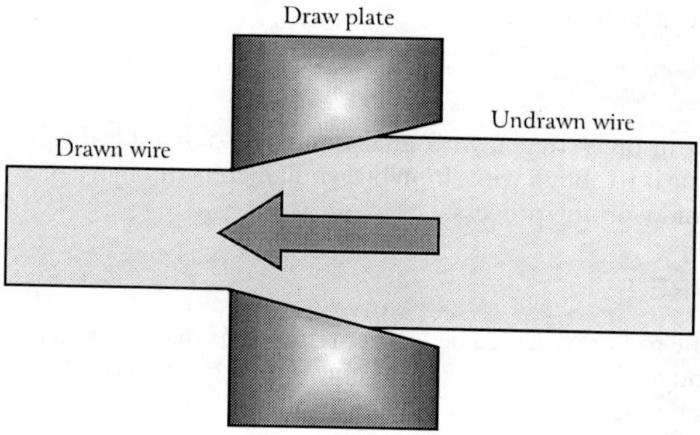

Fig. 2 Wire Drawing

PRECAUTIONS

(i) To make a safe distance form the machine.

E

EXPERIMENT-7

OBJECTIVE

To study about forging and power hammers in forging with its operation.

THEORY

Forging may be defined as a process in which metal or alloys plastically deformed to the desired shape by application of compressive force. Forging may be done by hot or cold working condition. In forging process pressure is applied intermittently. Dies are used to apply pressure to the materials is not cylindrical roll but usually a rectangular block of certain thickness and the dies have linear motion instead of rotation.

The forging equipment used may be grouped under two main categories:

1. **Hammer**: Compressive force is applied through impact load in vertical position.
2. **Forging machine**: Through this equipment the compressive force is applied continuously and the material is gradually pressed into shape, the forging machine operates in horizontal position.

Classification of forging:

(i) Open die forging

In open die forging workpiece is pressed between two flat surfaces. It is used when a large number of machine components to be forged is too small to justify the cost of impression die or sizes are too large and too irregular to be contained in usually impression die. The shapes most commonly used by this process are **bars**, **slab**, **billets**, etc.

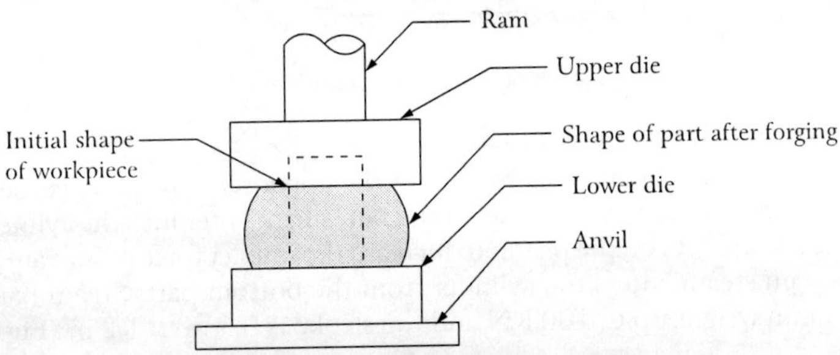

Fig. 1 Open die forging

(a) Hammer : **A hammer may consists of four main parts**

- Falling weight
- Frame or guide
- Base
- Lifting machine for ram.

The capacity of hammer is specified by weight of the falling part. The hammer is energy restricted equipment. The total kinetic energy attained by falling parts when the upper die touching the workpiece positioned on the lower die during a work stroke is dissipated by plastic deformation of the work piece material. The kinetic energy depends on weight of the falling and velocity attained at the end of downward stroke.

TYPES OF HAMMERS

(a)1. Single acting steam hammer

In this type of hammer steam or air is used to lift the ram which then falls by gravity. Steam or compressed air is delivered to the cylinder through inlet port, thus raising the piston and ram connected to it. The force necessary to ensure quick lift of ram can be 3 to 5 times the ram weight. When the ram has been lifted to a sufficient height the steam or is cut off which escapes through inlet port **(Fig. 2)**.

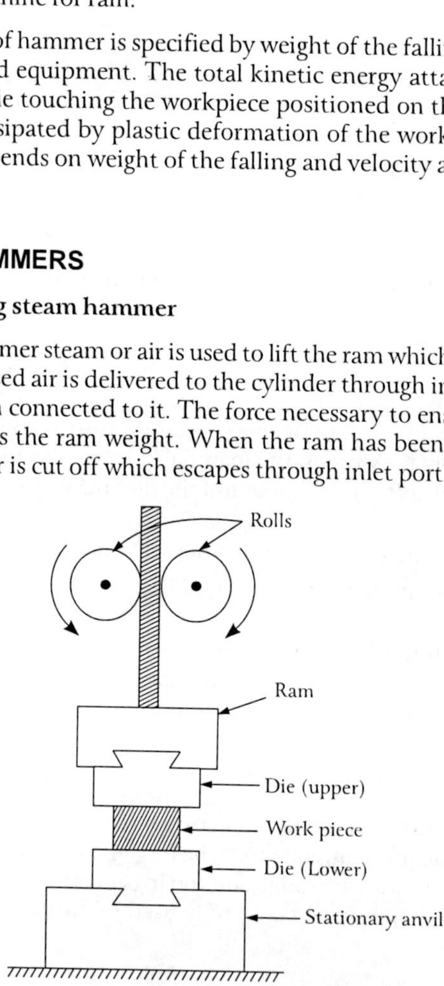

Fig. 2 Single acting Steam Hammer

(a)2. Double acting steam hammer

In this type of hammer, parts are provided at the upper section and lower section of the cylinder. On the downward stroke of the ram steam enter into the cylinder from the upper part and acts on the piston to increase the impact force of the ram. For the lifting the ram steam enters the cylinder from the bottom part. Steam hammer is built with ram weight up to 100 KN and the stroke is 0.9 m to 1.2 m **(Fig. 3)**.

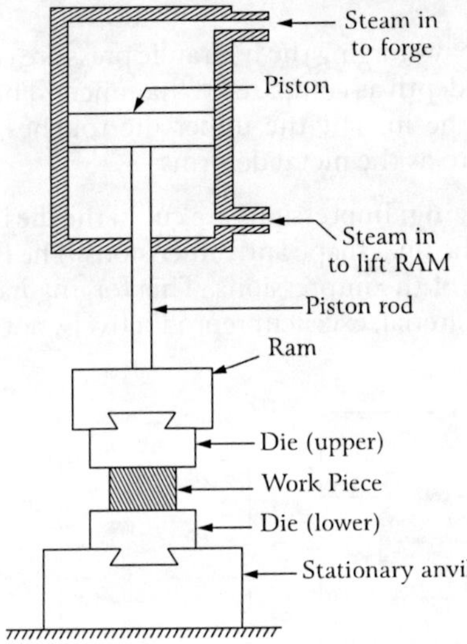

Fig. 3 Double acting Steam Hammer

(a)3. Pneumatic hammer

This type of hammer is used generally in smith forge for small parts. In this hammer the weight of anvil is varies from 15 to 20 times the falling weight. It has two cylinders, a compressor and a ram cylinder. Air is compressed in both stroke and delivered to ram cylinder where it actuates the ram, delivering the forging blow to the work. The weight of falling part varies from 0.5 to 10 KN. Such hammers operates at 80 to 200 blows per minute **(Fig. 4)**.

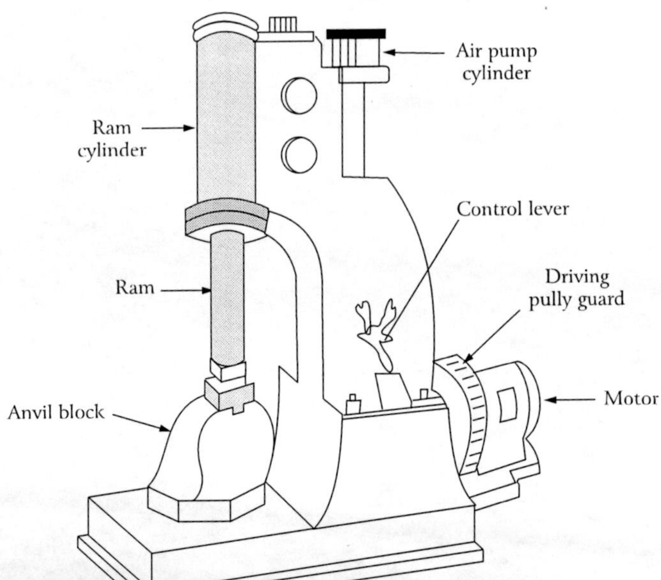

Fig. 4 Pneumatic Power Hammer

(b) Hydraulic press

For the fabrication of heavy forging the hydraulic press are employed. The material is worked up to greater depth as compared to hammers. The maximum amount of energy is available at the instant the upper die touches the workpiece and is gradually reduces to zero as the metal deforms.

(i) **Closed die forging:** Impressions are cut in the die block, in which metal is forced to take the final shape and dimensions. The flow of metals is limited by the surfaces of the impressions. This forging has saving of time, good utilization of material, excellent reproductively, not required highly skilled workers.

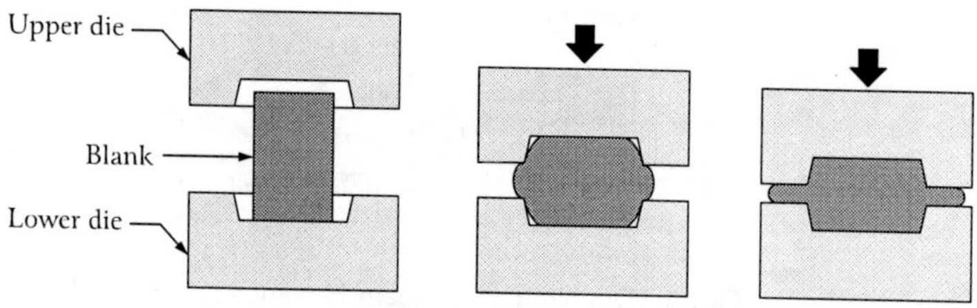

Fig. 5 Forging: Impression forging

(ii) **Drop forging:** In this process forging is made by hammering a heated bar or billet into aligned die cavity. It has added provisions of larger and more accurate guides for the vertical movement of the ram, so that the top die meets the bottom die in close alignment. Also the frame of the drop hammer is secured to the anvil block and total hammer structure is mounted on single structure.

E

EXPERIMENT-8

OBJECTIVE

To Prepare a hook (Nail) of the given dimension in the Black Smithy Shop.

MATERIAL USED

MS Rod of Diameter- 10 mm, Length -205 mm.

TOOLS AND EQUIPMENTS USED

Tong, Anvil, Hammer (cross peen and sledge), Round nose tong, Swage block, Hand hacksaw, steel rule.

MACHINE USED

Open hearth furnace.

STEPS OF OPERATION

1. First take the material of mild steel rod cut it into the required size.
2. Prepare the furnace for heating with the burning of fuel (coal), wooden pieces.
3. Now put one portion of rod in the furnace, heat one end of the job up to plastic stage.
4. When that portion gets red hot drawn it from furnace by tong.
5. With the help of anvil and sledge hammer draw out the portion and make it pointed and as given in the drawing.
6. Repeat this process of heating and drawing out will be nail portion has been completed.
7. The job is heated in the furnace and after heating bends this portion as given in the drawing.
8. Give the final finishing by hammer and cool in the water tank.

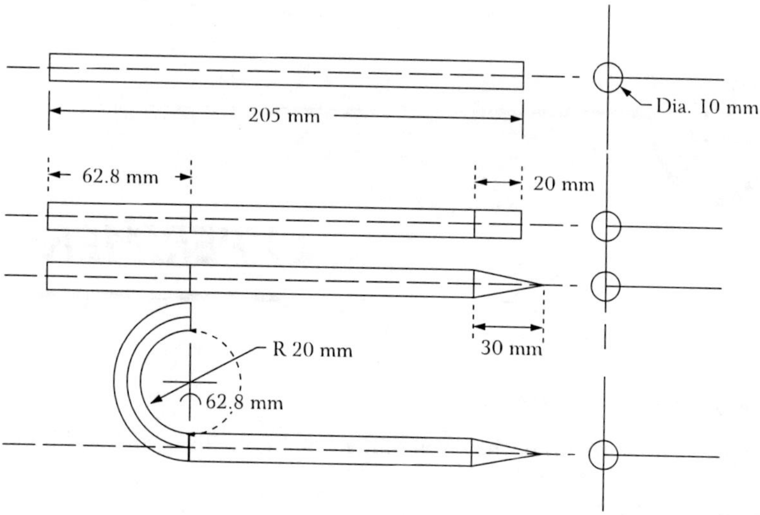

Fig. 1 Hook of Mild steel rod (Length: 205 mm, Diameter: 10 mm)

PRECAUTIONS

1. The workpiece should not be heated above its melting point.
2. Hammering should be done carefully and use the correct tools for forging.
3. Always wear leather apron so that hot pieces of scale flying around during forging do not burn your cloth.
4. Strike the hammer at right point and use the open hearth furnace carefully.

❑❑❑

B

BIBLIOGRAPHY

AFS, *"Cupola and its operations, American Foundrymen's Society"*, Des Plaines, 1965.

Avitzer, B., *"Metal Forming Processes and Analysis"*, Mc-Graw Hill, New York, 1991.

C.E. Pearson and R. N. Perkine, *"Extrusion of Metals"*, John Wiley & Sons inc, New York, 1960.

Chvorinov, N., *"Theory of Solidification of Casting"*, Geisserei Vol. 27, PP. 177-225, 1940.

Degarmo, E. Paul, *"Material and Processes in Manufacturing"*, The Macmillan Company, New Delhi, 1974.

Eary, D.F. and Reed, *"Techniques of Press Working Sheet Metal"*, Printic Hall, New Jersey, 1974.

Flinn R.A., *"Fundamental of Metal Casting"*, Addision-Wesley, Reading Massachusetts, 1963.

G.C. Spencer, *"Introduction to plasticity"*, Champan and Hall Ltd.

German. R.M., *"Powder Metallurgy Science"*, Princeton; Metal Powder Industries Federation, 1994.

Ghosh, Amitabh and Mallik, A.K., *"Manufacturing Science"*, East-West Press, New Delhi.

Heine, R.W., C.R. Loper and P.C. Rosenthal, *"Principal of Metal Casting"*, Mc-Graw Hill, New York, 1967.

Hoffman, & Sachs, G, *"Introduction to the theory of plasticity for engineers"*, Mc-Graw Hill book Co., New York, 1953.

http://Plastics.inwiki.org/

Jain, P.L., *"Principles of Foundry Technology"*, 2nd ed. Tata Mc Graw-Hill, New Delhi, 1986.

K.P. Sinhna and S.C. Prasad, *"Analysis of Metal Flow during Indentation Phase of Rotary Floring Proc. Indo. Pri. Conference Production Engineering"*, New Delhi, 1976.

Kalpak Jian, "*Manufacturing Engineering and Technology*", Addison-Wesley, 1995.

Kamenshikov. G., "*Forging Practice*", Moscow Foreign Language Publication House.

Kaushik J.P., "*Manufacturing Process*", Printic Hall of India, New Delhi.

Khachanor, L.M., "*Fundamental of the plasticity*", M. M. Publication, MOSCOOW USSR.

Khan, M.I., & Haque, S. "*Manufacturing Science*", PHI Learning Private Limited New Delhi, 2011.

Khanna, O.P., "*Foundry Technology*", Dhanpat Rai Publications (P) Ltd., New Delhi, 2005.

Minkoff. I., "*Materials Processes A short introduction*", Springer Verlag Berlin Heidelberg, New York.

Nagpal, G.R. "*Metal Forming Processes*". Khanna Publishers, New Delhi 2005.

Pandey, P.C., and C.K. Singh, "*A Textbook of Production Engineering*", Standard Publishers Distributors, Nai Sarak, New Delhi, 1972.

Prashant. P.D., "*Introduction to Manufacturing Technologies*" Jaico Publishing House, Mumbai.

Ranghuwanshi, B.S., "*Manufacturing Process*", Dhanpat Rai and Co., New Delhi.

Rao, P.N., "*Manufacturing Technology*", Tata Mc-Graw Hill, Publishing Company Limited, New Delhi.

Rizvi, S.A., Ali, Wajahat, "*Advanced Welding Technology*" S.K. Kataria & Sons, New Delhi.

Roway, G.W., "*Fundamental of Metal Forming Process*" PBS Publishers and Distributors, New Delhi.

Sharma, P.C., "*Production Engineering*", S. Chand and Company Ltd., New Delhi.

Singh, B. and Singh , C.K., "*A Textbook of Production Engineering*" Standard Publishers Distributors, Delhi.

Taylor, F.H. et al., "*Foundry Engineering*", Wiley Eastern (P) Ltd., New Delhi 1959.

Timoshenko, S & Goodier, J. N., "*Theory of Elasticity 2nd Edition*", Mc Graw Hill, New York.

❑❑❑

B. TECH.

(SEM IV) THEORY EXAMINATION, 2010-2011

MANUFACTURING SCIENCE-I

Time: 3 Hours *Maximum Marks : 100*

Note: *Attempt All questions:*

1. Attempt **any four** of the following: (4 × 5 = 20)
   (*a*) What is the role of mass-manufacture in raising the standard of living of human beings?
   (*b*) Write down the names of material and manufacturing processes used in manufacture of
      (*i*) Pistons for two wheelers (*ii*) Electrical conductors
      (*iii*) Plastic buckets (*iv*) Crockery
      (*v*) Plain washers.
   (*c*) Differentiate between hot working and cold working.
   A workpiece made of mild steel is heated to 400°C and mechanical work is being done on it. Is it an example of "hot working" or "cold working"?
   (*d*) State Tresca's and Von Mises criteria for plastic deformation. Show that according to Von Mises criterion, $K = \sigma_y/\sqrt{3}$; where σ_y is tensile yield strength of the material.
   (*e*) A metal workpiece has the size 200 mm × 100 mm × 150 mm ($b \times h \times w$). Assuming plane strain forging under sticking friction condition, calculate the peak pressure. The material has a yield stress in uniaxial tension of 160 MPa.
   (*f*) A metal workpiece having size of $b \times h \times w$ is undergoing open die forging under mixed friction condition. Size w does not change. Draw a graph showing pressure distribution Vs breadth b of the block. If the value of coeff. of friction increases, will the distance from centre of workpiece to where sticking ends increase or decrease? Explain why?

2. Attempt **any two** parts: (2 × 10 = 20)
   (*a*) From first principles, derive the formula

$$\frac{\sigma_{xb}}{2k} = \left(\frac{1+B}{B}\right)\left[\left(\frac{D_b}{D_a}\right)^{2b} - 1\right]$$

   for extrusion of a wire with friction; where σ_{xb} refers to the stress in wire at inlet to the die, D_b and D_a are the inlet and outlet diameters of the wire, $B = \mu \cot \alpha$ (μ is coeff. of friction and α is half die angle) and k is the critical shear stress.
   (*b*) (*i*) Write a note on rolling defects indicating the defects, their causes and remedies.
      (*ii*) Calculate the bite angle when rolling 15 mm thick plates using rolls of 400 mm diameter. Final thickness of plates 12 mm.
   (*c*) Describe the process of wire-drawing. What is the material of drawing dies? Why is "in-process" annealing done?
   Give an idea of wire drawing speeds.

3. Attempt **any two** parts: (2 × 10 = 20)
   (*a*) Show that during deep drawing of a cup, the radial stress σ_r at radius r is given by

$$\frac{\sigma_r}{2k} = \frac{\mu F_h}{2\pi K r_j t} + \log_e \frac{r_j}{r}, \text{ where}$$

F_h is the blank holding force, r_j is initial blank radius, t = plate thickness and μ = coeff. of friction K is the shear yield strength.

Also prove that to prevent tensile fracture at the bottom of the cup,

$$\left[\frac{\mu F_h}{2\pi K r_j t} + \log\frac{r_j}{r_d}\right] e^{\mu\pi/2} \le 1, \text{ where } r_d \text{ is die radius.}$$

(*b*) Explain "shearing process" in detail and explain the terms shearing force, penetration and clearance. Draw a shearing force vs thickness curve and explain how this can be used to determine the energy required in shearing. How can the maximum shearing force be reduced?

(*c*) (*i*) Describe the processes of "Air bending" and V-bending. What is meant by "spring back" and how is it compensated?

(*ii*) What precautions are necessary in bending a tube? How is the initial length required for making a bent tube component worked out? What kind of defects can occur while bending a tube?

4. Attempt **any two** parts: (2 × 10 = 20)

(*a*) Bring out the difference between a jig and a fixture. Describe with neat sketches, any three type of clamping devices used for clamping the workpiece on to a fixture. A fixture was designed for machining castings of a component sourced from Foundry 'A'. If the source for obtaining castings of the same component is changed to Foundry B, will any changes in the fixture be necessary? Component, why?

(*b*) How are powder-metallurgy components manufactured? Discuss various steps involved.

(*c*) (*i*) Describe the explosive forming method.

(*ii*) What is the difference between thermoplastics and thermosetting plastics? Describe three common methods of producing parts made of plastic. How can coloured plastic parts be manufactured?

5. Attempt **any four** parts: (4 × 5 = 20)

(*a*) What are the raw material fluxing agent and fuel used in a cupola to produce Grey C.I. castings?

(*b*) What is "dendritic structure? Explain briefly.

(*c*) Draw a sketch showing various elements of a gating system. Label all the elements shown.

(*d*) Show that for a bottom gating system, time taken to fill up the mould cavity is given by

$$t = \frac{A_m}{A_g} \cdot \sqrt{\frac{2}{g}}\left(\sqrt{H} - \sqrt{H - h_m}\right),$$ where A_m and A_g refer to the cross-section area of mould and gate respectively. H represents the height of liquid in pouring basin above the gate and h_m is the height of mould.

(*e*) Discuss CAINE's method for design of a riser.

(*f*) Describe some non-destructive inspection methods for castings.

❑❑❑

B. TECH.

(SEM IV) EVEN SEMESTER THEORY EXAMINATION, 2011-2012

MANUFACTURING SCIENCE-I

Time: 3 Hours *Maximum Marks : 100*

Note: *(1) Attempt **All** questions.*

(2) They carry marks shown against each.

1. Attempt any **two** parts of the following: (2 × 10 = 20)

(*a*) What is meant by "cold working" and "hot working"? Explain the difference between the two. What is meant by anisotropy ? Explain. Further explain how the effect of cold working of metals and alloys removed.

(*b*) Show that in forging a strip of cross-section $b \times h$ under mixed friction condition, x_s, the distance from the centre of this strip where sticking friction ends is given by:

$$x_s = \frac{b}{2} - \frac{h}{2\mu}\log_e \frac{1}{2\mu}$$

where μ is the coefficient of friction and h is the height of strip. State clearly, assumptions made if any in deriving the expression for x_s. What will be the value of forging pressure at x_s?

(*c*) What are the criteria commonly used for plastic deformation ? State these criteria and show that according to Tresca's criterion, critical shear stress is $\frac{\sigma_y}{2}$, where σ_y is the tensile yield stress of the material.

2. Attempt **any two** parts of the following: (2 × 10 = 20)

(*a*) In a wire drawing operation, initial wire diameter is 6 mm and final diameter is 5 mm. Die angle is 20°, find the drawing stress considering $\mu = 0.1$ and $k = 18$ N/mm^2.

Also calculate the maximum reduction possible in one pass. Assume that there is no back pull.

(*b*) A strip with a cross-section 150 mm × 45 mm is being rolled with 20% reduction of area using 450 mm diameter rolls. Find the angle subtended by the deformation zone at the roll centre, final strip thickness and the location of the neutral plane. Assume coefficient of friction to be 0.1.

(*c*) Describe briefly the hot and cold extrusion processes you know. What are the common extrusion defects and why are they caused?

3. Attempt **any two** parts: (2 × 10 = 20)

(*a*) Draw a sketch of a die-punch assembly. What is the function of the stripper plate?

A hole of 100 mm diameter is to be punched in a metal sheet of 6 mm thickness. The ultimate shear stress of the metal is 400 MPa. Normal radial clearance is 10% of sheet thickness and cutting completes at 40%

penetration. Find suitable punch and die diameters and the press capacity in metric tons, if a punch without shear is used.

(*b*) Show that in deep drawing of a cup, the limiting case for fracture at bottom is given by:

$$e^{-\mu\pi/2} = \frac{\mu F_h}{2\pi k r_j \cdot t} + \log_e \frac{r_j}{r_d}; \text{ where,}$$

F_h is the blank holding force, r_j is the initial blank radius, r_d is the radius of the die, k is yield shear strength, t is plate thickness and μ is coefficient of friction.

(*c*) Explain the processes of air bending and bottoming with the help of suitable sketches. What is "spring back" and how is it controlled?

A piece of metal 2.5 thickness is bent to an angle of 90° with and inner radius of 6.5 mm. What is the original length of metal which goes into the bend?

4. Answer **any four** parts of the following: (4 × 5 = 20)

(*a*) Why are jigs and fixtures used? Differentiate between a jig and a fixture.

(*b*) Explain the process of Electromagnetic Forming. What are its advantages? Indicate some typical applications.

(*c*) In powder-metallurgy process, some secondary operations are performed after sintering. What are these secondary operations and what is their object? Explain briefly.

(*d*) What characteristic properties of polymers display? Explain the meaning of "monomer molecules". What is a copolymer?

(*e*) Describe the general characteristics and typical applications of any three of the following plastics. Acrylics, cellulosics, fluorocarbons and polyethylenes.

(*f*) Explain "Injection moulding" and "blow moulding" methods of processing plastics.

5. Answer **any four** parts: (4 × 5 = 20)

(*a*) What is a pattern? Name the various allowances given on the pattern and why are they provided?

(*b*) Describe, step-by-step, the procedure for making a mould with a two piece split pattern.

(*c*) What are the requisite properties in a good foundry sand?

(*d*) Draw a properly labelled sketch of a cupola and write a brief account of its operation.

(*e*) Describe the effect of addition of ferrosilicon and rate of cooling on the properties of cast iron.

(*f*) Enumerate some common casting defects and their causes.

❑❑❑

B. TECH.

(SEM IV) THEORY EXAMINATION, 2013-2014

MANUFACTURING SCIENCE-I

Time: 3 Hours *Maximum Marks : 100*

Note: *Attempt* ***All*** *the questions as directed.*

SECTION-A

1. Answer **all** parts: (2 × 10 = 20)

   (a) Classify Manufacturing Process.

   (b) Define Recrystallisation.

   (c) Define forgeability. How can it be tested ?

   (d) Differentiate between Plane Stress and Plane Strain.

   (e) Define the terms, Angle of Bite and Neutral Plane in Rolling.

   (f) Write down different Pattern Allowances.

   (g) Define Notching and Nibbling operations.

   (h) What is Curing in Plastic?

   (i) Write the basic steps of Casting.

   (j) Differentiate the basic difference between Hot working and Cold working.

SECTION-B

2. Attempt any three of the following: (10 × 3 = 30)

   (a) Differentiate between open and closed die forging with their advantages and disadvantages. Also explain the defects in forging.

   (b) What is extrusion? Explain all the types with a neat sketch.

   (c) Differentiate between jigs and fixtures. Also explain the types of Jigs with neat sketch.

   (d) Differentiate between punching and blanking. Also derive the relation between force applied by punch having flat face and that having shear.

   (e) What are the design considerations of powder metallurgy? Explain.

SECTION-C

3. Attempt all parts with internal choice : (10 × 5 = 50)

   (a) What are different types of dies used in sheet metal forming. Explain any two with neat sketch.

   OR

   A metal component 25 mm × 25 mm × 150 mm long having a yield stress of 7 MPa in tension, is to be pressed between flat dies to a size 6 mm × 100 mm × 150 mm. If $\mu = 0.2$ then calculate the maximum forging load.

(b) Describe the Gating system in casting with neat sketch and prove the volume/area ratio for most compact economic side riser of height *h* diameter *d*/6 and *d*/*n* ratio 1.

OR

Explain Die Casting with neat sketch. Also explain all the casting defects briefly.

(c) Briefly explain principle and mechanism of rolling process. Also derive the relation for max draft obtained in rolling process.

OR

Explain the working principle, advantages and disadvantages of injection moulding process.

(d) What is a centrifugal casting? Explain about different types of centrifugal casting methods.

OR

(i) Explain briefly about electromagnetic forming, mention its advantages and its applications.

(ii) Distinguish between explosive forming and electro-hydraulic froming process.

(e) A hole 100 mm diameter is to be punched in a steel plate of 6 mm thickness. The material is cold rolled C_{40} steel for which maximum shear strength can be taken as 550 MPa. with normal clearance on the tools, cutting is competed at 40% penetration of punch, find diameters of punch and dies and shear angle on the punch in order to bring the work within the capacity of a 200 kN press available in shop.

OR

Explain all the type of moulding sand and discuss briefly the materials added to moulding sand to improve their moulding properties.

❑❑❑

TU-121

B. Tech. Examination May, 2014

M.E.

MANUFACTURING SCIENCE-I

CODE: BT-414(N)

Time: 3 Hours *Maximum Marks : 100*

Note: Attempt any **five** questions. **All** questions carry equal marks.

1. (a) What is manufacturing. Enumerate the importance of manufacturing? 10
   (b) Describe the following types of deformation in metals: 10
      (i) Elastic deformation
      (ii) Plastic deformation
2. (a) List the limitation of open die forging and closed die forging process? Explain closed die forging in brief. 10
   (b) Define forging? Explain any two types of forging with neat sketch. 10
3. (a) What do you mean by rolling? Explain briefly, the principle of rolling? 10
   (b) Give few examples of rolled products. 10
4. (a) Explain and differentiate: 10
      (i) Blanking and piercing
      (ii) Progressive and Compound die
   (b) Give the detail of various parts of power press 10
5. (a) Discuss the different methods of producing metal powders. 10
   (b) With the help of neat sketch, explain deep drawing operation 10
6. (a) Differentiate between Jig and Fixture.
   (b) What are the advantages and limitations of powder metallurgy 10
7. (a) Describe the process of compression moulding 10
   (b) What are the economic aspects of the use of Jigs and Fixtures? Differentiate between drill Jig and a Fixture 10
8. (a) Describe the various kinds of pattern in use? What are the allowances provided, when making a pattern? 10
   (b) What are the properties required for moulding sand? Also give the names of various types of moulding sand? 10
9. (a) Write short notes on the following: 10
      (i) welding of plastics
      (ii) future of plastics
   (b) Explain the Mechanism of solidification of casting of pure metals. 10

10. (a) Define the following Grating System: 10

(i) Parting line gate

(ii) Bottom gate

(iii) Top gate

(b) Give the causes and remedies of the following casting defects: 10

(i) Blow holes

(ii) Pin hole Porosity

(iii) Rat tails

(iv) Hot tears

❑❑❑

INDEX

INDEX

❑❑❑

Win Prizes!

ATTENTION READERS

We request you for your frank assessment regarding some of the aspects of the book given as under:

MANUFACTURING SCIENCE-I

By

S.A. Rizvi
Dr. S.P. Tewari
Wajahat Ali

Please fill up the given space in neat capital letters.

(i) What topics of your syllabus that are important from you examination point of view are not covered in the book ?

....
....
....
....

(ii) What are the chapters wherein the treatment of the subject matter is not systematic or organised or updated ?

....
....
....
....

(iii) Have you come across misprints/mistakes/factual inaccuracies in the book? Please specify the chapters and the page numbers.

....
....
....
....

(iv) What other books on the same subject have you found/heard better than the present book ? Please specify in terms of price and quality.

....
....
....
....

(v) Further suggestions and comments for the improvement of the book:

....
....
....

(PS: If need be please attach a separate sheet to write your views.)

Other Details:

(i) Who recommended you the book ? (Please tick in the box near the option relevant to you.)

☐ Teacher ☐ Friends ☐ Bookseller

(ii) Name of the recommending teacher, his designation and address.

....

....

....

(iii) Name and address of the bookseller you purchased the book from :

....

....

....

....

(iv) Name and address of your institution (Please mention the University or Board, as the case may be)

....

....

....

....

(v) Your name and complete postal address :

....

....

....

....

(vi) Course you are enrolled in :

....

....

....

....

The best assessment will be awarded each month. The award will be in the form of our publications as decided by the Editorial Board.

Please mail the filled up coupon at your earliest to:

S.K. Kataria & Sons®

4885/109, Prakash Mahal, Dr. Subhash Bhargav Lane,
Opposite Delhi Medical Association, Daryaganj, New Delhi-110002
Phone: +91-11-23243489, +91-11-43551243; Mobile: +91-9871775858
e-mail: katariabook@gmail.com, katariabooks@yahoo.com
Website: www.skkatariaandsons.com